Vertebrate Lectins

Van Nostrand Reinhold Advanced Cell Biology Series
Miguel M. Azar, Series Editor

CONTROL OF CELL GROWTH AND PROLIFERATION, edited by *Carlo M. Veneziale*
NEW METHODOLOGIES IN STUDIES OF PROTEIN CONFIGURATION, edited by *Tai Te Wu*
CHROMOSOME STRUCTURE AND FUNCTION, edited by *Michael S. Risley*
MITOCHONDRIAL PHYSIOLOGY AND PATHOLOGY, edited by *Gary Fiskum*
VETEBRATE LECTINS, edited by *Kenneth Olden* and *James Brian Parent*

VERTEBRATE LECTINS

Edited by

KENNETH OLDEN
Howard University Cancer Center

JAMES BRIAN PARENT
Howard University Cancer Center

Van Nostrand Reinhold
Advanced Cell Biology Series

VNR VAN NOSTRAND REINHOLD COMPANY
New York

Library of Congress Catalog Card Number 86-24580
ISBN 0-442-26988-9

Printed in the United States of America

Van Nostrand Reinhold Company Inc.
115 Fifth Avenue
New York, New York 10003

Van Nostrand Reinhold Company Limited
Molly Millars Lane
Wokingham, Berkshire RG11 2PY, England

Van Nostrand Reinhold
480 La Trobe Street
Melbourne, Victoria 3000, Australia

Macmillan of Canada
Division of Canada Publishing Corporation
164 Commander Boulevard
Agincourt, Ontario M1S 3C7, Canada

16 15 14 13 12 11 10 9 8 7 6 5 4 3 2 1

Library of Congress Cataloging-in-Publication Data
Vertebrate lectins.
(Van Nostrand Reinhold advanced cell biology series)
Includes index.
1. Lectins—Physiological effect. 2. Vertebrates—Physiology. I. Olden, Kenneth. II. Parent, James Brian. III. Series. [DNLM: 1. Lectins. QW 640 V567]
QP552.L42V47 1987 596′.019′245 86-24580
ISBN 0-442-26988-9

To Mack and Augusta Olden

Contents

Preface

Although lectins, or carbohydrate-binding proteins, have been of interest to biomedical researchers for more than a century, they became the focus of intense research by a large cross section of researchers only about a decade ago. The increased interest in lectins coincided with their discovery in mammalian cells.

The objective of this book is to summarize current knowledge on the function of lectins, with emphasis on the intracellular species. But, unfortunately, the functions of the various lectins have been considered primarily from the perspective in which they were discovered, and invariably this occurred as a consequence of pinocytosis. The intracellular localization of lectins suggests the possibility of their involvement in a biosynthetic process involving a progression from the endoplasmic reticulum to the Golgi and cell surface. Elucidation of the function of intracellular lectins has been slowed by failure to isolate tissue glycoconjugants to which endogenous lectins normally bind; to date only a few have been identified.

The first chapter deals with structural studies on the oligosaccharide moieties of representative glycoproteins, and it will become apparent in the succeeding chapters that these structural studies are essential for functional studies of the carbohydrate-binding proteins. The second chapter gives a historical perspective of lectins in terms of their discovery, characteristics, possible functions, and technological applications. Chapter 3 discusses the role of the mannose 6-phosphate receptor in intracellular sorting of newly synthesized lysosomal enzymes. This receptor is one of the few vertebrate lectins whose function has been firmly established. The author emphasizes the amino acid sequence, carbohydrate-binding site, and membrane orientation of the chicken and rat hepatic lectins. The galactosyl receptor protein is discussed in chapter 4 in terms of its recycling and ligand processing, with emphasis on recent results obtained in the author's laboratory. The liver mannan-binding protein is considered in chapter 5 in terms of its subcellular distribution and possible function in packaging of acid hydrolases into the lysosomal compartment. The possible role of α-glucose-1-phosphate-binding proteins is discussed in chapter 6 in terms of their possible participation in the subcellular localization of phosphodisester-linked glucose-bearing glycoproteins. This represents a second protein that binds oligosaccharides containing phosphate. Chapter 7 describes the metabolism of asialotransferrin and recycling of the Gal/GalNAc-specific transferrin receptor. The potential role of cell surface lectins in lymphocyte recirculation is discussed in chapter 8. The studies described suggest that a lectin on the surface of the lymphocyte and possibly a glycoconjugate on the surface of the post-capillary high endothelial venules contribute to this adhesive interaction. Chapter 9 is an extension of the previous chapter in that the role of lectins in mediation of the complex relationship between phagocytic and nonphagocytic immune cells is discussed. The author reviews findings indicating that interactions between lectin-like receptors and carbohydrate portions of surface glycoconjugates participate in immune recognition. The author suggests that additional studies of the type she has described will be required to answer the fundamental question of how immune cells communicate. A discussion of developmentally regulated soluble lectins is presented in chapter 10 in terms of their possible role in intracellular interactions during tissue differentiation and in the secretion or organization of extracellular glycoconjugates. Data summarized in chapter 11 implicate intracellular lectins in the intracellular transport of liver secretory glycoproteins. The evidence presented is indirect in that it involves the use of agents which specifically modify the structure of the oligosaccharide moieties of the glycoproteins. The final chapter deals with the use of a novel approach, the use of chimeric toxins, to investigate the function of endogenous lectins. This new technology involves the design and synthesis of chimeric or hybrid-protein toxins as tools for probing lectin function.

KENNETH OLDEN
JAMES BRIAN PARENT

Contributors

Nobuko Kawasaki
Department of Biological Chemistry, Faculty of Pharmaceutical Sciences, Kyoto University, Kyoto 606, Japan

Toshisuke Kawasaki
Department of Biological Chemistry, Faculty of Pharmaceutical Sciences, Kyoto University, Kyoto 606, Japan

David Kobiler
Department of Microbiology, Israel Institute for Biological Research, Ness Ziona 70450, Israel

Yasunori Kozutsumi
Department of Biological Chemistry, Faculty of Pharmaceutical Sciences, Kyoto University, Kyoto 606, Japan

Richard B. Marchase
Department of Anatomy, Duke University Medical Center, Durham, North Carolina 27710

Jean Montreuil
Universite des Sciences et Techniques de Lille, Laboratoire de Chimie Biologique et Laboratoire Associe au C.N.R.S. n 217, Villeneuve d'Ascq, France

Kazutoshi Mori
Department of Biological Chemistry, Faculty of Pharmaceutical Sciences, Kyoto University, Kyoto 606, Japan

Sheila A. Newton
Department of Oncology, Howard University Cancer Center, Howard University College of Medicine, Washington, D.C. 20060

Shogo Oka
Department of Biological Chemistry, Faculty of Pharmaceutical Sciences, Kyoto University, Kyoto 606, Japan

Kenneth Olden
Department of Oncology, Howard University Cancer Center, Howard University College of Medicine, Washington, D.C. 20060

James Brian Parent
Department of Oncology, Howard University Cancer Center, Howard University College of Medicine, Washington, D.C. 20060

Erwin Regoeczi
Department of Pathology, McMaster University Health Sciences Centre, Hamilton, Ontario, Canada L8N 3Z5

Steven D. Rosen
Department of Anatomy, University of California, San Francisco, California 94143

G. Gary Sahagian
Department of Physiology, Tufts University School of Medicine, 136 Harrison Avenue, Boston, Massachusetts 02111

Nathan Sharon
Department of Biophysics, The Weizmann Institute of Science, Rehovot, Israel

David L. Simpson
Department of Pathology, School of Dental Medicine, University of Pennsylvania, 4001 Spruce St., Philadelphia, Pennsylvania 19104

Lloyd M. Stoolman
Department of Pathology, University of Michigan, Ann Arbor, Michigan 48109

Paul H. Weigel
Division of Biochemistry, Department of Human Biological Chemistry and Genetics, The University of Texas Medical Branch, Galveston, Texas 77550

Sandra L. White
Department of Microbiology, Howard University Cancer Center, Howard University College of Medicine, Washington, D.C. 20060

Ikuo Yamashina
Department of Biological Chemistry, Faculty of Pharmaceutical Sciences, Kyoto University, Kyoto 606, Japan

Kiang-Teck Yeo
Department of Oncology, Howard University Cancer Center, Howard University College of Medicine, Washington, D.C. 20060

Tet-Kin Yeo
Department of Oncology, Howard University Cancer Center, Howard University College of Medicine, Washington, D.C. 20060

Vertebrate Lectins

1

Structure and Conformation of Glycoprotein Glycans

Jean Montreuil
Université des Sciences et Techniques de Lille Flandres-Artois

INTRODUCTION

Glycoproteins (for books and general reviews, see refs. 1-18), together with glycolipids, constitute the class of glycoconjugates; glycoproteins result from the association through covalent linkages of certain carbohydrate moieties, called *glycans,* with proteins. Glycosylation of proteins represents one of the most important, if not the most important, of the post-translational events, because of the universality of the phenomenon. In fact, most proteins are glycosylated, and the glycoproteins are widely distributed in animals, plants, microorganisms, and viruses. Moreover, a series of discoveries and observations have pointed out that glycans perform important biological roles and, in particular, that they may act as carriers of biological information. In fact, it appears more and more obvious that glycans act in the following roles:

1. the protection of the peptide chain against proteolytic attack
2. the induction and maintenance of the protein conformation in a biologically active form
3. the decrease of the immunogenicity of proteins
4. the mechanisms of recognition and association with viruses, with enzymes during glycan biosynthesis, and with lectins—particularly with membrane lectins

5. the control of glycoprotein uptake by cells and of the lifetime of circulatory glycoproteins and cells
6. intercellular recognition and adhesion, and in cell-contact inhibition.

Moreover, we know now that the glycan structure of the cell membrane glycoproteins is profoundly altered in cancer cells, as has been revealed by the use of lectins and by applying chemical and physical methods to determine the glycan primary structure. This molecular transformation could be related to the appearance of cell surface neoantigens and could be a factor in cancer induction and metastatic diffusion.

It then clearly appears that, according to Nathan Sharon (2), "the specificity of many natural polymers is written in terms of sugar residues, not of amino acids or nucleotides" and that the knowledge of the primary structure and conformation of glycans is necessary to lay the foundations of the molecular biology of glycoproteins and to understand the mechanisms of sugar-lectin recognition and interaction.

TYPES OF GLYCAN-PROTEIN LINKAGES: CLASSIFICATION OF GLYCOPROTEINS

Glycans are conjugated to peptide chains through two types of primary covalent linkages, N-glycosyl and O-glycosyl linkages, leading to the definition of two classes of glycoproteins, N-glycosylproteins and O-glycosylproteins (Table 1-1). Until now, the only N-glycosidic bond found in glycoproteins is the N-acetylglucosaminyl-asparagine (GlcNAc(β1-N) Asn), in contrast with the O-glycosidic type which presents a wide variety of linkages. The most distributed of these linkages are the following:

1. linkage between N-acetyl-D-galactosamine and L-serine or L-threonine (GalNAc(α-1, 3)Ser or GalNAc (α-1, 3)Thr), alkali-labile, found in very numerous glycoproteins that are said to be of the mucin type
2. linkage between D-xylose and L-serine (Xyl(β1-3)Ser), also alkali labile, involved in the acidic mucopolysaccharide-protein bond of proteoglycans
3. linkage between D-galactose and 5-hydroxy-L-lysine (Gal(β1-5)OH-Lys), alkali stable, characterized in collagens
4. linkage between L-arabofuranose and 4-hydroxy-L-proline (L-δAra(β1-4)OH-Pro), alkali stable, identified in plant glycoproteins.

Numerous investigations clearly demonstrate that the primary structure and conformation of the peptide chain plays an important role in its glycosylation. In fact, it is now well established that the existence of key

Table 1-1. Glycan-Peptide Linkages in Glycoproteins

	Amino Acid and Sugar Residues Generally Engaged in the Linkage	
Type of Glycoproteins	*Amino Acid*	*Corresponding Monosaccharide*
N-glycosylproteins	Asn	GlcNAc
O-glycosylproteins		
Mucin type	Ser, Thr	GalNAc
Proteoglycan type	Ser	Xyl
Collagen type	OH-Lys	Gal
Extensin type	OH-Pro	δAra

peptide sequences called *sequons* (19), around the glycan receptor amino acid, is a prerequisite for glycosylation of a protein. For example, the tripeptide sequence Asn-X-Ser(Thr), where X can be almost any amino acid except proline, is the prerequisite for N-glycosylation of proteins. Recently, a second requirement has been defined that is associated with a specific secondary structure of the polypeptide chain: the β-turn which consists of only four amino acids favoring hairpin structures. Asparagine, serine, and threonine are especially abundant in these regions. By applying predictive methods for determining the secondary structure of proteins, Aubert and co-workers (20, 21) and Beeley (22, 23) demonstrated that glycans are located in β-turns. Thus, the glycosylation may be favored, as turn conformations are generally located at the surface of globular proteins, making the acceptor amino acids readily accessible to glycosyltransferases.

PRIMARY STRUCTURE OF GLYCOPROTEIN GLYCANS

The problem of the determination of glycan primary structure could be considered as solved due to the improvements of sensitive and efficient methods. In particular, the development of a process based uniquely on permethylation and high-resolution nuclear magnetic resonance (NMR) spectroscopy allows the rapid determination of glycan structures, requiring amounts of glycans varying from 25 to 100 μg only (6, 11, 18).

The knowledge we now possess of the structure of numerous glycans raises a very exciting problem of comparative biochemistry. In fact, glycan structures are not randomly constructed. On the contrary, they may be divided into families, in which structures are very similar and contain common oligosaccharide sequences, whether they originate from animals, plants, microorganisms, or viruses. Consequently, a series of classes, concepts, and rules are firmly established.

CONCEPTS AND RULES

Concept of the Common Inner Core

The carbohydrate moieties on N- and O-glycosylproteins derive from the substitution of oligosaccharide structures common to all glycans of a given class of glycoproteins. These nonspecific and invariant structures are conjugated to the peptide chain and constitute the most internal part of glycans; namely, the *inner core*. The most current inner cores characterized until now in glycans are described in Figure 1-1. Cores A and B exist in the O-glycosylproteins of the mucin type. Core C is the terminal sequence of almost all of the glycosaminoglycans of proteoglycans. Core D is common to all N-glycosylproteins.

Concept of the Antenna

The glycan structures derive from the substitution of the invariant inner cores by a wide variety of oligosaccharide structures that carry the specificity of the glycans and constitute the variable fraction of the latter. On the basis of their morphology, their flexibility, and their property of being recognition signals, the term *antennae* has been proposed for the outer arms substituting the inner cores (1).

```
                  Gal(β1-4)GalNAc(α1-3)Ser or Thr          A

             GlcNAc(β1-3) GalNAc(α1-3)Ser or Thr          B

                  Gal(β1-3)Gal(β1-4)Xyl(β1-3)Ser          C

Man(α1-3)
         \
          Man(β1-4)GlcNAc(β1-4)GlcNAc(β1-N)Asn            D
         /
Man(α1-6)
```

Figure 1-1. Oligosaccharide inner cores of glycoprotein glycans.

Substitution Rules

The survey of the structures described until now shows that the substitution of monosaccharides as part of a glycan structure conforms to a certain orthodoxy and is subject to laws that are very restrictive; the basis of these laws is found in the specificity of glycosyltransferases and in a conservative evolution of these enzymes. In fact, the possibilities of substitution of a given sugar are limited to one, two, or three well-defined monosaccharides, generally conjugated by an unique type of glycosidic linkage. Therefore, the concept of one glycosidic bond-one glycosyltransferase appears to be established; however, this process is guided by substitution rules related to the primary structure and conformation of the nascent glycan. For example, the monosaccharide residues of the inner-core Gal(β1-13)GalNAc of numerous O-glycosylproteins are substituted as follows: GalNAc residue by sialic acid or by GlcNAc residues in the C-6 position only; Gal residue by sialic acid or by GlcNAc in the C-3 position, and by Fuc in the C-2 position. In the case of the N-glycosidically linked glycans, some rules relative to the substitution of the basic biantennary structure,

```
6           5           4
Gal(β1-4)GlcNAc(β1-2)Man(α1-3)\
                               \
                                Man(β1-4)GlcNAc(β1-4)GlcNAc(β1-N)Asn,
                               /  3         2           1
Gal(β1-4)GlcNAc(β1-2)Man(α1-6)/
6′          5′          4′
```

are as follows: GlcNAc-1 by α-L-fucose in the C-6 position; GlcNAc-5 and/or-5′ by α-L-fucose in the C-3 position, preventing the presence of N-acetyneuraminic acid (NeuAc) residues on Gal-6 and/or -6′ due to the fact that glycosylation by sialyl-and fucosyltransferases is mutually exclusive; Gal-6′ by α-NeuAc in the C-3 or C-6 position, by β-GlcNAc in the C-3 position, by α-L-Fuc in the C-2 or C-6 position. In addition, the structures containing five and nine mannose residues described in Figure 1-5 are the only ones synthesized by eukaryotic cells.

Microheterogeneity of Glycans

In addition to genetic variants expressed as variations in their polypeptide chains, almost all glycoproteins reveal another form of polymorphism associated with their carbohydrate residues that can be recognized by electrophoresis, by anion-exchange chromatography, or by the use of lectins (affinity chromatography or cross-immunoaffinity electrophoresis). In fact, a given glycan located at a given amino acid in a glycoprotein often presents a structural heterogeneity that is produced by partial substitution of sugar residues on a basically similar core structure. This type of diversity is termed *micro-heterogeneity* or *peripheral heterogeneity,* because it often involves the number and positions of the most external monosaccharides in the glycan residues. The microheterogeneity is related to variations in the level of sialylation or to more profound modifications, such as the number of antennae in N-glycosylproteins, as in the case of serum transferring that contains the bi- and triantennary structures 2, 5, and 6 of Figure 1-7 and of α_1-acid glycoproteins in which coexist the bi-, tri-, and tetraantennary glycans 2, 5, and 7 of Figure 1-7.

DESCRIPTION OF GLYCAN PRIMARY STRUCTURES

Glycan Structures Present in O-Glycosylproteins

Glycans Conjugated Through a GalNAc(α1-3)Ser or Thr Linkage. This group comprises mainly structures consisting of mucins, and these glycans are often designated as *mucin-like* structures, even if they are present in plasma cell membranes and in glycoproteins from biological fluids. Some of these structures are described in Figure 1-2. The main observation that could be made concerning these structures is the lack of structural specificity in several of them. In fact, the same structures can be found in glycoproteins of very different origins and roles. They are generally regarded as structures playing essentially a role in the physicochemical order and in stabilization of the conformation of the peptide chain. In this connection, it is worthwhile to note that this kind of glycan is often located in "strategic" domains of the protein moiety; for example, in the hinge region of immunoglobulins.

Glycans Conjugated Through a Xyl(β1-3)Ser Linkage. This very homogeneous family of glycans consists of the acid mucopolysaccharides or glycosamino glycans. All are linear polymers made up of disaccharide repeating units (Fig. 1-3 and Table 1-2) and O-glycosidically linked in the so-called proteoglycans to the peptide chain through the trisaccharide inner-core C or Figure 1-1, except in two cases: (a) that of hyaluronic acid, which does not seem to exist as a proteoglycan; (b) that of kerátan sulfate I, which is conjugated to the protein through an N-acetylglucosaminyl-asparagine linkage.

Glycan Structures Present in N-Glycosylproteins

The N-glycosylproteins are divided into three families, according to the nature of the carbohydrate moiety linked to the pentasaccharidic inner-core

```
                                   GalNAc(α1-3)Ser or Thr          1

                      NeuAc(α2-6)GalNAc(α1-3)Ser or Thr            2

                         Gal(β1-3)GalNAc(α1-3)Ser or Thr           3

          NeuAc(α2-3)Gal(β1-3)GalNAc(α1-3)Ser or Thr               4

                         Gal(β1-3)
                                  \
                                   GalNAc(α1-3)Ser or Thr          5
                                  /
                      NeuAc(α2-6)

          NeuAc(α2-3)Gal(β1-3)
                              \
                               GalNAc(α1-3)Ser or Thr              6
                              /
                  NeuAc(α2-6)

Fuc(α1-2)Gal(β1-4)GlcNAc(β1-3)
                              \
                               Gal(β1-3)GalNAc(α1-3)Ser or Thr     7
                              /
Fuc(α1-2)Gal(β1-4)GlcNAc(β1-6)

          Gal(β1-4)GlcNAc(β1-3)
                               \
                                GalNAc(α1-3)Ser or Thr             8
                               /
          Gal(β1-4)GlcNAc(β1-6)
```

Figure 1-2. Some examples of structures of glycans O-glycosidically conjugated to protein through the linkage GalNAc(α1-3)Ser or Thr. The described glycans are distributed in the following glycoproteins (nonexhaustive list). Structure 1: Tn-determinant of erythrocytes, porcine submaxillary mucin, human serum immunoglobulins IgA_1; structure 2: submaxillary mucins, human erythrocytes; structure 3: antifreeze glycoprotein from antarctic fish, human chorionic gonadotropin β-subunit, human serum IgA_1, epiglycanin of TA3-Ha cells, T-reactive erythrocytes, rat brain glycoproteins; structure 4: human glycophorin, bovine kappa-casein, fetuin, human milk secretory IgA, N blood group; structure 5: submaxillary mucins, rat brain glycoproteins, trout and herring eggs; structure 6: human glycophorin and gonadotropin β-subunit, fetuin, bovine kappa-casein, lymphocyte plasma membrane, M-blood group; structure 7: porcine blood group H substance; structure 8: bronchial mucin of patients suffering from cystic fibrosis (9, 14).

```
[ GalNAc(β1-4)GlcUA(β1-3) ]
[   | 4                   ]
[  SO3-                   ]n

              GalNAc(β1-4)GlcUA(β1-3)Gal(β1-3)Gal(β1-4)Xyl(β1-3)Ser
```

Figure 1-3. Primary structure of seryl-chondroitin 4-sulfate.

Table 1-2. Structure of Acid Mucopolysaccharides (Glycosaminoglycans)

Mucopolysaccharide	*Disaccharide Repeating Unit*	*Position of the Sulfate Group*	*Disaccharide-Disaccharide Linkage*	*Sugar-Protein Linkage*
Hyaluronic acid	GlcNAc(β1-4)GlcUA	None	(β1-3)	*(a)*
Chondroïtine sulfate A	GalNAc(β1-4)GlcUA	C-4 of GalNAc	(β1-3)	Xyl-Ser
Chondroïtine sulfate C	GalNAc(β1-4)GlcUA	C-6 of GalNAc	(β1-3)	Xyl-Ser
Keratan sulfate I	Gal(β1-4)GlcNAc	C-6 of Gal and GlcNAc	(β1-3)	GlcNAc-Asn

[a]The existence of hyaluronic acid in the tissues as a proteoglycan has not yet been demonstrated unequivocally.

D (Fig. 1-1). In the first family, the glycans contain mannose (man) and N-acetylglucosamine only (part 1 of Fig. 1-4). They are called glycans of the *oligomannosidic* type *or high-mannose* type glycans. In the second family, the sugar composition of glycans is more complex. In fact, these glycans contain galactose, fucose, and sialic acids in addition to mannose and N-acetylglucosamine. They derive fundamentally from the addition to the pentasaccharidic inner-core D (Fig. 1-1) of a variable number of N-acetyllactosamine residues (Gal(β1-4)GlcNAc). These structures have been called glycans of the N-*acetyllactosaminic* type or of the *complex* type (part 2 of 1-4). In the third family, the glycans simultaneously have structures of the oligomannosidic and of the N-acetyllactosaminic type. They belong to the *oligomannosido*-N-*acetyllactosaminic* type or *hybrid* type glycans.

Glycans of the Oligomannosidic Type (High-Mannose Type). The glycans of the oligomannosidic type constitute a very homogeneous group. Figure 1-5 shows the structures of two glycans of this type, the most widely distributed in plants and animals. These structures are common to numerous glycoproteins of different origins and roles, and thus do not present any structural specificity. On the basis of the findings concerning the biosynthesis of the N-glycosidically linked glycans (7, 9, 10, 12, 14, 15, 16, and 24), they could be regarded as metabolic intermediates in the maturation of the glycans of the N-acetylactosaminic type. However, this cannot be considered as their only role. In fact, some observations show that oligomannosidic structures act as recognition signals. For example, *Escherichia coli* binds to human epithelial cells, a prerequisite for colonization and infection, by means of a lectin present on the *pili* of the bacteria with a specificity for oligo- and polymannosides (25).

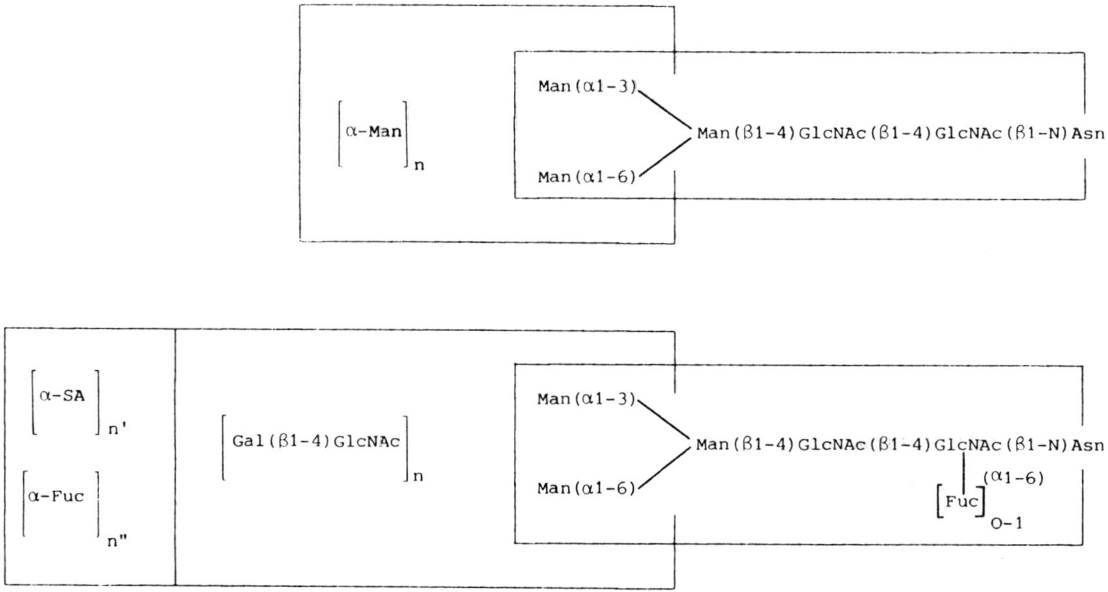

1

2

Figure 1-4. General structural schemes of N-glycosidically linked glycans. Structure 1: oligomannosidic type or high-mannose type; structure 2: N-acetyllactosamine type or complex type; SA: sialic acid.

Moreover, phosphorylated glycans of the oligomannosidic type target newly synthesized lysosomal enzymes to lysosomes by specific recognition of receptors present in the Golgi apparatus and plasma membrane (26, 27). Evidence that the recognition marker was mannose-6-phosphate was brought by Kaplan and associates (28). The structure of such a phosphorylated glycan is given in Figure 1-6 (29).

Glycans of the N-Acetyllactosaminic Type (Complex Type). In contrast to the previously described glycans, glycans of the N-acetyllactosaminic type present a wide variety of structures, and this observation favors the hypothesis that they act essentially as recognition signals. In this connection, the structures described in Figure 1-7 underline the particular importance of two monosaccharides whose systematically external position designates them as recognition signal; that is, sialic acids (30, 31) and fucose (32). Occasionally, the branchings are incomplete and the formation of the

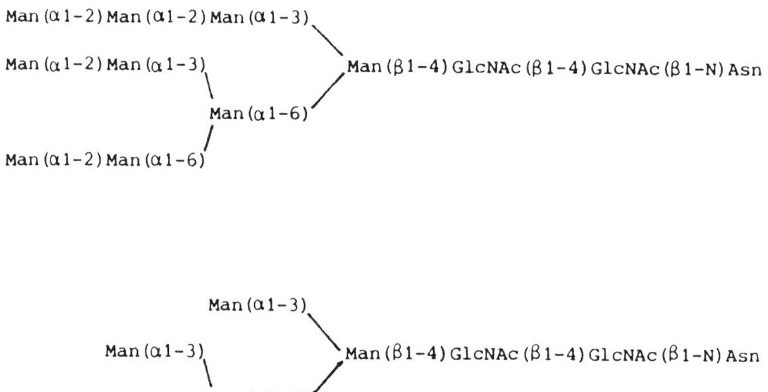

1

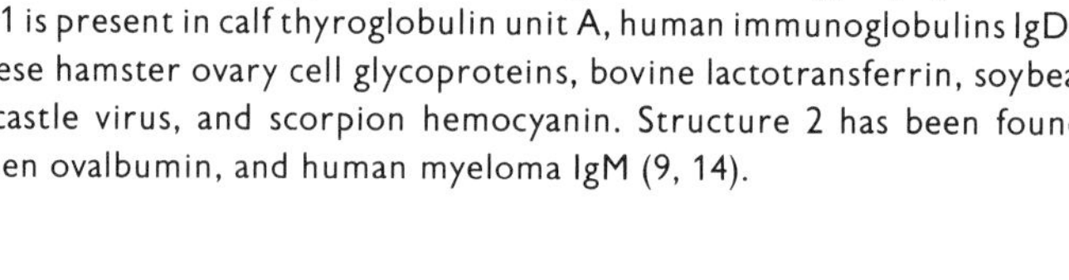

2

Figure 1-5. Structure of glycans of the oligomannosidic type (high-mannose type). Structure 1 is present in calf thyroglobulin unit A, human immunoglobulins IgD, myeloma IgM, Chinese hamster ovary cell glycoproteins, bovine lactotransferrin, soybean agglutinin, Newcastle virus, and scorpion hemocyanin. Structure 2 has been found in Taka-amylase, hen ovalbumin, and human myeloma IgM (9, 14).

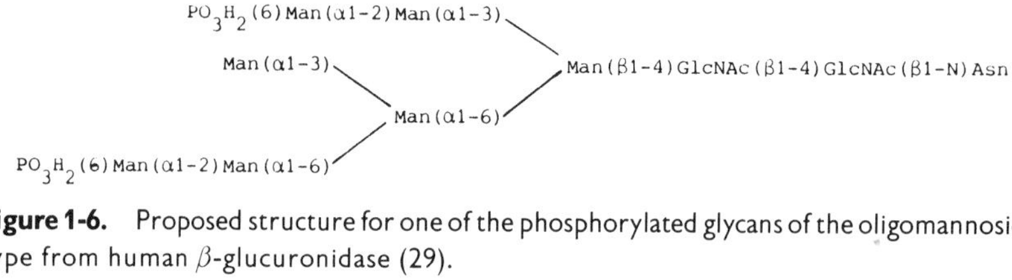

Figure 1-6. Proposed structure for one of the phosphorylated glycans of the oligomannosidic type from human β-glucuronidase (29).

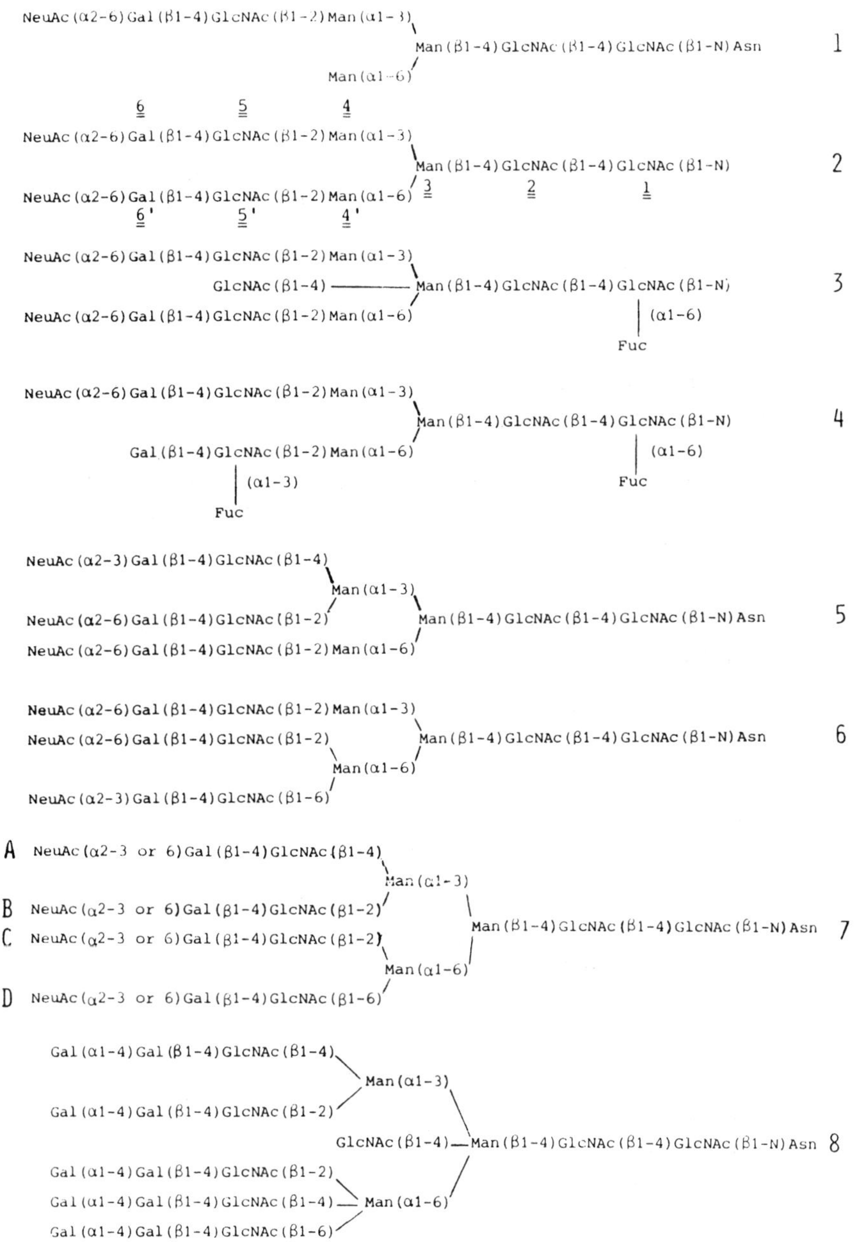

Figure 1-7. Structure of glycans of the N-acetyllactosaminic type (complex type) (nonexhaustive list). Structure 1: monoantennary glycan of secretory component from human milk and human chorionic gonadotropin; structure 2: biantennary glycan of human serum transferrin; structure 3: monofucosylated and "bisected" (presence of a bisecting N-acetylglucosamine residue) biantennary glycan of human IgG; structure 4: difucosylated biantennary glycan of human lactotransferrin; structures 5 and 6: triantennary glycans of human serum transferrin; structure 7: tetraantennary glycan of human α_1-acid glycoprotein; structure 8: pentaantennary glycan of turtle dove ovomucoid possessing a P_1 serologic activity (9, 14).

N-acetyllactosamine residues is started only in outline, as in the glycans of hen ovotransferrin (Fig. 1-8); this incompleteness probably due to a lack of galactosyltransferases or to a signal of nongalactosylation. Moreover, structures containing linear oligo- or poly-N-acetyllactosaminyl sequences (Fig. 1-9), named poly(glycosyl)peptides, have been characterized in erythrocyte membrane (33), human lactotransferrin (34), and Chinese hamster ovary cell glycoprotein (35).

Glycans of the Hybrid Type. Findings on ovalbumin glycopeptides led to the definition of a third group of N-glycosylproteins in which the glycans simultaneously present structures of the oligomannosidic and of the N-acetyllactosaminic type (Figure 1-10), and thus belong to the *hybrid* type. These structures are considered as intermediates in the biosynthesis of glycans of the N-acetyllactosaminic type.

SPATIAL CONFORMATION OF GLYCANS

In the last few years, our knowledge of the spatial conformation of glycans and glycoproteins, for a long time restricted to the human immunoglobulin G (IgG) crystallizable fragment (Fc fragment) (36–40), progressed rapidly, thanks to the development of physical methodologies; that is, X-ray diffraction, electron paramagnetic resonance, nuclear magnetic resonance (NMR), including two-dimensional ^{1}H-NMR and one-dimensional ^{1}H-nuclear Overhauser effect (NOE) experiments, neutron scattering, and computerized molecular modeling. The information obtained in this field has been decisive in the understanding of the mechanism of glycoprotein biosynthesis and of the biological roles played by the glycan moiety.

Glycosaminoglycans and Proteoglycans

X-ray diffraction studies have shown that the mucopolysaccharides are in an helical rod conformation. In the intercellular matrix of cartilage (41, 42), most of the proteoglycans exist in the form of aggregates that are formed by

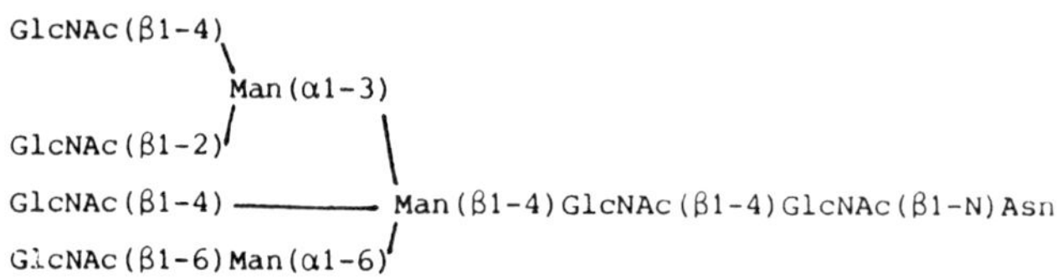

Figure 1-8. Structure of hen ovotransferrin glycan.

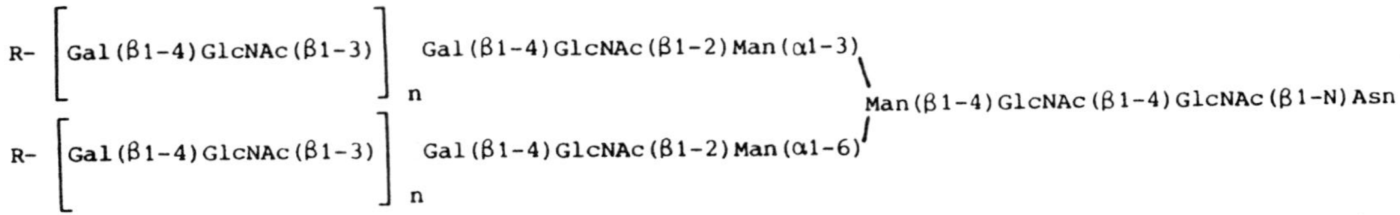

Figure 1-9. General basic structure of the biantennary glycans of the poly-(glycosyl) peptides in which n varies from 1 to 15. In human erythrocyte membrane, R = NeuAc(α2-3)-Gal(β1-4)GlcNAc, NeuAc(α2-6)Gal β1-4)GlcNAc, Fuc(α1-2)Gal(β1-4)GlcNAc (Blood group O), or GalNAc(α1-3) [Fuc(α1-2)] Gal(β1-4) GlcNAc (Blood group A), or Gal(α1-3) [Fuc(α1-2)] Gal(β1-4)GlcNAc (Blood group B).

the association of proteoglycan monomers with hyaluronic acid. The structure of the monomer molecules is that of a central protein core to which glycosaminoglycans are covalently bound (Fig. 1-11). A low-molecular-weight (43-48 kilodaltons [Kd]) glycoprotein, called *link protein,* is responsible for promoting proteoglycan-hyaluronic acid (HA) aggregation leading to the molecular model of Figure 1-12. Whether this kind of architecture could be

```
GlcNAc(β1-2)Man(α1-3)
                     \
                      Man(β1-4)GlcNAc(β1-4)GlcNAc(β1-N)Asn          1
                     /
          Man(α1-6)
```

```
GlcNAc(β1-2)Man(α1-3)
                     \
  Man(α1-3)           Man(β1-4)GlcNAc(β1-4)GlcNAc(β1-N)Asn          2
           \         /
            Man(α1-6)
           /
  Man(α1-6)
```

```
Gal(β1-4)GlcNAc(β1-4)
                     \
                      Man(α1-3)
                     /         \
        GlcNAc(β1-2)            \
        GlcNAc(β1-4)------------ Man(β1-4)GlcNAc(β1-4)GlcNAc(β1-N)Asn   3
                                /
          Man(α1-3)            /
                   \          /
                    Man(α1-6)
                   /
          Man(α1-6)
```

Figure 1-10. Structure of glycans of the hybrid type characterized. Structure 1: bovine rhodopsin; structure 2: bovine rhodopsin and human myeloma IgM; structure 3: in hen ovalbumin (9, 14).

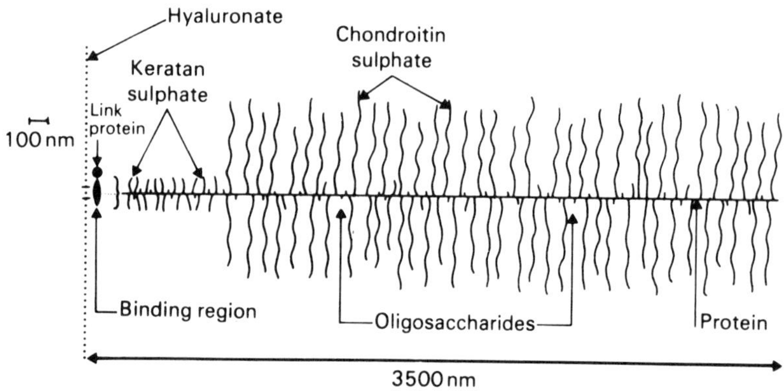

Figure 1-11. Structural model of the major cartilage proteoglycan monomer (41, 42).

generalized and could exist in the intercellular matrix of any tissue remains to be demonstrated.

Mucin-Like Glycans and Glycoproteins

Very little is known about the conformation of O-glycosidically linked glycans and of O-glycosylproteins, except in the case of submaxillary mucins. In fact, mucins have an elevated axial ratio and are in a rod conformation maintained as a result of repulsive electronegative charges carried by the sialic acid residues present in the 800 disaccharide moieties NeuAc(α2-6) GalNAc(α1-3)Ser or Thr(NeuAc-N acetylneuraminic acid); these moieties are distributed almost regularly along the long peptide chain (43). This extended conformation is responsible for the high viscosity of mucin solutions and of secretions of mucous cells. Removal of sialic acid makes the glycoprotein globular, and the solutions become weakly viscous. This example clearly illustrates the role the glycans could play in maintaining the protein in a biologically active conformation.

Glycans of N-Glycosylproteins

From the Y- to the "Broken-Wing" Conformation of Biantennary Glycans. The studies shown in Figure 1-13 illustrate the evolution of ideas

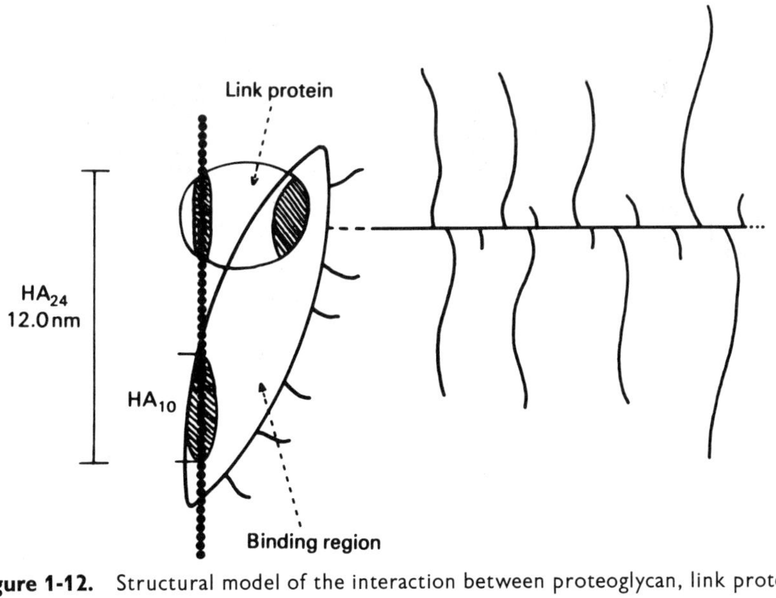

Figure 1-12. Structural model of the interaction between proteoglycan, link protein, and hyaluronic acid (41, 42).

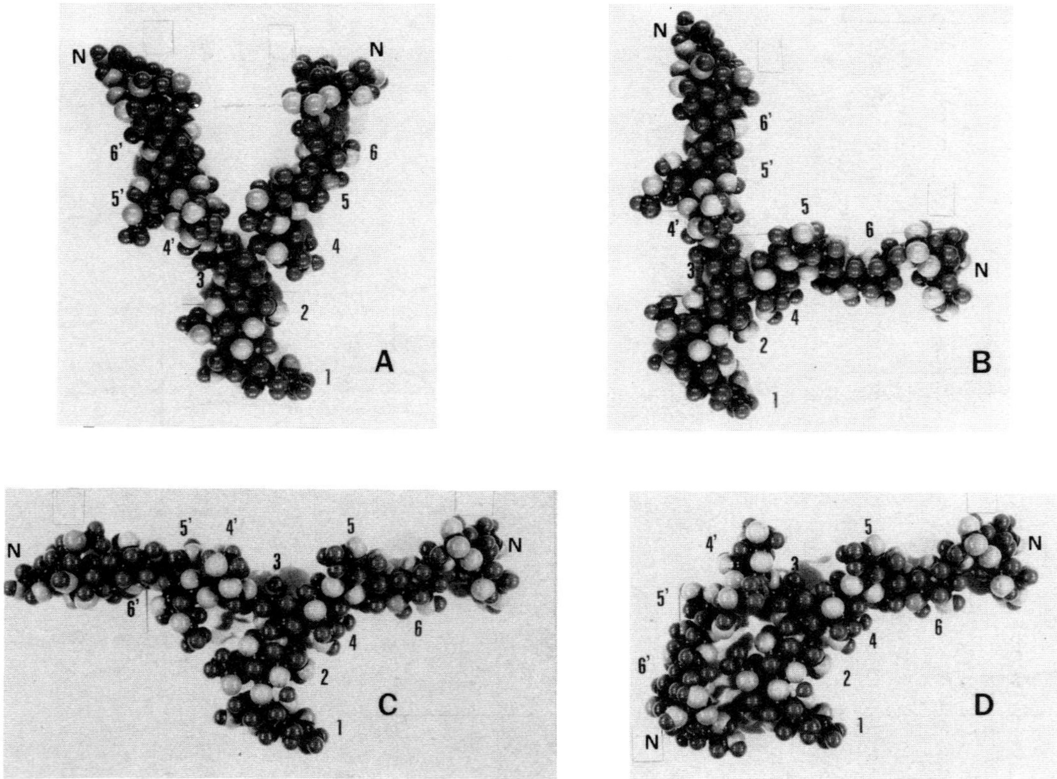

Figure 1-13. Spatial conformation of a biantennary glycan of the N-acetyllactosaminic type. *A:* Y conformation (1); *B:* T conformation (46); *C:* bird conformation (8, 14, 44, 54); *D:* "broken-wing" conformation (8, 54). Numbers correspond to the numbering used in structure 2 of Figure 1-7. N: N-acetyl-neuraminic acid.

concerning the glycan conformation, with the biantennary glycan of the human serum transferrin (Fig. 1-7, structure 2) being a typical example (1, 9, 14, 17, 44). The construction of the molecular model of this glycan, in creating thermodynamically possible hydrogen bonds, led to the first proposal: that of the *Y conformation* (1). Later, on the basis of X-ray diffraction studies (45), the *T conformation* was proposed (46), in which the α-1, 3 antenna is disposed perpendicularly in relation to the α-1, 6 antenna and in an invariant orientation (47-53). Then, the *bird conformation* was suggested (8, 14, 44, 54) from data obtained by a large rotation around the glycosidic bond of the α-1, 6-linked mannose residue on the basis of low-angle X-ray scattering studies (55-57), computer calculations (47, 58), crystallographic data (36-41, 50, 51), and NMR (47) data. More recently, NMR experiments showed that the α-1, 6 antenna is able to fold back towards the protein backbone and to set along the inner core (11, 49, 59, 60), leading to the *broken-wing* conformation.

Generalization of the Concept of the Extended Conformation of Glycans: The Umbrella Conformation. The generalization of the concept of the bird conformation to additional antennae leads to the *umbrella conformation* (8, 14, 44) of the human α_1-acid glycoprotein tetraantennary glycan (Fig. 1-7, structure 7), view of which is shown in Figure 1-14. This speculative view has been confirmed by Bock and co-workers (47), who used high-resolution NMR, and by Li and co-workers (61) on the basis of small-angle neutron scattering.

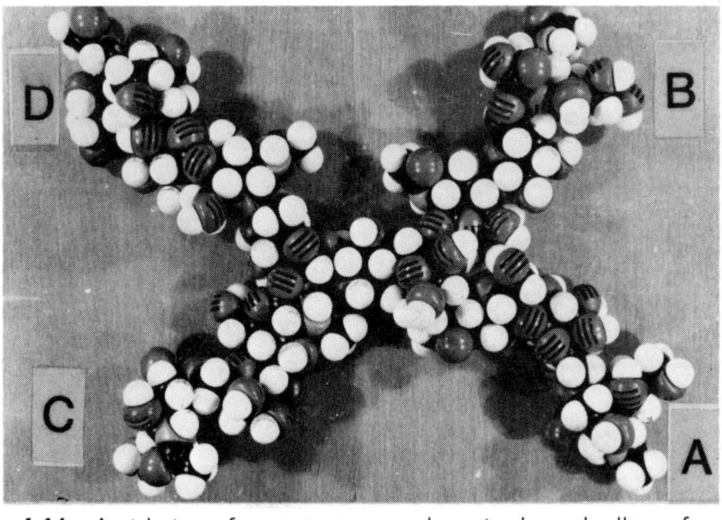

Figure 1-14. Aerial view of a tetraantennary glycan in the umbrella conformation. Letters refer to the four antennae of structure 7 of Figure 1-7.

Are Extended Conformations Compatible with the Properties and the Biological Behavior of Glycans and Glycoproteins? Interaction with Lectins. We know now that lectins could recognize and bind laterally, and not only terminally, mono- or oligosaccharides in an internal position. For example, as has been demonstrated by Kornfeld and Ferris (62), the most Concanavalin A (Con-A)-active part of an N-glycosidically linked glycan is not an α-mannose residue in a terminal position, but the disaccharide GlcNAc(β1-2)Man(α1-3 or 6). Thus the glycans in an extended conformation provide good targets for recognition by other proteins in general and by lectins in particular.

The nonreactivity with Con-A of glycans containing a bisecting N-acetylglucosamine residue, as in human IgG (Fig. 1-7; structure 3) can be explained by a steric hindrance due to the fact that this residue interacts with the two α-mannose residues and covers a wide area of the surface of the GlcNAc(β1-2)Man(α1-3)Man(β1-4) unit (53, 59).

Substitution by fucose residues of C-6 of N-acetylglucosamine-1 and of C-3 of N-acetylglucosamine-5′, as in human lactotransferrin (Fig. 1-7, structure 4), leads to a compact structure (Fig. 1-15) with four confluent methyl groups (two from the fucose residues and two from the acetamido groups of N-acetylglucosamine residues) that can interact through hydrophobic means. Thus, the glycan molecule becomes rigid as a whole, because of the invariance of the α-1, 3 antennae. This rigidity accounts for the fact that the desialylated glycan of human lactotransferrin has the highest inhibitory effect on a galactose-lectin released from guinea pig colonic epithelial cells. The particularly rigid spatial conformation of the glycan renders the terminal galactose of the α-1, 3 antenna more accessible to the lectin (63).

Figure 1-15. Molecular model of a difucosylated biantennary glycan (structure 4 of Fig. 1-7). Arrow indicates the confluence of four methyl groups.

Protective Effects of Glycans Toward the Protein Moiety. Considering the structures shown in Figures 1-13 and 1-14, the question arises as to the size and surface dimension of such glycans in comparison with those of the protein moiety. In the case of a sialylated biantennary glycan such as that of human serum transferrin (Fig. 1-7, structure 2), the dimensions are as follows: length, 5.5 nm; height, 2.4 nm; thickness, 0.5 nm; area, 2.75 nm^2. These dimensions are important compared with those of a globular protein such as the human serum transferrin, which is an oblate spheroid with semiaxes of lengths 4.66 ± 0.14 nm, 4.66 ± 0.14 nm, and 1.58 ± 0.38 nm (64). In the same way, a tetraantennary glycan, in adopting an umbrella conformation, covers an area of about 20-25 nm^2. In the case of glycoproteins such as the human α_1-acid glycoprotein (MW 37.000), which contains five glycan moieties essentially of the tri- and tetraantennary type, the protein could be almost completely enveloped by the glycans (Fig. 1-16). Such glycans acting as protective shields could explain the carbohydrate stabilization and weak antigenicity of the polypeptide component of certain glycoproteins. Moreover, the umbrella conformation of glycans is solidly maintained by ionic bonds between the electronegative charges of sialic acid residues and electropositive charges of basic amino acids, as has been demonstrated by the small-angle neutron scattering study of Li and associates (61). Removal of sialic acid residues renders the antennae free and consequently mobile, and abolishes the protective effect of glycans with respect to protease accessibility and polypeptide antigenicity.

If the structures shown in Figures 1-13 and 1-14 are present in viral envelopes, one can appreciate and understand how the glycans may play a role in masking or altering the nature of viral protein antigenic sites and the role that they may play in viral escape in immune recognition. The possibility

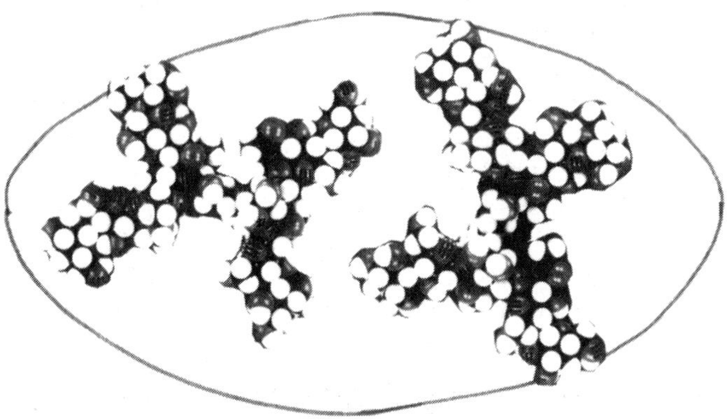

Figure 1-16. Area covered by two tetraantennary glycans associated with a protein having the same size as that of human serum transferrin (8, 14, 44, 54).

that glycans modulate recognition of viral proteins by the immune system is discussed in a paper by Wiley and associates (65). Also, the protective effect that glycans exhibit towards the protein moieties could be responsible for the peculiar behavior and for the resistance to the immune system of metastatic cancerous cells, since we now know that the membrane glycoproteins of cancerous cells are significantly enriched in tri- and tetraantennary glycans, as has been demonstrated by several authors (54, 66–72). Moreover, the new glycan structures expressed on the surface of metastatic cells could explain the organ-specific localization of the latter by specific lectin recognition. For example, Delannoy and co-workers (73) recently observed that fibronectin from cancerous cells in culture is enriched in tri- and tetraantennary glycans, leading to a complete inversion of the ratio of biantennary to tri- plus tetraantennary structures (Table 1-3). The enhancement in branch chain formation is accompanied by a similar increase in the degree of sialylation. The profound modifications observed could be related to (a) the loss of reactivity of the protein moiety of the fibronectin by masking the lectin sites; (b) the disappearance of fibronectin from cancerous tissues; and (c) the mechanism of metastasis itself.

Glycan and Protein Conformation. The concept of the extended conformation of glycans provides a better understanding of the role the carbohydrate moiety plays in the conformation of proteins through specific protein-carbohydrate and carbohydrate-carbohydrate interactions. A very fine demonstration using X-ray diffraction was recently reported by Sutton and Phillips (50). This demonstration concerns the Fc fragment of the rabbit IgG molecule, which is derived from the classic association of two half-molecules, each carrying a glycan residue of the biantennary N-acetyllactosaminic type. The latter is in an internal position in the C_H2 domain, and its integrity is indispensable for the maintenance of all properties of immunoglobulins. The authors have demonstrated that (a) the two antennae of each glycan are in an extended conformation, the orientation of the α-1, 3 antenna being T-shaped and invariant; (b) the α-1, 6 antennae of each glycan make several contacts, mainly hydrophobic, with amino acid residues on the surface of the protein; and (c) the α-1, 3 antennae have no contact with the surface of the C_H2 domain but do interact with each other between the GlcNAc-5 residue of the one chain and the GlcNAc-2 and Man-3 residues of the other chain. It therefore appears that they are responsible for maintaining the disposition of the two C_H2 domains.

Conformational Changes of Glycans. Mobility of the Antennae. Glycan structures must not be considered as rigid conformations, because they have

Table 1-3. Glycans and Sialylation Associated with Malignant Transformation of BHK $21/C_{13}$ Fibroblasts[a]

Branching Type of Glycans		*Biantennary*		*Triantennary and Tetra-antennary*	
Normal BHK $21/C_{13}$		60		40	
HSV-transformed BHK $21/C_{13}$		28		72	
Sialic Acid Residues per Glycan	*0*	*1*	*2*	*3*	*4*
Normal BHK $21/C_{13}$	36	45	14	4	1
HSV-transformed BHK $21/C_{13}$	12	25	33	23	7

[a]Increase in proportions of tri- and tetra-antennary fibronectin glycans and of the degree of sialylation associated with the viral malignant transformation of BHK $21/C_{13}$ fibroblasts.

a certain freedom of motion or rotation in solution, which is expressed during their interactions with other carbohydrate or protein groups. In fact, glycans represent points of flexibility leading to the relative mobility of the antennae, due to the rotation of the rigid pyranose rings around the two bonds attached to the glycosidic oxygen (rotation angles ϕ and ψ). The rotation is limited in the case of $1 \rightarrow 2$, $1 \rightarrow 3$, and $1 \rightarrow 4$ linkages, the $1 \rightarrow 3$ linkage being considered the most rigid one. On the contrary, the $1 \rightarrow 6$ linkages are more flexible because of the additional rotation angle ω. In the case of biantennary structures of the N-acetyllactosaminic type (Fig. 1-7, structure 2), the most important point of flexibility is located at the α-1,6 bond linking the mannose-4′ residue to the mannose-3 residue. This characteristic explains that the Y-, T-, bird-, and broken-wing conformers are interconvertible, as has been demonstrated by low-angle X-ray scattering of artificial glycoconjugates in aqueous solutions (55, 57) and by electron paramagnetic resonance of spin-labeled bi- and triantennary glycans (74). Small-angle neutron scattering of human α_1-acid glycoprotein has also shown that in the case of tetraantennary structures all antennae are mobile and able to adopt an umbrella conformation or a reversed umbrella conformation (61).

The above data suggest that the mannotriose cores play a fundamental role in the conformation of glycans of N-glycosylproteins. In this connection, it is amazing to note that all "strategic" branching points systematically associate a rigid 1,3 linkage with a mobile 1,6 linkage, as is shown in Figure 1-17. Why are T-shaped 1,3 bonds associated in a relatively invariant orientation with a highly flexible and mobile 1,6 bond? This structural organization is probably important for the extended conformation of the antennae, which itself favors (a) the occupation of their "vital space" by the glycans; (b) the interaction of the glycans with their own protein; (c) the glycan-glycan interactions; (d) the action of enzymes; and/or (e) the recognition of glycans by receptors and lectins.

CONCLUSIONS

The development of chemical, physical, and enzymic methods for exploring the primary structure of carbohydrates has, during the past 15 years, allowed the determination of the structure of several hundred glycans. Moreover, the procedures have been miniaturized and are applicable to the study of cell membrane glycoproteins. In this field, the introduction of high-resolution NMR has been a decisive event. In fact, this procedure is rapid, nondestructive, and very sensitive.

In the same way, the intervention of the biophysicists led to the definition of the spatial conformation of glycans, particularly of N-glycosylprotein glycans, and to the concept of their steric conformational changes, consistent with the role they have to play. In fact, experimental results obtained in the past two years from exploration of the spatial conformation of glycans through physical methods—X-ray diffraction, nuclear magnetic resonance, electron paramagnetic resonance, and small-angle neutron scattering—have led to the fundamental finding that the antennae, which support the biologically active part of the glycans, are not rigid formations at all. On the

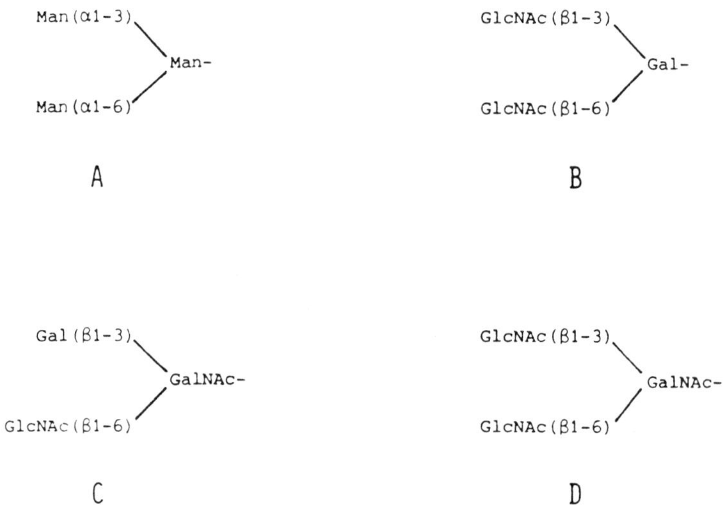

Figure 1-17. "Strategic branchings" associating 1,3 and 1,6 linkages. *A:* inner core of N-glycosidically linked glycans; *B* and *C:* branchings currently found in O-glycosylprotein glycans of the mucin-type; *D:* inner core of tissue antigen group Ii active glycans from sheep gastric mucin (75).

contrary, they must be considered as flexible and mobile parts of glycan molecules, solidly planted on the rigid arm comprising the terminal trisaccharide, Man(β1-4)GlcNAc(β1-4)GlcNAc(β1-N), linked to the peptide chain. In fact, this arm probably accounts for the conformation of the glycan, preventing the twist of the whole molecule due to the relative rigidity of the GlcNAc-Asn linkage (76).

The concept of the antennae changing their conformation according to the environment fits perfectly with the concept of glycans acting as recognition signals. In fact, because of their unimpeded mobility, the antennae could rapidly adapt themselves to the receptor sites and interact with them.

ACKNOWLEDGMENT

We wish to thank Miss Brigitte Mahieu and Mrs. Jocelyne Celen for excellent technical assistance in the preparation of this manuscript.

REFERENCES

1. J. Montreuil, 1975, Recent data on the structure of the carbohydrate moiety of glycoproteins. Metabolic and biological implications, *Pure Appl. Chem.* **42:**431-437.
2. N. Sharon, 1975, *Complex Carbohydrates, Their Chemistry, Biosynthesis and Functions,* Addison-Wesley, Reading, Mass., 466p.
3. M. Horowitz, and W. Pigman, 1977, *The Glycoconjugates,* vol. 1, Academic Press, New York, 560 p.
4. M. Horowitz, and W. Pigman, 1978, *The Glycoconjugates,* vol. 2, Academic Press, New York, 464p.
5. E. F. Walborg, 1978, Glycoproteins and glycolipids in disease processes, *ACS Symposium n° 80,* American Chemical Society, Washington, D.C.
6. J. Montreuil, and J. F. G. Vliegenthart, 1979, Primary structure and conformation of glycans N-glycosidically linked to peptide chains, in *Glycoconjugates Res., Proc. Int. Symp. Glycoconjugates,* vol. 1, J. D. Gregory and R. W. Jeanloz, eds., Academic Press, New York, pp. 35-78.
7. W. J. Lennartz, 1980, *The Biochemistry of Glycoproteins and Proteoglycans,* Plenum Press, New York, 381p.
8. J. Montreuil, 1980, Present knowledge on the conformation of glycoproteins, *Proc. Aharon Katzir-Katchalski Conf. Carbohydr.-Protein Interact.,* Kibbutz Kiryat Anavim, **9:**3-4.
9. J. Montreuil, 1980, Primary structure of glycoprotein glycans. Basis for the molecular biology of glycoproteins, *Adv. Carbohydr. Chem. Biochem.* **37:**157-223.
10. N. Sharon, and H. Lis, 1981, The chemistry and biology of glycoproteins, *Chem. Eng. News* **59:**21-44.
11. J. F. G. Vliegenthart, H. Van Halbeek, and L. Dorland, 1981, The applicability of 500-MHz high-resolution ^{1}H-NMR spectroscopy for the structure determination of carbohydrates derived from glycoproteins, *Pure Appl. Chem.* **53:**45-77.
12. E. G. Berger, E. Buddecke, J. P. Kamerling, A. Kobata, J. C. Paulson, and J. F. G. Vliegenthart, 1982, Structure, biosynthesis and functions of glycoprotein glycans, *Experientia* **38:**1129-1258.
13. M. Horowitz, 1982, *The Glycoconjugates,* vol. 3, part A, and vol. 4, part B, Academic Press, New York, 369p. and 362p.

14. J. Montreuil, 1982, Glycoproteins, in *Comprehensive Biochemistry,* 19B, part II, A. Neuberger and L. L. M. Van Deenen, eds., Elsevier, Amsterdam, pp. 1-188.
15. N. Sharon, and H. Lis, 1982, Glycoproteins, in *The Proteins,* H. Neurath and R. L. Hill, eds., Academic Press, New York, **5:**1-144.
16. R. C. Hughes, 1983, *Glycoproteins,* Chapman and Hill, London, 9p.
17. H. Popper, W. Reutter, F. Gudat, and E. Köttgen, 1983, Structural carbohydrates in the liver, *Falk. Symp. N° 34,* MTP Press Ltd., Lancaster, 707p.
18. J. F. G. Vliegenthart, L. Dorland, and H. Van Halbeek, 1983, High-resolution, ^{1}H-nuclear magnetic resonance spectroscopy as a tool in the structural analysis of carbohydrates related to glycoproteins, *Adv. Carbohydr. Chem. Biochem.* **41:**209-374.
19. R. D. Marshall, and A. Neuberger, 1968, The metabolism of glycoproteins and blood-group substances, in *Carbohydrate Metabolism and Its Disorders,* F. Dickens, P. J. Randle, and W. J. Wheelan, eds., Academic Press, New York, pp. 213-258.
20. J.-P. Aubert, and M.-H. Loucheux-Lefebvre, 1976, Conformational study of α_1-acid glycoproteins, *Arch. Biochem. Biophys.* **175:**400-409.
21. J.-P. Aubert, G. Biserte, and M.-H. Loucheux-Lefebvre, 1976, Carbohydrate peptide linkage in glycoproteins, *Arch. Biochem. Biophys.* **175:**410-418.
22. J. G. Beeley, 1976, Location of the carbohydrate groups of ovomucoid, *Biochem. J.* **159:**335-345.
23. J. G. Beeley, 1977, Peptide chain conformation and the glycosylation of glycoproteins, *Biochem. Biophys. Res. Commun.* **76:**1051-1055.
24. H. Schachter, 1984, Glycoproteins: Their structure, biosynthesis and possible clinical implications, *Clin. Biochem.* **17:**3-14.
25. I. Ofek, D. Mirelman, and N. Sharon, 1977, Adherence of *E. coli* to human mucosal cells mediated by mannose receptors, *Nature* **265:**623-625.
26. E. F. Neufeld, and G. Ashwell, 1980, Carbohydrate recognition systems for receptor-mediated pinocytosis, in *The Biochemistry of Glycoproteins and Proteoglycans,* W. J. Lennartz, ed., Plenum Press, New York, pp. 241-266.
27. W. S. Sly, 1980, Saccharide traffic signals in receptor-mediated endocytosis and transport of acid hydrolases, in *Structure and Function of Gangliosides,* L. Svennerholm, P. Mandel, H. Dreyfus, and P. F. Urban, eds., Plenum Press, New York, pp. 433-451.
28. A. Kaplan, D. T. Achord, and W. S. Sly, 1977, Phosphohexosyl components of a lysosomal enzyme are recognized by pinocytosis receptors of human fibroblasts, *Natl. Acad. Sci. Proc. (USA)* **74:**2026-2030.
29. M. Natowicz, J. U. Baenziger, and W. S. Sly, 1982, Structural studies of the phosphorylated high-mannose type oligosaccharides of human β-glucuronosidase, *J. Biol. Chem.* **257:**4412-4420.
30. A. Rosenberg, and C. L. Schengrund, 1976, *Biological Role of Sialic Acid,* Plenum Press, New York, 375p.
31. R. Schauer, 1982, *Sialic Acids. Chemistry, Metabolism and Functions,* Springer, Vienna, 344p.
32. M. C. Glick, 1978, Fucosylation. A role in cell function, *ACS Symposium n° 80,* American Chemical Society, Washington, D.C., pp. 404-411.
33. T. Krusius, J. Finne, and H. Rauvala, 1978, The poly(glycosyl) chains of glycoproteins. Characterization of a novel type of glycoprotein saccharides from human erythrocyte membrane, *Eur. J. Biochem.* **92:**289-300.
34. G. Spik, G. Strecker, B. Fournet, S. Bouquelet, L. Dorland, J. F. G. Vliengenthart, and J. Montreuil, 1982, Primary structure of the glycans from human lactotransferrin, *Eur. J. Biochem.* **121:**413-419.
35. E. Li, R. Gibson, and S. Kornfeld, 1980, Structure of an unusual complex-type oligosaccharide isolated from Chinese hamster ovary cells, *Arch. Biochem. Biophys.* **199:**393-399.

36. J. Deisenhofer, P. M. Colman, O. Epp, and R. Huber, 1976, Crystallographic structural studies of human Fc fragment, II. A complete model based on a Fourier map at 3.5 Å resolution, *A. Physiol. Chem.* **357:**1421-1434.
37. R. Huber, 1976, Antibody structure, *Trends Biochem. Sci.* **1:**174-178.
38. R. Huber, J. Deisenhofer, P. M. Colman, M. Matsushima, and W. Palm, 1976, Crystallographic structure studies of an IgG molecule and an Fc fragment, *Nature* **264:**415-420.
39. R. Huber, 1980, Spatial structure of immunoglobulin molecules, *Klin. Wochenschr.* **58:**1217-1231.
40. J. Deisenhofer, 1981, Crystallographic refinement and atomic models of a human Fc fragment and its complex with fragment B of protein A from *Staphylococcus aureus* at 2.9- and 2.8-Å resolution, *Biochemistry* **20:**2361-2370.
41. T. Hardingham, 1981, Proteoglycans: Their structure, interactions and molecular organization in cartilage, *Biochem. Soc. Trans.* **9:**489-497.
42. H. Muir, 1983, Proteoglycans as organizers of the intracellular matrix, *Biochem. Soc. Trans.* **11:**613-622.
43. A. Gottschalk, 1960, Correlation between composition, structure, shape and function of a salivary mucoprotein, *Nature* **186:**949-951.
44. J. Montreuil, 1983, Conformation of the glycoprotein glycans of the N-acetyllactosaminic type (complex type), *Biochem. Soc. Trans.* **11:**134-136.
45. V. Warin, G. Basert, R. Fouret, G. Strecker, G. Spik, B. Fournet, and J. Montreuil, 1979, The crystal and molecular structure of O-α-D-mannopyranosyl-(1 $\rightarrow$ 3)-O-β-D-mannopyranosyl-(1 $\rightarrow$ 4)-2-acetamido-2-deoxy-α-D-glucopyranose, *Carbohydr. Res.* **76:**11-22.
46. J. Montreuil, B. Fournet, G. Spik, and G. Strecker, 1978, Etude théorique de la conformation spatiale des glycannes de la sérotransferrine humaine, *C. R. Acad. Sci. Paris* **287D:** 837-840.
47. K. Bock, J. Arnarp, and J. Lönngren, 1982, The preferred conformation of oligosaccharides derived from the complex-type carbohydrate portions of glycoproteins, *Eur. J. Biochem* **129:**171-178.
48. S. W. Homans, R. A. Dwek, D. L. Fernandes, and T. W. Rademacher, 1982, Solution conformation of the biantennary N-linked oligosaccharide of human serotransferrin using ^{1}H-NMR nuclear Overhauser effect measurements, *FEBS Lett.* **150:**503-506.
49. J. R. Brisson, and J. P. Carver, 1983, Solution conformation of asparagine-linked oligosaccharides: α(1-6)-linked moiety, *Biochemistry* **22:**3680-3686.
50. B. J. Sutton, and D. C. Phillips, 1983, The three-dimensional structure of the carbohydrate within the Fc fragment of immunoglobulin G, *Biochem. Soc. Trans.* **11:**130-132.
51. T. W. Rademacher, S. W. Homans, D. L. Fernandes, R. A. Dwek, T. Mizuochi, T. Taniguchi, and A. Kobata, 1983, Structural and conformational analysis of immunoglobulin-derivated N-linked oligosaccharides, *Biochem. Soc. Trans.* **11:**132-134.
52. J. R. Brisson, and J. P. Carver, 1983, Solution conformation of αD(1-3)- and αD(1-6)-linked oligomannosides using proton nuclear magnetic resonance, *Biochemistry* **22:**1362-1368.
53. J. R. Brisson, and J. P. Carver, 1983, Solution conformation of asparagine-linked oligosaccharides: α(1-2)-, α(1-3)-, β(1-2)- and β(1-4)-linked units, *Biochemistry* **22:**3671-3680.
54. J. Montreuil, H. Debray, P. Debeire, and P. Delannoy, 1983, Lectins as oligosaccharide receptors, *Falk Symp. N° 34,* MTP Press Ltd., Lancaster, pp. 239-258.
55. A. Douy, and B. Gallot, 1980, Synthesis and ordered structure of amphipatic block copolymers with a saccharide and a peptide block, *Biopolymers* **19:**493-507.
56. A. Douy, M. Gervais, and B. Gallot, 1980, Synthesis and mesomorphic structures of new amphipathic liposaccharides, *Makromol. Chem.* **181:**1199-1208.
57. M. Gervais, and B. Gallot, 1982, Glycan conformational change in mesomorphic structures of ternary systems: Liposaccharide/phospholipid/water, *Biochim. Biophys. Acta* **688:**586-596.

58. V. Warin, S. Perez, and J. Montreuil, unpublished results.
59. J. R. Brisson, and J. P. Carver, 1983, The relation of three-dimensional structure to biosynthesis in the N-linked oligosaccharides, *Can. J. Biochem. Cell Biol.* **61:**1067-1078.
60. B. Meyer, 1983, Structural analysis of the core oligosaccharides of glycoproteins by HSEA-calculations, *Proc. 7th Int. Symp. Glycoconjugates,* Lund-Ronneby, Rahms, Lund, p. 126.
61. Z. Q. Li, S. J. Perkins, and M. H. Loucheux-Lefebvre, 1983, α_1-Acid glycoprotein: A small-angle neutron scattering study of a human plasma glycoprotein, *Eur. J. Biochem.* **130:**275-279.
62. R. Kornfeld, and C. Ferris, 1975, Interaction of immunoglobulin glycopeptides with concanavalin A, *J. Biol. Chem.* **250:**2614-2619.
63. A. Pierce-Crétel, M. Izhar, Y. Nuchamowitz, G. Strecker, J. Montreuil, G. Spik, and D. Mirelman, 1983, Oligosaccharide structural specificity of the soluble agglutinin released from guinea pig colonic epithelial cells, *FEMS Lett.* **20:**237-242.
64. B. Gorinski, C. Horsburgh, P. F. Lindley, D. S. Moss, M. Parkar, and J. L. Watson, 1979, Evidence for the bilobal nature of diferric rabbit plasma transferrin, *Nature* **281:**157-158.
65. D. C. Wiley, I. A. Wilson, and J. J. Skehel, 1981, Structural identification of the antibody-binding sites of Hong Kong influenza haemagglutination and their involvement in antigenic variation, *Nature* **289:**373-378.
66. S. Ogata, T. Muramatsu, and A. Kobata, 1976, New structural characteristics of the large glycopeptides from transformed cells, *Nature* **259:**580-582.
67. P. Emmelot, W. P. Van Beek, and L. A. Smets, 1977, Cell surface carbohydrates and cell transformation: A general change signifying tumorigenicity, in *Membrane Alterations as Basis of Liver Injury,* H. P. Popper, L. Bianchi, and W. Reutter, eds., MTP Press Ltd., Lancaster, pp. 179-185.
68. L. Warren, C. A. Buck, and G. P. Tuszynski, 1978, Glycopeptide changes and malignant transformation. A possible role for carbohydrate in malignant behavior, *Biochim. Biophys. Acta* **516:**97-127.
69. D. L. Blithe, C. A. Buck, and L. Warren, 1980, Comparison of glycopeptides from control and virus-transformed baby hamster kidney fibroblasts, *Biochemistry* **19:**3386-3395.
70. S. Takasaki, H. Ikehira, and A. Kobata, 1980, Increase of asparagine-linked oligosaccharides with branched outer chains caused by cell transformation, *Biochem. Biophys. Res. Commun.* **92:**735-742.
71. L. A. Smets, and W. P. Van Beek, 1983, Glycoprotein modification after transformation with oncogenic viruses, *Falk Symp. N° 34,* MTP Press Ltd., Lancaster, pp. 325-332.
72. H. Debray, Zhu Qin, P. Delannoy, J. Montreuil, D. Dus, C. Radzikowski, B. Christensen, and J. Kieler, 1986, Altered glycosylation of membrane glycoproteins in human uroepithelial cell lines, *Int. J. Cancer* (in press).
73. P. Delannoy, H. Debray, and J. Montreuil, 1985, Structural modifications of fibronectin glycans related to the transformation of BHK 21/C13 cells by the hamster sarcoma virus. Hypothesis concerning the metastatic diffusion based on the masking of lectin-sites of fibronectin, *C. R. Acad. Sci. Paris* **301:**767-772.
74. J. Davoust, V. Michel, G. Spik, J. Montreuil, and P. Devaux, 1981, Flexibility of bi- and triantennary glycans of the N-acetyllactosaminic type, *FEBS Lett.* **125:**271-275.
75. E. F. Hounsell, M. Fukuda, M. E. Powell, T. Feizi, and S. I. Hakomori, 1980, A new O-glycosidically linked tri-hexosamine core structure in sheep gastric mucin: A preliminary note, *Biochem. Biophys. Res. Commun.* **92:**1143-1150.
76. B. M. Austen, and R. Marshall, 1974, Some stereochemical aspects of the carbohydrate moieties of glycoproteins, in *Méthodologies de la Structure et du Métabolismedes Glycoconjugués,* vol. 1, J. Montreuil, ed., Collogue n° 221 du CNRS, CNRS, Paris, pp. 465-482.

2

Lectins: An Overview

Nathan Sharon
The Weizmann Institute of Science
National Institutes of Health

Lectins are operationally defined as carbohydrate-binding proteins of nonimmune origin that agglutinate cells or precipitate polysaccharides and glycoproteins (1, 2). This definition implies that lectins are multivalent. Typically, lectins are oligomeric proteins with several sugar-binding sites per molecule, usually with one site per subunit. The emphasis on nonimmune origin is in order to distinguish lectins from anti-carbohydrate antibodies that may act as cell agglutinins. Furthermore, many lectins are found in plants and bacteria, which do not synthesize immunoglobulins. In addition, antibodies are structurally similar to each other, whereas lectins are structurally diverse. Lectins vary in molecular size, amino acid composition, metal requirements, and three-dimensional structure. In their structural diversity, lectins are akin to enzymes, but they are devoid of catalytic activity.

Although lectins have been known for nearly a century, they have become the focus of intense interest only during the last decade or two. This interest is evidenced by the rapid growth of the literature on the subject, which since the 1960s has increased nearly 20-fold. The total number of publications that deal with various aspects of lectins is fast approaching the 2,000/year mark. Included in these are numerous reviews on lectins, and several recent books (3, 4, 5).

There are many reasons for the current interest in lectins. Prominent among these is their usefulness in detecting and studying carbohydrates in solution and on cell surfaces. As a result of this characteristic, lectins have become indispensable tools in many areas of biological research. Another reason for the interest in lectins is the belief that they may serve as recognition determinants in a variety of biological systems in microorganisms, plants, and animals. Bacterial surface lectins mediate the sugar-specific adherence of the bacteria to epithelial cells, which is an essential prerequisite for infection. Plant lectins may act, on the one hand, as protective agents against fungal phytopathogens. On the other hand, they may mediate the attachment of nitrogen-fixing bacteria to the roots of leguminous plants. In animals, membrane-associated lectins appear to function in the clearance of glycoproteins from the circulatory system and in the intracellular translocation and targeting of glycoproteins.

With the expansion of our knowledge about lectins, it became apparent that they deserve attention in their own right: Some of them exhibit unusual structural properties, they constitute a convenient source of well-defined plant glycoproteins, and they are excellent models for studying protein-carbohydrate interactions. Furthermore, the availability of lectins from different species and tribes of the same family (of plants or bacteria) makes them suitable objects for taxonomic and phylogenetic studies, as well as for evolutionary correlations.

Contributing to the increasing popularity of lectins is not only their ease of purification (mainly by affinity chromatography on immobilized carbohydrates), but also their increased availability from commercial sources. Over 100 lectins have been purified to date, and more than 40 of these are available commercially.

This chapter will serve as an introduction to lectins, and as an aid for the reader of the subsequent chapters of this book, which focuses on vertebrate lectins. Examples will, however, be given mainly from studies with plant lectins. This is justified, since plant lectins are very common, easily purified, well characterized, and widely employed in many laboratories.

Only selected references will be given, primarily to reviews and books, to provide access to the voluminous literature on the subject. The reviews and monographs either deal with general aspects of lectins (3, 6-10) and the possible roles of lectins in nature (11-16), or emphasize specific topics such as lectins from plants (3, 17, 18, 19), microorganisms (5, 20, 21), slime molds and animals (22, 23), and membranes (24, 25, 26). Other reviews deal with applications of lectins to the study of glycoconjugates in solution and on cell surfaces (27-29), to immunology (10, 30, 31), to blood typing (32, 33), to microbiology (34, 35), to cell identification and separation (27, 31), and as histochemical probes (36, 37), or they discuss lectin-resistant animal cell mutants (38, 39). A monograph on concanavalin A describes the many applications of this lectin (40), but the procedures described are also applicable to other lectins. Techniques for lectin purification and some of their applications can be found in several volumes of *Methods of Enzymology* (41-44).

HISTORY

A survey of the history of lectins (6, 45) is of interest because it gives insight into the development of an area of research that has wide implications and increasing importance. It also provides information on some of the outstanding properties of lectins, and how we came to know about them.

Lectins were first designated as *hemagglutinins,* or more commonly as *phytohemagglutinins,* because they were found almost exclusively in plants. The earliest report on the occurrence in plants of proteins that agglutinate human or animal erythrocytes was published in 1888 by H. Stillmark, who described the hemagglutination caused by extracts of the toxic *Ricinus communis* seeds (castor bean). Soon after, hemagglutinins were found in extracts of the Jequirity bean, *Abrus precatorius.*

Lectins were first employed by Paul Ehrlich in the early 1890s as antigens. Ehrlich realized that the easily obtainable ricin and abrin would be more useful antigens than the bacterial toxins, such as that of diphtheria, which were popular research tools at the time. Although his preparations were very crude by present-day criteria (we now know that both the ricin and abrin were mixtures of a weakly agglutinating protein toxin and nontoxic agglutinin), he was able to establish with them some of the fundamental principles of immunology. Ehrlich found that immunization of animals with ricin and abrin resulted in the formation of serum proteins capable of specifically precipitating and neutralizing these substances, and that there was a quantitative relationship between the amount of antiserum and the amount of antigen it could neutralize. Moreover, anti-ricin did not protect the animals against the toxic effects of abrin, and vice versa. This lack of cross-reactivity provided clear evidence for the specificity of the immune response. Ehrlich also demonstrated that immunity to the toxins is transferred from the mother to the offspring by the blood during pregnancy and by the milk after birth.

The early results obtained by Stillmark indicated some selectivity in the lectin-mediated agglutination of red cells from different animals. This observation was corroborated and further extended by K. Landsteiner and H. Raubitschek in 1908. They established that the relative hemagglutinating activities of various seed extracts were quite different when tested with red blood cells from different animals. For instance, pea extracts were very effective in agglutinating rabbit erythrocytes but far less effective with sheep or pigeon erythrocytes, whereas human red blood cells were strongly agglutinated by bean extracts and weakly agglutinated by pea or lentil extracts.

The sugar specificity of lectins was discovered by J. B. Sumner in 1936. Nearly two decades before, in 1919, he crystallized concanavalin A from the jack bean and thus obtained for the first time a pure lectin. Together with S. F. Howell, he reported in 1936 that in addition to its ability to agglutinate cells such as erythrocytes and yeasts, concanavalin A precipitated glycogen from solution. They further showed that hemagglutination by concanavalin A was inhibited by cane sugar and suggested that this might be the consequence of a reaction of the plant protein with carbohydrates on the surface of the red cells.

In spite of the demonstration of the species specificity of lectins by Landsteiner and Raubitschek, it was presumed for several decades that these agglutinins were nonspecific. It was not until the late 1940s that W. C. Boyd and R. M. Reguera and, independently, K. O. Renkonen reported that certain seeds contain agglutinins specific for some human blood group antigens. Thus, a crude extract of lima bean *(Phaseolus limensis)* and of the seeds of tufted vetch *(Vicia cracca),* both specific for α-N-acetylgalactosamine (sugars are of the D configuration unless otherwise noted), agglutinated only blood type A erythrocytes; the lectin from winged pea *(Lotus tetragonolobus),* specific for L-fucose, agglutinated only type O(H) erythrocytes. These studies provided a further demonstration of the carbohydrate specificity of lectins as well as the first evidence that sugars are the immunodeterminants of human blood types. Additional lectins specific for blood types A, B, O(H), M, N, and other blood groups are now known (32, 33). In 1954 Boyd and Elizabeth Shapleigh proposed to name the various agglutinins *lectins* (from the Latin *legere,* to select or choose) on the basis of the ability of these substances to distinguish between different blood groups.

A major development in lectin research occurred in 1960, when P. C. Nowell discovered that PHA, the phytohemagglutinin of the red kidney bean *(Phaseolus vulgaris)* is mitogenic; that is, it possesses the ability to stimulate lymphocytes to undergo mitosis. This discovery has had a revolutionary impact on immunology in that it shattered the view, held until the early 1960s, that lymphocytes are dead-end cells that could neither divide nor differentiate further.

Much excitement has been created in the biomedical community by experiments, originally from the laboratory of J. C. Aub, on the ability of lectins to preferentially agglutinate malignant cells. It was initially believed that high susceptibility to agglutination by lectins was a property common to all malignant cells and that it was correlated with loss of contact inhibition and with tumorigenicity. This is now known not to be generally true. Furthermore, the molecular basis for increased agglutinability of certain malignant cells is far from being understood. Mainly for these reasons, the interest in this phenomenon has waned.

The discovery by Y. Reisner, in our laboratory in 1976, that peanut agglutinin binds to cortical mouse thymocytes and not to the medullary cells provided a starting point for the application of lectins to the separation of cells, especially lymphocyte subpopulations (31). This use of lectins has culminated in the application of soybean agglutinin for removal from human bone marrow of cells responsible for the lethal graft-versus-host disease and the isolation of stem cells suitable for transplantation across histocompatability barriers. By now, more than 50 children born with severe combined immunodeficiency disease have been successfully treated with bone marrow transplants prepared by this technique (46, 47). Studies on the application of the soybean agglutinin method to the treatment of leukemic patients are in progress.

Two recent advances have provided insights into the roles of lectins in nature. One of these was the discovery in 1974 by G. Ashwell and A. G. Morell and their coworkers of the first mammalian lectin, the hepatic binding protein specific for galactose, and the demonstration that it may be involved in the clearance of glycoproteins from the circulatory system. This advance was followed by the finding in animal cells of lectins with other sugar specificities. Some of these lectins function similarly to the hepatic binding protein, while others such as the mannose-6-phosphate binding protein are responsible for the intracellular routing of glycoproteins.

Another advance came from studies of bacterial adherence and agglutination reactions. Since the early 1950s, it has been known that *Escherichia coli* and related organisms adhere to epithelial cells and agglutinate erythrocytes in a reaction that is mannose-specific. However, only in the 1970s was it realized that this agglutination is mediated by bacterial surface lectins, in the form of pili or fimbriae, which also mediate the adherence of the bacteria to epithelial cells (20, 21). Direct evidence for the involvement of these lectins in the initiation of infection was first obtained in 1979, when we showed that urinary tract infection in mice by mannose-specific *E. coli* could be prevented by methyl α-mannoside. These studies raise the possibility that sugar inhibitors of lectin-mediated bacteria adherence may prove to be useful in protection against natural infection.

LECTIN DETECTION AND PURIFICATION

The simplest way to detect the presence of a lectin in a crude preparation is by showing that the latter agglutinates erythrocytes or causes precipitation of a polysaccharide or glycoprotein, and that these reactions are specifically inhibited by mono- or oligosaccharides; that is, they are sugar-specific (8, 27, 48).

Hemagglutination is usually determined by the serial dilution technique using erythrocytes from humans or rabbits. Occasionally, erythrocytes that have been treated with trypsin or sialidase are employed, since such cells are often more sensitive to agglutination than untreated cells. The cells may also be treated with glutaraldehyde or formaldehyde to stabilize them and to provide a standard preparation that can be used for long periods of time. The same methods are also used to monitor the activity of lectins during purification.

Most of the lectins that have been studied in detail are easily soluble in aqueous solution. In such cases, dilute buffers are sufficient for their extraction. For solubilization of lectins associated with membranes, harsher treatments are required, such as the use of detergents, which may damage the carbohydrate-binding activity of the lectins. Lectins bound to endogenous carbohydrates may be solubilized by treatment with appropriate sugars.

In the past, lectins were purified by traditional methods of protein fractionation. Currently, affinity chromatography is the method of choice (27, 49). Knowledge of the sugar specificity of a lectin, which can be obtained by using a crude lectin preparation, permits the design of a suitable affinity chromatography procedure. Whenever possible, commercially available supports, such as Sephadex and Sepharose, are employed. In cases of lectins with sugar specificities for which no ready-made adsorbents are available, affinity columns are prepared by coupling the appropriate carbohydrate to an insoluble support. The carbohydrate can be introduced in the form of a glycoprotein, a glycopeptide, or a suitable synthetic derivative of the specific sugar.

Most sources contain one lectin, but examples are known of plant seeds that contain two (or more) lectins that differ in their sugar specificities. Thus, two lectins have been isolated from the seeds of gorse *(Ulex europeus),* one specific for N-acetylglucosamine and the other specific for L-fucose. Two lectins with distinct sugar specificities have also been isolated from *V. cracca,* one specific for N-acetylgalactosamine and the other specific for mannose. Out of the complex mixture of lectins present in seeds of *Griffonia simplicifolia* (also known as *Bandeiraea simplicifolia*), several lectins have been isolated, which are specific for N-acetylgalactosamine, galactose, and N-acetylglucosamine.

Individual lectins frequently occur as a group of closely related proteins or isolectins that are separable by ion exchange chromatography, for example, as was first demonstrated for soybean agglutinin. Multiple forms of wheat germ agglutinin and of peanut agglutinin have been described. The occurrence of isolectins is under genetic control.

CARBOHYDRATE SPECIFICITY

Studies of the carbohydrate specificity of lectins are usually done by the hapten inhibition technique (8, 27, 48). Currently, carbohydrate specificity is being examined to an increasing extent by physicochemical methods, such as spectrophotometry, fluorimetry, and nuclear magnetic resonance, as well as by rapid reaction techniques. These methods provide a deeper insight into the nature of the combining sites of lectins, and may also lead to a better understanding of the binding process.

The sugar specificity of a lectin is defined in terms of the monosaccharide that is the best inhibitor of lectin-induced hemagglutination or of glycoconjugate precipitation. Lectins can be grouped according to their monosaccharide specificity (Table 2-1). All of the monosaccharides listed in the table are common constituents of animal glycoconjugates. This may be a reflection of the method commonly employed for the detection of lectins; that is, hemagglutination. As a result, lectins specific for sugars other than those present on erythrocytes might be overlooked.

Lectins belonging to the same group may differ in their fine specificity towards substituted monosaccharides, oligosaccharides, or cells. For instance, most lectins specific for mannose combine well with glucose and to some extent with N-acetylglucosamine. Exceptions are the mannose-specific lectins from *E. coli,* which do not react with glucose or N-acetylglucosamine. Although concanavalin A and pea lectins are both classified as mannose-specific, concanavalin A is poorly inhibited by 3-O-methylglucose, whereas pea lectin is strongly inhibited by this sugar. Many lectins specific for galactose also interact well with N-acetylgalactosamine, while others (e.g., peanut agglutinin) exhibit strict specificity for the former monosaccharide. More important, many lectins exhibit greater specificity for oligosaccharides than for monosaccharides. Thus, although peanut agglutinin and *Erythrina cristagalli* agglutinin are both galactose-specific, they have considerable greater affinity for the disaccharides Galβ3GalNAc and Galβ4GlcNAc, respectively. These disaccharides are common constituents of the carbohydrate units of glycoproteins. The relative affinity of Manα3(Manα6)Man to concanavalin A is 130-fold greater than that of methyl α-mannoside (50). The mannose-specific lectin of *E. coli* is inhibited strongly by oligosaccharides such as Manα3Manβ4GlcNAc (21). Both trisaccharides are glycoprotein constituents.

For some lectins no efficient monosaccharide inhibitors are known, and they are inhibited only by oligosaccharides. The minimal structural unit that inhibits L-PHA, the leukoagglutinating isolectin of PHA is the disaccharide GlcNAcβ2Man; the most active inhibitor is the pentasaccharide Galβ4GlcNAcβ2(Galβ4GlcNAcβ6)Man. The lectin from the mushroom *(Agaricus bisporus)* is inhibited by glycopeptides from erythrocyte membranes, fetuin, and immunoglobulin A. It is of interest that this lectin, the blood group

Table 2-1. Specificity and Properties of Purified Lectins

Group	Monosaccharide Specificity	Lectin Source	Molecular Weight	Oligosaccharide Specificity
1	Mannose	Jack bean[a]	108,000	Manα(1→3)[Manα(1→6)]Man
		Fava bean	52,500	
		Lentil	46,500	
		Pea	46,000	
		Escherichia coli	[17,000]N	Manα(1→3)Manβ(1→4)GlcNAc
2	Galactose	Castor seed	120,000	
		Erythrina cristagelli	58,000	Galβ(1→4)GlcNAc
		Peanut	110,000	Galβ(1→3)GalNAc
3	N-Acetyl-galactosamine	Dolichos biflorus	110,000	GalNAcα(1→3)GalNAc
		Helix pomatia	79,000	
		Soybean	120,000	
		Vicia villosa	140,000	GalNAcα(1→3)Gal
4	N-Acetyl-glucosamine	Wheat germ	36,000	GlcNAcβ(1→4)GlcNAcβ(1→4)GlcNAc
		Griffonia simplicifolia[b]	113,000	
		Ulex europeus[b]		
		Potato	100,000	GlcNAcβ(1→4)GlcNAcβ(1→4)GlcNAc β(1→4)GlcNAc
5	L-Fucose	*Lotus tetragonolobus*[c]		
		Ulex europeus[c]		
6	Sialic acid	Horseshoe crab	400,000	

[a]Lectin known as concanavalin A.
[b]Lectin II.
[c]Lectin I.

N-specific lectin of *Vicia graminea*, and the lectin of *Vicia villosa*, appear to recognize carbohydrate sequences together with the amino acid or peptide to which the latter are linked. The most efficient inhibitor of the potato lectin is the β1-4-linked pentasaccharide of N-acetylglucosamine, which is 2.5, 25, and 50 times more active than the corresponding tetramer, trimer, and dimer, respectively; free N-acetylglucosamine is not inhibitory at concentrations up to 0.2 M. The lectin from *Datura stramonium* is also inhibited by oligomers of N-acetylglucosamine and not by the monomer.

MOLECULAR PROPERTIES

Lectins comprise a heterogeneous group of proteins with a wide spectrum of molecular properties (8, 27, 48). Their molecular weights (Mr) range from over 30,000 daltons (32,000 Da for the electric eel lectin and 36,000 Da for wheat germ agglutinin) to 400,000 Da (horseshoe crab lectin). The molecular weights of the mannose-specific lectins of *E. coli* are even higher, since they consist of long arrays of pilin monomers, each of Mr 17,000 Da.

Lectins usually consist of two or four subunits, although the number of subunits may be as high as 20 (for the horseshoe crab lectin) or more. The subunits are in most cases identical, but lectins composed of nonidentical subunits are also known. In some lectins (e.g., from soybean and lima bean) the subunits differ only slightly, but in others (such as those from pea, lentil, and fava bean) the subunits differ markedly in size and other properties.

Usually each subunit has a single sugar-binding site, which in most cases is identical in all subunits of the same lectin. However, cases are known in which subunits of the same lectin differ in sugar specificity. The group I lectins from *G. simplicifolia* consist of a family of five closely related proteins, each of which is a tetramer of one or two types of subunit, A and B. The structure of the five isolectins can be represented as A_4, A_3B_1, A_2B_2, A_1B_3, and B_4. Subunit A is specific for α-N-acetylgalactosamine (but also reacts with α-galactose), whereas subunit B is specific for α-galactose only. The structure of *G. simplicifolia* isolectins is analogous to that of PHA isolectins, which consist of two classes of subunits: E and L. The PHA isolectins differ in sugar specificity and biological properties: E_4 is the most potent erythroagglutinin, L_4 is the most potent lymphocyte mitogen, and intermediate forms possess both properties. There is a striking similarity in the carbohydrate and overall amino acid composition of the E and L subunits and in their tryptic peptide maps. The differences in biological properties of the erythroagglutinating (E_4) and the mitogenic (L_4) isolectins are probably due to small differences in the primary sequences of their subunits.

With a few exceptions (e.g., concanavalin A, wheat germ agglutinin, and peanut agglutinin, which are devoid of covalently bound sugar), most lectins are glycoproteins, some of which contain up to 50% carbohydrate. On the basis of carbohydrate composition, two types of glycoprotein lectins can be discerned: (a) those containing mainly N-acetylglucosamine and mannose (with or without L-fucose and xylose—many legume lectins belong to this group); and (b) those containing L-arabinose or galactose, as is the case with potato and tomato lectins. Of the former type, soybean agglutinin has been most thoroughly investigated, and the structure of its carbohydrate chain has been fully established. This structure is identical to that of the N-linked $Man_9GlcNAc_2$ oligosaccharide found in animal glycoproteins. It appears that the carbohydrate is not required for the sugar binding or biological activities of lectins, since wherever it was modified or removed from the protein these activities were retained.

In spite of their structural diversity, lectins can be grouped into families with similar features (51). Lectins from plants of the Solanaceae family (e.g., potato and tomato) are characterized by an unusually high carbohydrate content and the presence of hydroxyproline residues. Many of the legume lectins are rich in aspartic acid and serine, and contain little or no methionine or cysteine. In contrast, the cereal lectins examined to date (e.g., from wheat or barley) are basic proteins and have a large number of cysteine residues (up to 20%).

Although the most abundant lectins appear to be localized in the same tissue in different plants of the same family, this tissue is not necessarily identical in different families: The legume lectins are found in the cotyledons, the Solanaceae lectins are found in modified stems, and the cereal lectins are found in the embryos. It is thus possible that in different organisms, whether plants or animals, as well as in different tissues within the same organism, distinct types of lectins are present that have been adapted to perform different functions (51).

LECTINS AND ANTIBODIES

As was mentioned earlier, lectins differ from antibodies in their overall structure as well as in their origin. There are, however, remarkable similarities between the carbohydrate-binding sites of lectins and antibodies. Both lectins and antibodies react with mono- and oligosaccharides in a reversible manner, with similar association constants—usually in the range of from 10^3 to 10^5 M^{-1}. In both systems, binding is due primarily to hydrophobic interactions and hydrogen bonding, but the relative contribution of the two types of bond to the total energy of interaction is not known. Electrostatic forces do not seem to contribute to the lectin-carbohydrate or antibody-carbohydrate interactions, since most of the ligands are neutral.

Precipitation of glycoproteins and polysaccharides by lectins is similar in many respects to the well-known precipitation reaction between antibody and antigen. It is specific, it depends on the concentration of the reactants, it exhibits a maximum at an optimal ratio of the reactants, and it may be specifically inhibited by low-molecular-weight haptens. Thus it is clear why many of the methods used in lectin research are based on immunochemical techniques.

Table 2-2. Effects of Lectins on Cells and Organisms

Agglutination of erythrocytes and other types of cells
Mitogenic stimulation of lymphocytes
Induction of γ interferon formation by human leukocytes
Generation of suppressor cells
Mediation of killing of target cells by lymphocytes and macrophages
Enhancement of phagocytosis of yeasts and bacteria by macrophages
Insulin-like effects on fat cells
Toxicity to cells and animals
Promotion of cell adhesion and spreading
Induction of histamine release from basophil and mast cells
Inhibition of fungal growth

EFFECTS ON CELLS

Lectins exert a variety of effects on cells (Table 2-2; see also ref. 52). Among these, agglutination is the most easily detectable. For agglutination to occur, the bound lectin must form multiple cross-bridges between opposing cells. This seemingly simple process is affected by many factors (30), such as molecular properties of the lectin (e.g., number of saccharide-binding sites, molecular size), cell surface properties (e.g., number and accessibility of receptor sites, membrane fluidity), and the metabolic state of the cells. In addition, agglutination is affected by such assay conditions as temperature, cell concentration, and mixing, so lectins may bind to cells without causing agglutination. This limits the value of agglutination by lectins as a means for obtaining information about the structure and architecture of cell surfaces. When agglutination does occur and is inhibited by an appropriate sugar, it can be taken as an indication that carbohydrate structures for which the lectin is specific are present on the surface of the cells. It is also useful for assessing the effect of various treatments (e.g., with enzymes or sugar-modifying reagents) on the lectin receptors. More important selective agglutination for a particular cell population in a heterogeneous cell preparation provides a simple and efficient means for cell fractionation (31).

A dramatic effect of the interaction of many lectins with cells is mitogenic stimulation; that is, the triggering of quiescent, nondividing lymphocytes into a state of growth and proliferation. PHA, the first mitogenic agent found, is still widely used for this purpose. An interesting mitogen is the hepatic (carbohydrate)-binding protein (HBP), a lectin isolated from rabbit liver cell membranes. Most mitogenic lectins stimulate only the thymus-dependent population of lymphocytes (T cells) and are inactive or inhibitory for mitosis of the other class of lymphocytes, the immunoglobulin-producing (or B) cells. The hepatic binding protein is an example of a lectin that stimulates only sialidase-treated T lymphocytes. Lectins also differ in their ability to stimulate lymphocytes of different animal species.

In contrast to stimulation by antigens, in which specific clones of lymphocytes comprising less than 0.1% of the cells are induced to proliferate, lectins activate multiple lymphocyte clones irrespective of their antigen specificity. Therefore, the percentage of responding cells is rather high—up to 70-80% of an appropriate lymphocyte population. This is the main basis for the wide use of mitogenic lectins for studies of the mechanism of lymphocyte stimulation. Mitogenic stimulation of lymphocytes leads also to the release of many lymphokines. These include interleukin 2, γ-interferon, macrophage inhibitory factor, and lymphotoxin. Lentil lectin is a particularly good inducer of γ-interferon production by human peripheral blood mononuclear cells.

Another effect of mitogenic lectins on lymphocytes is the induction of the generation of suppressor cells capable of inhibiting the activities of T and B cells in vitro. In addition, in the presence of mitogenic lectins, lymphocytes and macrophages will lyse unrelated target cells. This phenomenon is known as lectin-dependent cytotoxicity (LDCC). Lectins that form bridges between bacteria and macrophages may also induce phagocytosis of the target cells.

Several lectins, such as concanavalin A and wheat germ agglutinin, mimic the effects of insulin on adipocytes and other target tissues. Insulinomimetic activities were also observed in vivo. It is possible that the lectins exert their insulinomimetic action by binding to the carbohydrate units of the insulin receptor, itself a membrane glycoprotein.

Several lectins are toxic to mammalian cells. These include concanavalin A, wheat germ agglutinin, PHA, and the lectin from *Robinia pseudoacacia*. Their toxicities are, however, much lower (1,000-2,000 times in vivo) than those of the toxins ricin and abrin, which are weak agglutinins. Ricin and abrin consist of two chains joined by disulfide bonds. One chain (B) possesses the carbohydrate-binding site specific for galactose or N-acetylgalactosamine, whereas the other chain (A) inhibits protein synthesis in cell-free systems, thus representing the toxic moiety of the molecule; only the intact molecule is active on cells. Subsequent to binding to cell surface sugars via the B chain, the toxins are taken up by the cell where the A chain inhibits protein synthesis by interfering with peptide chain elongation on polyribosomes.

ROLE IN NATURE

Many different biological roles have been proposed for lectins, several of which are discussed in detail in other chapters of this book. Some of the more likely possibilities are summarized in Table 2-3. The common assump-

Table 2-3. Possible Roles of Lectins

Possible Roles of Lectins
Plants
Attachment of nitrogen-fixing bacteria to legumes
Protection against phytopathogenic fungi
Regulation of cell wall extension
Bacteria
Adherence to host cells to initiate infection
Mediators of nonimmune phagocytosis
Slime molds
Determinants of cell adhesion
Animals
Endocytosis and intracellular translocation of glycoproteins
Regulation of cell migration and adhesion
Mediators of phagocytosis

tion is that lectins function as recognition molecules, both in intercellular systems (homotypic and heterotypic) and intracellular ones. In most cases, however, the evidence for the proposed roles is inconclusive. For example, in the slime mold (*Dictyostelium discoideum*), synthesis of lectins occurs as the single cells form aggregates and develop into slugs, spores, and stalks (22, 23). It has therefore been proposed that the lectins function as cell adhesion molecules, by binding to glycoconjugates on adjacent cells. However, the facts that antibodies to the lectins do not block adhesion in an in vitro assay, and that slime mold mutants defective in the synthesis of the lectins are able to undergo aggregation, argues against this possibility.

The proposal that lectins are involved in the first step of the association between legumes and nitrogen-fixing bacteria is based primarily on the observation that a lectin from a particular legume binds only to the rhizobial species that normally infects this legume, and not to bacteria that infect other legumes (12, 13, 15). It has been found, however, that many *Rhizobium* strains nodulate soybean roots but do not bind detectable amounts of soybean lectin, and that soybean lines that lack the lectin can be nodulated by symbiotic *Rhizobium japonicum.* Thus, although some legume lectins fulfill many criteria that are expected of a cell recognition molecule involved in the nitrogen-fixing *Rhizobium*-legume symbiosis, the evidence for the role of lectins in this process is inconclusive (53).

Undoubtedly, the best evidence for a role played by lectins comes from studies of the animal membrane lectins involved in endocytosis and intracellular transport and of lectin-mediated bacterial adherence to mammalian cells (20, 21). As has been mentioned, the latter lectins most likely serve as recognition determinants in infection, since experimental infection in mice by lectin-carrying bacteria could be prevented by suitable sugars.

SELECTED APPLICATIONS

Table 2-4 lists the major applications of lectins in biomedical research (27, 54). Lectins serve for the facile and sensitive detection of individual glycoproteins in complex mixtures that have been separated by polyacrylamide gel electrophoresis, either by incubating the lectin directly with the gels, or with gel blots. They can similarly be used as an aid for the identification of glycolipids separated by thin-layer chromatography. For preparative purposes, affinity chromatography of glycoproteins on immobilized lectins is widely employed. Since lectins may bind sugars in the presence of mild detergents, they are particularly useful for the isolation and purification of membrane glycoproteins. The advantage of lectins over other more specific molecules, such as antibodies, for the purification of glycoproteins is due to the fact that materials bound specifically to lectin columns can be eluted with simple sugars under mild conditions. For example, wheat germ agglutinin is routinely employed for the partial purification of the insulin receptor from human placenta and cultured lymphocytes (29). Occasionally, a single glycoprotein can be obtained in purified form by fractionation on an immobilized lectin. A case in point is laminin, a noncollagenous glycoprotein that is a major component of basement membrane and the intracellular matrix. This glycoprotein has been purified from extracts of EHS sarcoma cells by affinity chromatography on immobilized *G. simplicifolia* lectin, which is specific for α-galactosyl residues.

Affinity chromatography on lectins permits the separation of glycoproteins that differ only slightly in their carbohydrate composition or the structure of their carbohydrate units. Ovalbumin, a glycoprotein containing a single carbohydrate attachment site, has recently been separated into eight subfractions that differ in their carbohydrate chains (55). Three of the subfractions were homogeneous with respect to the structure of their carbohydrate, and did not exhibit the microheterogeneity characteristic of glycoproteins. In a related application, two biosynthetic intermediates of ovalbumin were purified by affinity chromatography on immobilized concanavalin A (56). Glycopeptides or oligosaccharides obtained by cleavage of purified glycoproteins can be similarly separated into individual species.

In addition to molecular separations, lectins are employed to an increasing extent for the identification, enumeration, and fractionation of cells into biologically distinct subpopulations (31). Prominent examples are the fractionation of mouse and human thymocytes by peanut agglutinin into cortical (immature) and medullary (mature) cells, the separation of T and B cells from mouse spleen by soybean agglutinin, and the isolation of a fraction of human bone marrow enriched in stem cells for transplantation from immunologically nonidentical donors into immunodeficient and leukemic

Table 2-4. Applications of Lectins

Applications
Isolation of glycoproteins and glycopeptides
Structural studies of carbohydrates
Investigation of protein-carbohydrate interactions
Identification of cell-surface sugars
Characterization of membrane organization and dynamics
Detection of cell surface changes in differentiation, development, and malignancy
Blood typing and detection of secretors
Cell identification and separation
Selection of lectin-resistant mutants

children. Some lectins are routinely employed in blood banks as an aid in blood typing, and for the identification of *secretors,* individuals who secrete blood group substances in their saliva or other body fluids (32, 33).

Hybrid molecules made by linking the toxic lectin ricin to antibodies are being used to eliminate T cells and tumor cells from bone marrow. These reagents bind via the antibody and can kill antigen-positive cells without killing nontarget cells. Initial clinical results using these reagents to treat bone marrow in HLA-matched, allogeneic transplants indicate that T cells can be eliminated and the stem cells will reconstitute the immune system of the recipient (57).

There are numerous reports on the application of lectins as histochemical probes for specific carbohydrate residues and for characterization of cells (36, 37). For this purpose, individual lectins are conjugated to fluorescent dyes, enzymes, or electron-dense substances. In laboratory medicine and pathology, such derivatives have served as markers to identify different cell types in histologic sections, to predict the clinical course of patients with transitional cell carcinomas of the urinary bladder, to distinguish between hormone-sensitive and -insensitive breast carcinomas, and to subtype hemopoietic cells. Lectins can also be used on tissue sections to distinguish between several carbohydrate storage diseases by identifying abnormally stored oligosaccharides. Thus, concanavalin A may be used for the diagnosis of mannosidosis, *U. europeus* agglutinin I for fucosidosis, and wheat germ agglutinin for sialidosis.

CONCLUDING REMARKS

We shall soon be celebrating the 100th anniversary of the discovery of the first known lectin—*R. communis* agglutinin. Research on lectins has enriched our knowledge in different areas of biochemistry, plant physiology, microbiology, and immunology. It has also provided us with important tools for biological and medical investigations in areas as diverse as bacterial typing and bone marrow transplantation. As we learn more about the properties of lectins, new and unexpected applications can be anticipated. We can also look forward to definitive answers to the puzzling question of their role in nature.

ACKNOWLEDGMENT

Thanks are due to Dr. Steven L. Spitalnik for his suggestions and criticism.

REFERENCES

1. I. J. Goldstein, R. C. Hughes, M. Monsigny, T. Osawa, and N. Sharon, 1980, What should be called a lectin? *Nature* **285:**66.
2. N. Sharon, and H. Lis, 1972, Lectins: Cell-agglutinating and sugar-specific proteins, *Science* **177:**949-959.
3. I. J. Goldstein, and M. E. Etzler, eds., 1983, Chemical taxonomy, molecular biology, and function of plant lectins, *Progress in Clinical and Biological Research,* vol. 138, Alan R. Liss, New York.
4. I. E. Liener, N. Sharon, and I. J. Goldstein, eds., 1986, *The Lectins: Properties, Functions and Applications in Biology and Medicine,* Academic Press, New York, in press.
5. D. Mirelman, ed., 1986, *Microbial Lectins and Agglutinins,* John Wiley, New York.
6. E. R. Gold, and P. Balding, 1975, *Receptor-Specific Proteins: Plant and Animal Lectins,* Excerpta Medica, Amsterdam.
7. N. Sharon, 1977, Lectins, *Sci. Am.* **236**(6):108-119.
8. I. J. Goldstein, and C. E. Hayes, 1978, The lectins: Carbohydrate-binding proteins of plants and animals, *Adv. Carbohydr. Chem. Biochem.* **35:**127-340.
9. J. Roth, 1978, The lectins, molecular probes in cell biology and membrane research, *Exp. Pathol. Suppl.* **3:**1-186.
10. M. E. A. Pereira, and E. A. Kabat, 1979, Immunochemical studies on lectins and their application to the fractionation of blood group substances and cells, *CRC Crit. Rev. Immunol.* **1:**33-78.
11. N. Sharon, 1979, Possible functions of lectins in microorganisms, plants and animals, in *Glycoconjugate Research,* R. W. Jeanloz and J. D. Gregory, eds., Academic Press, New York, pp. 459-491.
12. E. L. Schmidt, 1979, Initiation of plant root-microbe interactions, *Annu. Rev. Microbiol.* **33:**355-376.
13. E. L. Schmidt, and B. B. Bohlool, 1981, The role of lectins in symbiotic plant-microbe interactions, in *Plant Carbohydrates,* vol. II, W. Tanner and F. A. Loewus, eds., Encyclopedia Plant Physiology 13B, Springer-Verlag, New York, pp. 627-657.
14. L. Sequeira, 1978, Lectins and their role in host-pathogen specificity, *Annu. Rev. Phytopathol.* **316:**453-481.
15. F. B. Dazzo, and J. E. Sherwood, 1983, Trifoliin A: A *Rhizobium*-recognition lectin in white clover, in *Chemical Taxonomy, Molecular Biology, and Function of Plant Lectins,* I. J. Goldstein and M. E. Etzler, eds., Progress in Clinical and Biological Research, vol. 138, Alan R. Liss, New York, pp. 209-223.

16. N. Sharon, 1984, Surface carbohydrates and surface lectins are recognition determinants in phagocytosis, *Immunol. Today* **5:**143-147.
17. I. E. Liener, 1976, Phytohemagglutinins (phytolectins), *Annu. Rev. Plant Physiol.* **27:**291-319.
18. H. Lis, and N. Sharon, 1981, Lectins in higher plants in *Proteins and Nucleic Acids,* A. Marcus, ed., The Biochemistry of Plants: A Comprehensive Treatise, vol. 6, Academic Press, New York, pp. 371-447.
19. H. Kauss, 1981, Lectins and their physiological role in slime molds and in higher plants, in *Plant Carbohydrates,* vol. 2, W. Tanner, and F. A. Loewus, eds., Encyclopedia Plant Physiology 13B, Springer-Verlag, New York, pp. 627-657.
20. N. Sharon, Y. Eshdat, F. J. Silverblatt, and I. Ofek, 1981, Bacterial adherence to cell surface sugars, in *Adhesion and Micro-Organism Pathogenicity,* K. Elliott, M. O'Connor, and J. Whelan, eds., Ciba Symposium Series 80, Pitman Medical, Tunbridge Wells, United Kingdom, pp. 119-135.
21. N. Sharon, 1984, Lectin-like bacterial adherence to animal cells, in *Attachment of Organisms to the Gut Mucosa,* vol. 1, E. C. Boedeker, ed., CRC Press, Boca Raton, Fla., pp. 129-147.
22. S. H. Barondes, 1981, Lectins: Their multiple endogenous cellular functions, *Annu. Rev. Biochem.* **50:**207-231.
23. S. H. Barondes, 1984, Soluble lectins: A new class of extracellular proteins, *Science* **223:**1259-1264.
24. G. Ashwell, and J. Harford, 1982, Carbohydrate-specific receptors of the liver, *Annu. Rev. Biochem.* **51:**531-554.
25. M. Monsigny, C. Kieda, and A. C. Roche, 1983, Membrane glycoproteins/glycolipids and membrane lectins as recognition signals in normal and malignant cells, *Biol. Cell* **47:**95-110.
26. M. Monsigny, ed., 1984, The role of carbohydrates in cell recognition: Endogenous lectins, *Biol. Cell* **51**(2):113-294.
27. H. Lis, and N. Sharon, 1984, Lectins: Properties and applications to the study of complex carbohydrates in solution and on cell surfaces, in *Biology of Carbohydrates,* vol. 2, V. Ginsburg and P. W. Robbins, eds., John Wiley, New York, pp. 1-85.
28. R. Lotan, and G. L. Nicolson, 1979, Purification of cell membrane components by lectin affinity chromatography, *Biochim. Biophys. Acta* **559:**329-376.
29. J. A. Hedo, 1984, Lectins as tools for the purification of membrane receptors, in *Receptor Purification Procedures,* vol. 1, C. Venter and L. Harrison, eds., Alan R. Liss, New York, pp. 45-60.
30. H. Lis, and N. Sharon, 1977, Lectins: Their chemistry and application to immunology, in *The Antigens,* vol. 4, M. Sela, ed., Academic Press, New York, pp. 429-529.
31. N. Sharon, 1983, Lectin receptors as lymphocyte surface markers, in *Advances in Immunology,* vol. 34, F. J. Dixon and H. G. Kunkel, eds., Academic Press, New York, pp. 213-298.
32. G. W. G. Bird, 1978, Significant advances in lectins and polyagglutinable red cells, in *Proceedings of XII Congress of the International Society of Haematology and XV Congress of the International Society of Blood Transfusion.* Paris, pp. 87-95.
33. W. J. Judd, 1980, The role of lectins in blood group serology, *CRC Crit. Rev. Clin. Lab. Sci.* **1:**171-214.
34. T. G. Pistole, 1981, Interaction of bacteria and fungi with lectins and lectin-like substances, *Annu. Rev. Microbiol.* **35:**85-112.
35. R. J. Doyle, and K. Keller, 1984, Lectins in diagnostic microbiology: A review, *Eur. J. Clin. Microbiol.* **3:**4-9.
36. J. Schrevel, D. Gros, and M. Monsigny, eds., 1981, Cytochemistry of cell glycoconjugates, *Prog. Histochem. Cytochem.* **14**(2).

37. A. J. C. Leathem, and N. J. Atkins, 1983, Lectin binding to paraffin sections, in *Techniques in Immunocytochemistry,* vol. 2, G. R. Bullock and P. Petrusz, eds., Academic Press, London, pp. 39-70.
38. E. B. Briles, 1982, Lectin-resistant cell surface variants of eukaryotic cells, *Int. Rev. Cytol.* **75:**101-165.
39. P. Stanley, 1980, Surface carbohydrate alterations of mutant mammalian cells selected for resistance to plant lectins, in *The Biochemistry of Glycoproteins and Proteoglycans,* W. J. Lennarz, ed., Plenum, New York, pp. 161-189.
40. H. Bittiger, and H. P. Schnebli, eds., 1976, *Concanavalin A as a Tool,* Wiley, London.
41. V. Ginsburg, ed., 1972, Complex carbohydrates, part B, in *Methods in Enzymology 28,* Academic Press, New York.
42. V. Ginsburg, ed., 1978, Complex carbohydrates, part C, in *Methods in Enzymology 50,* Academic Press, New York.
43. V. Ginsburg, ed., 1982, Complex carbohydrates, part D, in *Methods in Enzymology 83,* Academic Press, New York.
44. W. B. Jakoby, and M. Wilchek, eds., 1974, Affinity techniques: Enzyme purification, part B, in *Methods in Enzymology 34,* Academic Press, New York.
45. J. Kocourek, 1986, Historical background, in *The Lectins: Properties, Functions and Applications in Biology and Medicine,* I. E. Liener, N. Sharon, and I. J. Goldstein, eds., Academic Press, New York, pp. 1-32.
46. R. A. Good, N. Kapoor, and Y. Reisner, 1983, Bone marrow transplantation—an expanding approach to treatment of many diseases, *Cell. Immunol.* **82:**36-54.
47. W. Friedrich, S. F. Goldmann, W. Ebell, R. Blütters-Sawatzki, G. Gaedicke, A. Raghavachar, H. H. Peter, B. Belohradsky, W. Kreth, B. Kubanek, and E. Kleihauer, 1985, Severe combined immunodeficiency: Treatment by bone marrow transplantation in 15 infants using HLA-haploidentical donors, *Eur. J. Pediatr.* **144:**125-130.
48. I. J. Goldstein, and R. D. Poretz, 1986, Isolation, physicochemical characterization, and carbohydrate-binding specificity of lectins, in *The Lectins: Properties, Functions and Applications in Biology and Medicine* I. E. Liener, N. Sharon, and I. J. Goldstein, eds., Academic Press, New York, pp. 33-248.
49. H. Lis, and N. Sharon, 1981, Affinity chromatography for the purification of lectins, *J. Chromatogr.* **215:**361-372.
50. F. Brewer, L. Bhattacharyya, R. D. Brown, III, and S. H. Koenig, 1985, Interactions of concanavalin A with a trimannosyl oligosaccharide fragment of complex and high mannose type glycopeptides, *Biochem. Biophys. Res. Commun.* **127:**1066-1071.
51. M. E. Etzler, 1983, Introduction, in *Chemical Taxonomy, Molecular Biology, and Function of Plant Lectins,* I. J. Goldstein and M. E. Etzler, eds., Progress in Clinical and Biological Research vol. 138, Alan R. Liss, New York, pp. 1-7.
52. H. Lis, and N. Sharon, 1986, Biological properties of lectins, in *The Lectins: Properties, Functions and Applications in Biology and Medicine,* I. E. Liener, N. Sharon, and I. J. Goldstein, eds., Academic Press, New York, pp. 265-291.
53. S. G. Peuppke, 1983, Soybean Lectin: Does it have an essential role in the Rhizobium-soybean symbiosis?, in *Chemical Taxonomy, Molecular Biology, and Function of Plant Lectins,* I. J. Goldstein, and M. E. Etzler, eds., Progress in Clinical and Biological Research, vol. 138, Alan R. Liss, New York, pp. 225-236.
54. H. Lis, and N. Sharon, 1986, Applications of lectins, in *The Lectins: Properties, Functions and Applications in Biology and Medicine,* I. E. Liener, N. Sharon, and I. J. Goldstein, eds., Academic Press, New York, pp. 293-370.

55. Y. Kato, H. Iwase, and K. Hotta, 1984, Preparation of eight ovalbumin subfractions by combined lectin affinity chromatography, *Anal. Biochem.* **138:**437–441.
56. Y. Kato, H. Iwase, and K. Hotta, 1984, Purification and characterization of two biosynthetic intermediates of ovalbumin, *J. Biochem. (Tokyo)* **95:**455–463.
57. A. H. Filipovich, D. A. Vallera, R. J. Youle, R. R. Quinones, D. M. Neville, and J. H. Kersey, 1984, Ex-vivo treatment of donor bone marrow with anti-T-cell immunotoxins for prevention of graft-versus-host disease, *Lancet* **1:**469–472.

3

The Mannose 6-Phosphate Receptor and Its Role in Lysosomal Enzyme Transport

G. Gary Sahagian
Tufts University Schools of Medicine and Veterinary Medicine

Of the many vertebrate lectins that have been identified, the manose 6-phosphate (Man-6-P) receptor is among the few whose function has been firmly established. While the receptor was initially discovered in the context of endocytosis, it is now widely accepted that its primary role is to act inside the cell to target lysosomal proteins to lysosomes. The receptor interacts specifically with lysosomal proteins by virtue of a common recognition marker composed of Man-6-P residues that are added to the enzymes during their biosynthesis.

This chapter focuses on the physical and chemical properties of the receptor, its interaction with lysosomal enzymes, and the cellular mechanism by which the receptor carries out its functions of endocytosis and intracellular transport. For other recent reviews on these and related subjects, see references 1, 2, and 3.

PURIFICATION AND PROPERTIES

The existence of the Man-6-P receptor was initially inferred from the specificity and saturability of lysosomal enzyme uptake by human fibroblasts deficient in particular lysosomal hydrolases (for a review of these studies, see references 4 and 5). The receptor was subsequently identified in detergent extracts of fibroblasts and several other cell types and purified from bovine liver membranes by co-immunoprecipitation with lysosomal β-galactosidase and affinity chromatography on β-galactosidase-Sepharose 4B (6). Several methods have since been described for large-scale purification of the receptor. Nearly homogeneous receptor can be obtained in milligram amounts from acetone powders of liver of bovine and other species (7, 8) and from microsomes of rat chondrosarcoma (9) by using a phosphomannan affinity matrix. Pentamannosyl monophosphate (10) and crude lysosomal enzyme preparations from *Dictyostelium discoideum* (11, 12) have also been used successfully as affinity ligands.

The binding characteristics of the purified receptor are identical in all respects to those of the cellular binding sites that mediate transport of lysosomal enzymes in vivo (6). Binding constants (Kd) for Man-6-P-containing lysosomal enzymes range from 10^{-7} to 10^{-9} M. Ligands with the highest affinity are lysosomal enzymes secreted from *D. discoideum* (13) and amine-treated fibroblasts (14, 15) and the core fragment of the phosphomannan secreted from *Hansenula holstii* (16). Binding of the ligands is maximal at neutral pH but drops sharply at pH values below 6 (6, 17). This unique dependence on pH is characteristic of many receptors that mediate protein transport and is thought to play a key role in the ability of such receptors to be reutilized. The only known inhibitors of binding to the receptor are analogs of Man-6-P; chelators of divalent cations have no effect on binding (6, 18).

The affinity-purified receptor appears to be a glycoprotein comprised of a single 215,000-dalton (Da) subunit (4, 7), although one group has obtained evidence suggesting that it may be composed of several smaller subunits (see below). Two forms of the receptor that differ in isoelectric point (pI 5.5 and 6.2) have been resolved by two-dimensional electrophoresis (9); only the major form (pI 5.5) is made more basic by treatment with neuraminidase. Solubilized receptor exists as multimers of up to four 215,000-Da subunits, as has been determined by electrophoresis in nondenaturing acrylamide gradient gels (unpublished data). Nothing, however, is known about the functional sizes of the receptor or its aggregation state in the lipid bilayer.

Amino acid analyses (8, 9, 19) have indicated that the receptor is an acidic protein that is rich in hydrophobic and half-cystine residues. The large decrease in electrophoretic mobility of the receptor upon reduction suggests that the receptor polypeptide is extensively cross-linked by disulfide bonds (20). Extensive disulfide cross-linking appears to be a characteristic shared with other receptors, most notably the receptor for low-density lipoprotein (21). The Man-6-P receptor is continually exposed to pH changes and proteolytic enzymes. The stability conferred by disulfide bonds may help prevent denaturation and/or degradation that might otherwise occur in such harsh environments.

Considerable information concerning the structure of the Man-6-P receptor has been obtained from biosynthetic studies. Receptor synthesized by Chinese hamster ovary (CHO) cells was shown to contain asparagine-linked oligosaccharides and phosphorylated serine residues (20). The structure of oligosaccharides of the receptor made by mouse lymphoma cells has been studied by Goldberg and co-workers (22). Typical complex-type structures containing fucose and sialic acid, as well as some high-mannose-type structures, were found in the fully mature glycoprotein. Asparagine-linked oligosaccharides are not required for ligand binding but may be important for proper protein folding and stability (20). The function of phosphoserine found in this and other receptors involved in transport of macromolecules (23) is not yet known.

There is some question about the nature of the newly synthesized receptor. Biosynthetic labeling studies with CHO cells revealed an exceptionally long lag in appearance of the 215,000-Da receptor polypeptide (20). Lag times of approximately 1 hour were observed regardless of whether the receptor was isolated by immunoprecipitation or by binding to an insolubilized ligand. Similar but shorter lags were apparent in studies with human skin fibroblasts (24) and established mouse cell lines (22). These results suggest that nascent receptor lacks binding activity and antigenic determinants required for recognition by the antibody and that assembly of functional receptor requires major post-translational modification of the receptor polypeptide (20, 25). The size of the nascent receptor and the nature of the modifications are not known. Jourdian's group (26, 27) has proposed that the 215,000-Da receptor polypeptide is actually an aggregate composed of smaller subunits and that association of the subunits might be responsible for the lag in appearance of the receptor in biosynthetic labeling studies. An alternative explanation is that conversion of the nascent receptor to the active immunoprecipitable form is dependent on slow formation of intra-chain disulfide bonds; intact disulfide bonds are known to be required for recognition of the receptor by at least one of the antibody preparations used in these studies (20). Further studies with antibodies that recognize nascent receptor are needed to determine the nature of the newly synthesized receptor and early events in the biosynthesis of the receptor.

Clathrin-coated vesicles isolated from rat liver and bovine brain have been used to examine the membrane topography of the Man-6-P receptor. Campbell and associates (28) showed that these vesicles are highly enriched in cryptic Man-6-P receptors. Subsequent studies with bovine brain-coated vesicles indicated that the receptor is a transmembrane protein (29). The lysosomal enzyme-binding site was shown to be located at the lumenal surface of the vesicle, and 12,000 Da of the C-terminal end of the protein was exposed at the vesicle's cytoplasmic surface. A similar transmembrane orientation has been reported for Man-6-P receptor present in microsomal fractions of human fibroblasts (30).

DETERMINANTS FOR BINDING OF LYSOSOMAL ENZYMES

The major determinant for binding of lysosomal enzymes to the receptor is Man-6-P located on asparagine-linked oligosaccharides of the enzymes (8, 31, 32). The receptor interacts strongly with the phosphate moiety and at the 2 and 4 positions of the phosphorylated sugar, as has been indicated by inhibition studies with monosaccharides and their derivatives (8, 18, 33, 34).

While it is clear that Man-6-P is necessary for binding, other factors also contribute, since the affinity of the receptor for lysosomal enzymes is three to four orders of magnitude greater than that for Man-6-P. This observation has led to suggestions that multivalent ligands are required for high-affinity binding (35) or that, in addition to Man-6-P, the receptor recognizes other structural features of the enzymes, either protein (36–38) or carbohydrate (8, 39) in nature.

An experiment to examine the contribution of the protein and carbohydrate portions of β-galactosidase to binding to the receptor is shown in Table 3-1. Extensive treatment of unlabeled enzyme with trypsin or endohexosaminidase H reduced its ability to inhibit binding of ^{125}I-labeled enzyme by four- to fivefold. This reduction was much less the 40-fold reduction observed when β-galactosidase was treated with alkaline phosphatase, which suggested a significant but minor role for the polypeptide. The inhibitory activity remaining after endo-H treatment was in fact in the released oligosaccharides. These oligosaccharides were 100 times more inhibitory than Man-6-P. The results suggest that most of the determinants for binding of β-galactosidase and other lysosomal enzymes are contained on their oligosaccharides and that structural features of the oligosaccharides, in addition to Man-6-P, are important.

Several factors could account for the apparent minor involvement of the polypeptide portion of lysosomal enzymes. Protein determinants could interact directly with the receptor or could act indirectly by affecting the conformation of phosphorylated oligosaccharides or their accessibility to the receptor. The protein might also contribute by producing a multivalent ligand that contains several phosphorylated oligosaccharides; such ligands would be expected to have a higher apparent affinity for the receptor than individual oligosaccharides of similar structure. While protein contributions may increase the affinity of many lysosomal enzyme for the receptor, they are not required, since phosphorylated oligosaccharides released from lysosomal

Table 3-1. Inhibition of Binding of ^{125}I-β-Galacatosidase to Purified Man-6-P Receptor

Inhibitor	K_i ($\mu M\ PO_4$)
β-Galactosidase	0.06
+trypsin	0.23
+endohexosaminidase H	0.28
+alkaline phosphatase	2.61
Oligosaccharide fraction	0.30
Mannose 6-phosphate	29.0

proteins with endo-H can be efficiently bound and internalized by the receptor (see below).

Further characterization of the interaction of the receptor with lysosomal enzymes has followed determination of the structure of phosphorylated oligosaccharides. Tabas and Kornfeld (40) showed that phosphorylated oligosaccharides of newly synthesized β-glucuronidase contained one or two Man-6-P residues, of which most were present in phosphodiester linkage with α-N-acetylglucosamine (GlcNAc) "covering" groups. At about the same time, Hasilik and associates (41) obtained similar results for β-hexosaminidase and cathepsin D secreted by human fibroblasts. The authors of both studies proposed that phosphodiester moieties were intermediates in the formation of Man-6-P residues on the oligosaccharides. Varki and Kornfeld (42) studied the structure of phosphorylated oligosaccharides synthesized by mouse lymphoma cells. Phosphorylation was shown to occur at five separate mannose residues of the high-mannose-type oligosaccharides, with individual molecules containing one or two phosphate residues in monoester or diester linkage. Similar results with some minor differences have been obtained by Natowicz and co-workers (43) for phosphorylated oligosaccharides of human β-glucuronidase.

Sly's group has studied the uptake of phosphorylated oligosaccharides by human fibroblasts (44) and their binding to Man-6-P receptor immobilized on agarose beads (11). Oligosaccharides with two phosphomonoesters were preferred to those with one phosphomonoester for both binding and uptake. Oligosaccharides with one or two phosphodiesters were neither bound nor internalized, indicating that the GlcNAc "covering" group interferes with binding of the oligosaccharides to the receptor. Similar conclusions were drawn from inhibition studies with phosphorylated oligosaccharides of β-galactosidase (8). In spite of their lower affinity for the receptor, certain oligosaccharides with one phosphomonoester may bind well enough in vivo to target hydrolases to lysosomes (15, 45).

Natowicz and associates (46) determined kinetic constants for uptake of phosphorylated oligosaccharides by human fibroblasts. Oligosacchorides bearing one and two phosphomonoesters were internalized with uptake constants (K uptake) of 3.2×10^{-7} and 3.9×10^{-8} M, respectively. The authors concluded that a phosphorylated high-manose-type oligosaccharide is the minimum structure required for recognition and uptake by the Man-6-P receptor.

A detailed analysis of anionic oligosaccharides of P388D_1 mouse macrophage-like cells has been performed by Varki and Kornfeld (12). In addition to the usual phosphorylated oligosaccharides, hybrid-type molecules containing both sialic acid and phosphate were identified. Analysis of the interaction of the isolated oligosaccharides with immobilized Man-6-P receptor confirmed the results of previous studies and demonstrated that the location of phosphomonoesters on the oligosaccharide and the presence or absence of certain nonphosphorylated sugar residues such as galactose and sialic acid can affect the interaction of the oligosaccharide with the receptor.

SYNTHESIS AND TRANSPORT OF LYSOSOMAL ENZYMES

In many respects, Man-6-P-containing lysosomal enzymes can be considered modified secretory proteins. Like secretory proteins, the enzymes are soluble proteins that are synthesized on membrane-bound polysomes at the endoplasmic reticulum (ER) (47, 48, and 49). The nascent polypeptide of cathepsin D has been shown to possess an amino terminal signal sequence (47) that interacts with the same signal recognition particle that mediates insertion of secretory and membrane proteins into the ER (50); indirect evidence has been obtained for the presence of signal sequences of several other nascent lysosomal enzymes (48, 49, 51, 52). Within the lumen of the ER, the signal sequences are cleaved and the polypeptides are glycosylated at asparagine residues in a manner analogous to that for other glycoproteins (47, 48, 49, 53). Studies with the glucosidase inhibitor deoxynojirimycin (54) and glucosylation-defective CHO cells (55) suggest that, as for some secretory glycoproteins (56), transport of lysosomal enzymes out of the ER may require removal of glucose residues from asparagine-linked oligosaccharides.

Unlike secretory glycoproteins, lysosomal enzymes are modified by addition of the Man-6-P recognition marker to one or more of their N-linked oligosaccharides. This modification is initiated after removal of glucose (Glc) residues from the oligosaccharides (53) but probably before addition of terminal sugars (2), placing the site of synthesis of the marker in the late ER, or early Golgi. Construction of the marker involves transfer of GlcNAc-1-PO_4 from UDPGlcNAc to the 6-position of mannose residues on the oligosaccharides to form phosphodiester moieties with the structure Man-6-P-GlcNAc, and subsequent removal of the covering GlcNAc residues to expose the underlying Man-6-P moieties (40, 41, 53). Further oligosaccharide processing, such as removal of terminal mannose residues linked $\alpha 1,2$ to phosphomannose residues, may also be required for completion of a fully functional marker (12). Although the N-linked oligosaccharides of newly synthesized lysosomal enzymes and secretory proteins are initially identical, only lysosomal enzymes are efficiently phosphorylated. This preferential phosphorylation has been attributed to the existence of an as yet undefined structural determinant in the protein sequence or conformation of lysosomal enzymes that is required for recognition by the GlcNAc-1-PO_4 transferase (57, 58). The GlcNAc-1-P transferase (57, 58) and the GlcNAc phosphodiesterase (59, 60) have been partially purified and characterized. The subcellular distributions of both enzymes have been examined in rat liver (60, 61, 62) and cultured cells (63); the results suggest that both activities act on lysosomal enzymes in the *cis* compartment of the Golgi apparatus.

In normal human fibroblasts, binding of lysosomal enzymes to the receptor and subsequent delivery of the enzymes to lysosomes occurs predominantly within the cell (64, 65, 66). Binding of the enzymes to the receptor and their segregation from secretory proteins is thought to occur in the Golgi apparatus (see below). However, significant portions of many enzymes, up to 35% in some cases, escape lysosomal delivery and are secreted. Much of the secreted enzyme carries the Man-6-P recognition marker and so can bind to Man-6-P receptor on the cell surface and be transported to lysosomes by the process of receptor-mediated endocytosis.

During the later stages of their maturation, most if not all lysosomal enzymes undergo discrete reductions in size that can vary from a few to greater than 20 kilodaltons (kDa; see references 3 and 67). In most cases such changes have been attributed to limited proteolysis in lysosomes. While proteolytic processing may be important for activation of some lysosomal enzyme—for example, cathepsin B (68) and cathepsin D (51, 69)—its significance for most enzymes is not yet known. It may be that the excised fragments are important for early events in enzyme maturation and are dispensable once those events are completed.

In most cultured cells binding of phosphorylated enzymes to the Man-6-P receptor is necessary for efficient sorting of the enzymes from secretory proteins and in delivery of the enzymes to lysosomes. When this interaction is prevented, the majority of most lysosomal enzymes is secreted. This situation exists in I-cell (ML II) and pseudo-Hurler (ML III) fibroblasts that are unable to phosphorylate lysosomal enzymes (70-72) because of a deficiency in the GlcNAc-1-P transferase (73, 74, 75), in several receptor-defective (76) and receptor-deficient (77) cells, and in normal cells when receptor function is blocked with lysosomotropic amines (17, 78) or anti-receptor antibodies (79). The only known exception is MOPC 315 cells that retain most of their lysosomal enzymes in spite of a very low level of Man-6-P receptor (77, 80).

Man-6-P-mediated transport is thought to be a general mechanism for packaging of lysosomal enzymes. However, three pieces of evidence suggest that other mechanisms also exist. First, I-cell fibroblasts, in spite of an inability to phosphorylate lysosomal enzymes, have near-normal levels of a few acid hydrolases (81). Among the retained hydrolases are β-glucocerebrosidase, a membrane-bound enzyme, acid phosphatase and α-glucosidase. Second, liver (82, 83) and certain other tissues (83) of I-cell patients that also lack GlcNAc-1-P transferase activity have normal levels of many hydrolases. And finally, lysosomes of I-cell fibroblasts (84) and receptor-deficient mouse cells (77) contain low but significant levels of lysosomal enzymes. Based on these observations, the existence of "alternate pathways" for packaging lysosomal proteins has been proposed (82–84, 77).

Recently, Hoflack and Kornfeld (85, 86) have purified a second Man-6-P binding protein of molecular weight (M_r) = 46,000. This binding protein appears to be distinct from the 215,000-Da receptor described here. The new binding protein has a similar specificity for phosphorylated lysosomal proteins except that divalent cations are required for binding and the protein does not bind lysosomal enzymes secreted by *D. discoideum;* phosphorylated oligosaccharides of these enzymes contain Man-6-P in a methylated diester form (Man-P-CH_3) rather than in phosphomonester form (87). While the function of this binding protein is not known, it has been suggested that it may provide an alternate pathway for lysosomal delivery of Man-6-P-containing proteins and may be responsible for targeting lysosomal proteins in cell lines that lack the 215,000-Da receptor (86).

ENDOCYTOSIS OF LYSOSOMAL ENZYMES AND RECEPTOR RECYCLING

The pathway for endocytosis of lysosomal enzymes by the Man-6-P receptor has been studied quite extensively and has been found to be similar to that for other macromolecules. Morphologic studies on the endocytosis of β-galactosidase by CHO cells (88) showed that this process involves binding of the enzyme to Man-6-P receptors on the cell surface, internalization at clathrin-coated pits, transport through various types of nonlysosomal structures, and accumulation in lysosomes. Among the nonlysosomal structures are endocytic vesicles that have been called receptosomes (89), endosomes (90), or CURL (compartment for the uncoupling of receptor and ligand) (91), and smaller clathrin-coated and uncoated structures, many of which were located in the general vicinity of the Golgi apparatus. The involvement of prelysosomal structures in the endocytosis of lysosomal enzymes has also been demonstrated by Percoll gradient centrifugation (20, 92).

Like many other receptors that mediate endocytosis, the Man-6-P receptor can be reutilized for several rounds of ligand internalization; the amount of ligand internalized by cells exceeds the number of Man-6-P receptor binding sites, even when synthesis of new receptor is blocked with cycloheximide (17, 18). Receptor reutilization generally involves internalization of the receptor with its ligand, dissociation of the complex within the cell, and cycling of the unoccupied receptor back to the cell surface (23, 93). Inhibition of recycling of cell surface Man-6-P receptors by lysosomotropic amines (17) and in CHO cell mutants defective in acidification (15, 94, 95) suggests that dissociation is caused by a drop in intravesicular pH, a condition known to cause release of lysosomal enzymes from the receptor (6, 17). Dissociation occurs prior to delivery of the ligand to lysosomes, since endocytic vesicles are acidified almost immediately after their formation at the cell surface (96, 97).

Less is known about how Man-6-P receptors are segregated from the dissociated ligands and returned to the cell surface. Morphologic studies of Geuze and co-workers (91) on the endocytosis of asialoglycoproteins suggest that segregation occurs in CURL endocytic vesicles (91). Their electron micrographs suggest that after dissociation of the ligand, receptors concentrate in tubular domains of the CURL membrane. These domains presumably bud off from CURL and fuse with the plasma membrane to deliver the receptors to the cell surface for reutilization. Recently, Man-6-P receptors have also been shown to be concentrated in tubular protrusions of CURL (98).

Several recent observations suggest that CURL-derived vesicles carrying receptors to be reutilized may not take a direct route to the cell surface or, alternatively, that some portions of these vesicles carry receptors to other subcellular destinations. Incubation of cells in culture medium containing antibodies directed against Man-6-P receptor results in rapid labeling (99) and inactivation (79) of intracellular receptor, indicating that cellular pools of receptor mix rapidly. Since a major portion of the intracellular receptor is located in the Golgi apparatus, the results suggest the existence of a route for exchange of membrane between the Golgi apparatus and the cell surface. Evidence for recycling of transferrin and asialoglycoprotein receptors through the Golgi apparatus has also been obtained. Snider and Rogers (100) have shown that following internalization, transferrin receptors can be modified by glycosyl transferases in the Golgi apparatus. The kinetics for the modification suggest that transport through the Golgi is not an obligatory process but occurs approximately once in every five to ten rounds of internalization. Internalized asialoglycoprotein receptors accumulate in the Golgi when receptor recycling is blocked with lysosomotropic amines (A. L. Schwartz, personal communication). Collectively, these observations provide suggestive, though indirect, evidence for transport of receptors between the endocytic compartment and Golgi apparatus.

RECEPTOR LOCALIZATION AND INTRACELLULAR ENZYME TRANSPORT

Hypotheses concerning the pathway for intracellular transport of lysosomal enzymes were formulated several years before the discovery of the phosphomannosyl recognition system. Studies using electron microscopy and cytochemistry to localize acid phosphatase led to proposals that lysosomes are formed either from the Golgi apparatus (101, 102) or from GERL (103), a smooth membrane network at the *trans* face of the Golgi. These hypotheses differed in that it was proposed that GERL receives lysosomal enzyme directly from the ER without ever entering the Golgi apparatus. The presence of acid phosphatase in coated vesicles that appeared to arise from some component of the Golgi apparatus or from GERL suggested that these structures are transport vesicles for lysosomal enzymes (101, 104). This notion has since been supported by biochemical evidence demonstrating that purified coated vesicles are enriched in Man-6-P receptor (28) and newly synthesized lysosomal enzymes (105).

In order to further define the intracellular pathway, several investigators have examined the subcellular location of the Man-6-P receptor. Using ultrastructural immunocytochemistry and antibodies against the purified bovine liver receptor, Willingham and associates (106) localized the receptor in CHO cells. Consistent with its role in endocytosis, the receptor was found in coated pits on the cell surface and in endocytic vesicles (receptosomes). However, most of the receptor was located in Golgi cisternae, in coated and uncoated structures in the Golgi region, and in the ER. Little or no receptor was detected in lysosomes, a finding confirmed by subcellular fractionation (20). The results were consistent with previous studies implicating coated vesicles and specialized regions of the Golgi apparatus in lysosome formation. The substantial ER localization in these cells has not been observed in most other cell types examined. Studies on the biosynthesis of the receptor in CHO cells (20) suggest that this localization represents a large biosynthetic pool of receptor that exists because of exceptionally slow transport of the receptor out of the ER following its synthesis (99).

Subsequently, Brown and Farquhar (10) used immunocytochemical methods to determine the subcellular location of the Man-6-P receptor in several secretory and adsorptive cells. In all cells examined, receptor was detected in Golgi cisternae and associated coated vesicles and in large vesicles designated endosomes and/or lysosomes. Most cell types showed little or no receptor in the ER. Within the Golgi region, the receptor was concentrated in cisternae and vesicles on one side of the Golgi stacks. In secretory cells where the orientation of the stacks could be determined, the receptor was detected in the *cis*-most cisternae, not in medial or *trans* cisternae or GERL. The authors proposed that lysosomal enzymes bind to Man-6-P receptor in the *cis* Golgi and are delivered from this location to lysosomes and/or endosomes via coated pits and vesicles. It has been difficult, however, to reconcile this hypothesis and the GERL hypothesis with the finding that receptor-bound enzymes (107) and enzymes in the lysosomes of normal (108) or sialidase-deficient (109) fibroblasts contain galactose and sialic acid, sugars added to glycoproteins in the *trans* Golgi (110).

In a more recent morphologic study by Geuze and co-workers (111) on rat liver, Man-6-P receptor and the lysosomal enzyme cathepsin D were both detected throughout the Golgi stacks. Double-label experiments showed that most of the enzyme in the Golgi was closely associated with the receptor, indicating that, at least in rat liver, enzyme and receptor traverse the entire Golgi complex on the way to lysosomes. Receptor was also found on the plasma membrane, in coated pits and vesicles, and in CURL, but not in lysosomes. Based on these findings, the authors have raised the possibility that in addition to its role in endocytosis, CURL might also be involved in intracellular lysosomal enzyme transport (98).

While little is known about the molecular mechanism for intracellular transport of lysosomal proteins, several similarities with the mechanism for endocytosis have become apparent. Because both processes involve clathrin-coated pits and vesicles, it is likely that a similar mechanism for concentration of receptor-enzyme complexes in the lipid bilayer and for vesicle formation is used. In addition, both processes appear to use the same mechanism for receptor recycling: lysosomotropic amines that inhibit recycling of cell surface Man-6-P receptors by raising the pH of prelysosomal endocytic vesicles also inhibit recycling of Man-6-P receptors functioning in the Golgi apparatus (17). Furthermore, intracellular transport of lysosomal proteins is inhibited in CHO cell mutants that are defective in endosome acidification (15, 95). These results, taken together with the absence of Man-6-P receptor in lysosomes, strongly suggest that intracellular lysosomal enzyme transport involves acidification-dependent recycling of the receptor and that segregation of receptor and ligand occurs in a prelysosomal compartment that is the same as or analogous to that involved in endocytosis (99).

COMMON MECHANISM FOR ENDOCYTOSIS AND INTRACELLULAR TRANSPORT OF LYSOSOMAL ENZYMES

At present, endocytosis and intracellular transport of lysosomal enzymes are viewed by most as separate processes. However, the apparent similarities in the cellular mechanisms for these processes and the observation that pools of Man-6-P receptors functioning in endocytosis and intracellular transport readily mix have suggested a more connected relationship (98, 99). Figure 3-1 provides a working hypothesis (99) for transport of Man-6-P-containing lysosomal enzymes from both intracellular and extracellular compartments.

Following their synthesis in the ER, lysosomal enzymes are transported to the Golgi apparatus, where they acquire the Man-6-P recognition marker. As the enzymes pass through the Golgi, they bind to Man-6-P receptors, which then concentrate in clathrin-coated pits located in the Golgi apparatus or the region called the trans-Golgi network (112) or GERL. The precise location in the Golgi where the enzymes bind to the receptor and the place where the complexes concentrate and leave the Golgi apparatus are not yet clear. Some enzyme escapes through the secretory pathway and is released

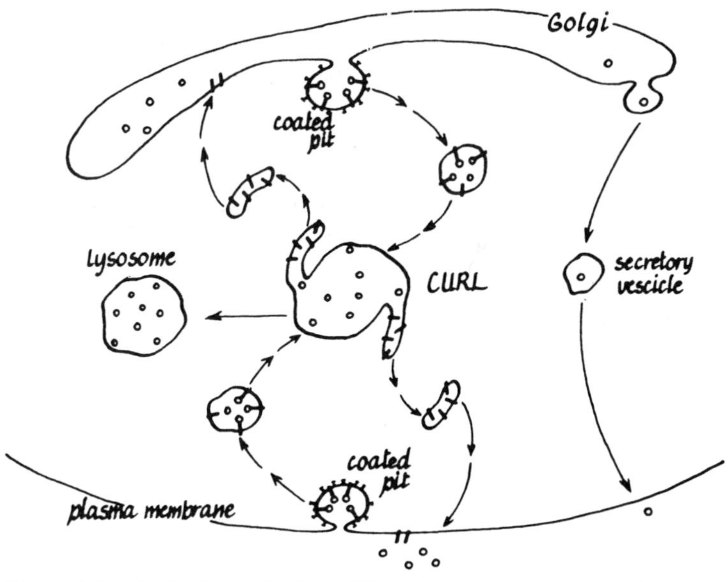

Figure 3-1. Model for packaging of Man-6-P-containing lysosomal enzymes.

from the cell. Secreted enzyme that carries a functional Man-6-P recognition marker can bind to receptors on the plasma membrane and be concentrated into coated pits on the cell surface.

Receptor-enzyme complexes in coated pits on the cell surface or Golgi are then packaged into vesicles that mature into or fuse with CURL. In CURL the receptors are segregated from their ligands and recycled, with dissociation occurring in the vesicles or CURL as a result of acidification. The receptor leaves CURL by budding of the membrane and is transported back to the cell surface or Golgi apparatus via recycling vesicles. Since different receptors display varying propensities for transport to the Golgi or plasma membrane, some sorting of receptors into appropriate recycling vesicles would be required. Mixing of receptor pools can be accounted for by direct mixing of receptors in CURL. Additional mixing might also occur by transport of the receptor from the Golgi apparatus to the cell surface via secretory vesicles.

This scheme suggests a relatively simple mechanism for Man-6-P receptor function. Whether on the cell surface or in the Golgi apparatus, the receptor would serve as a vehicle for the transport of lysosomal proteins from coated pits to CURL. At least two recognition signals on the receptor are suggested: one for directing enzyme-receptor complexes to coated pits and another for concentrating the receptor in appropriate domains of the CURL membrane to allow efficient segregation and recycling. Other receptors should possess one or both of these signals, depending on whether they are recycled to the plasma membrane or Golgi, or transported to lysosomes for degradation. Detailed information concerning the structure of the receptor will be required to identify these signals and the cellular constituents with which they interact.

ACKNOWLEDGMENTS

I would like to thank Dr. E. F. Neufeld for helpful discussions and critical reading of the manuscript and Dr. A. D'Azzo for illustrations.

REFERENCES

1. K. E. Creek, and W. S. Sly, 1984, the role of the phosphomannosyl receptor in the transport of acid hydrolases to lysosomes, in *Lysosomes in Biology and Pathology*. J. T. Dingle, R. T. Dean, and W. S. Sly, eds., Elsevier/North Holland, Amsterdam, pp. 63-82.
2. D. Goldberg, C. Gabel, and S. Kornfeld, 1984, Processing of lysosomal enzyme oligosaccharide units, in *Lysosomes in Biology and Pathology*, J. T. Dingle, R. T. Dean, and W. S. Sly, eds., Elsevier/North Holland, Amsterdam, pp. 45-62.
3. A. Hasilik, and K. von Figura, 1984, Processing of lysosomal enzymes in fibroblasts, in *Lysosomes in Biology and Pathology*, J. T. Dingle, R. T. Dean, and W. S. Sly, eds., Elsevier/North Holland, Amsterdam, pp. 3-16.

4. E. F. Neufeld, and G. Ashwell, 1980, Carbohydrate recognition systems for receptor-mediated pinocytosis, in *The Biochemistry of Glycoproteins and Proteoglycans,* W. J. Lennarz, ed., Plenum Publishing Corp., pp. 241-266.
5. W. S. Sly, H. D. Fischer, A. Gonzalez-Noriega, J. H. Grubb, and M. Natowicz, 1981, Role of the 6-phosphomannosyl-enzyme receptor in intracellular transport and adsorptive pinocytosis of lysosomal enzymes, *Methods Cell Biol.* **23:**191-214.
6. G. G. Sahagian, J. Distler, and G. W. Jourdian, 1981, Characterization of a membrane-associated receptor from bovine liver that binds phosphomannosyl residues of bovine testicular β-galactosidase, *Proc. Natl. Acad. Sci. (USA)* **78:**4289-4293.
7. G. G. Sahagian, J. Distler, and G. W. Jourdian, 1982, Membrane receptor for phosphomannosyl residues, *Methods Enzymol.* **83:**392-396.
8. G. G. Sahagian, 1981, Receptor-Mediated Uptake of Lysosomal Enzymes, Ph.D. dissertation, University of Michigan.
9. A. W. Steiner, and L. H. Rome, 1982, Assay and purification of a solubilized membrane receptor that binds the lysosomal enzyme α-*L*-iduronidase, *Arch. Biochem. Biophys.* **214:**681-687.
10. W. J. Brown, and M. G. Farquhar, 1984, The mannose-6-phosphate receptor for lysosomal enzymes is concentrated in cis Golgi cisternae, *Cell* **36:**295-307.
11. H. D. Fischer, K. E. Creek, and W. S. Sly, 1982, Binding of phosphorylated oligosaccharides to immobilized phosphomannosyl receptors, *J. Biol. Chem.* **257:**9938-9943.
12. J. Varki, and S. Kornfeld, 1983, The spectrum of anionic oligosaccharides released by endo-β-N-acetylglucosaminidase H from glycoproteins: Structural studies and interactions with the phosphomannosyl receptor, *J. Biol. Chem.* **258:**2808-2818.
13. H. H. Freeze, A. L. Miller, and A. Kaplan, 1980, Acid hydrolases from *Dictyostelium discoideum* contain phosphomannosyl recognition markers, *J. Biol. Chem.* **255:**11081-11084.
14. H. D. Fischer, A. Gonzalez-Noriega, W. S. Sly, and D. J. Morre, 1980, Phosphomannosyl-enzyme receptors in rat liver: Subcellular distribution and role in intracellular transport of lysosomal enzymes, *J. Biol. Chem.* **255:**9608-9615.
15. A. R. Robbins, S. S. Peng, and J. L. Marshall, 1983, Mutant Chinese hamster ovary cells pleiotropically defective in receptor-mediated endocytosis, *J. Cell Biol.* **96:**1064-1071.
16. A. Kaplan, D. Fischer, and W. S. Sly, 1978, Correlation of structural features of phosphomannans with their ability to inhibit pinocytosis of human β-glucuronidase by human fibroblasts, *J. Biol. Chem.* **253:**647-650.
17. A. Gonzalez-Noriega, J. H. Grubb, V. Talkad, and W. S. Sly, 1980, Chloroquine inhibits lysosomal enzyme pinocytosis and enhances lysosomal enzyme secretion by impairing receptor recycling, *J. Cell Biol.* **85:**839-852.
18. L. H. Rome, B. Weissmann, and E. F. Neufeld, 1979, Direct demonstration of binding of a lysosomal enzyme, α-*L*-iduronidase, to receptors on cultured fibroblasts, *Proc. Natl. Acad. Sci. (USA)* **76:**2331-2334.
19. D. C. Mitchell, T. Maler, and G. W. Jourdian, 1984, Detergent dissociation of bovine liver phosphomannosyl binding protein, *J. Cell. Biochem.* **24:**319-330.
20. G. G. Sahagian, and E. F. Neufeld, 1983, Biosynthesis and turnover of the mannose 6-phosphate receptor in cultured Chinese hamster ovary cells, *J. Biol. Chem.* **258:**7121-7128.
21. J. L. Goldstein, M. S. Brown, R. G. W. Anderson, D. W. Russell, and W. J. Schneider, 1985, Receptor-mediated endocytosis: Concepts emerging from the LDL receptor system, *Annu. Rev. Cell Biol.* **1:**1-39.
22. D. E. Goldberg, C. A. Gabel, and S. Kornfeld, 1983, Studies of the biosynthesis of the mannose 6-phosphate receptor in receptor-positive and -deficient cell lines, *J. Cell Biol.* **97:**1700-1706.

23. M. S. Brown, R. G. W. Anderson, and J. L. Goldstein, 1983, Recycling receptors: The round-trip itinerary of migrant membrane proteins, *Cell* **32:**663-667.
24. K. E. Creek, and W. S. Sly, 1983, Biosynthesis and turnover of the phosphomannosyl receptor in human fibroblasts, *Biochem. J.* **214:**2628-2633.
25. G. G. Sahagian, 1984, The mannose 6-phosphate receptor: Function, biosynthesis and translocation, *Biol. Cell* **51:**207-214.
26. D. C. Mitchell, T. Maler, and G. W. Jourdian, 1984, Detergent dissociation of bovine liver phosphomanosyl binding protein, *J. Cell. Biochem.* **24:**319-330.
27. T. Maler, B. B. Rosenblum, and G. W. Jourdian, 1985, Properties of the Syrian hamster phosphomannosyl receptor: An aggregate of low molecular weight proteins, *Proc. Natl. Acad. Sci. (USA)* **82:**8379-8383.
28. C. H. Campbell, R. E. Fine, J. Squicciarini, and L. H. Rome, 1983, Coated vesicles from rat liver and calf brain contain cryptic mannose 6-phosphate receptors, *J. Biol. Chem.* **258:**2628-2633.
29. G. G. Sahagian, and C. J. Steer, 1985, Transmembrane orientation of the mannose 6-phosphate receptor in isolated clathrin-coated vesicles, *J. Biol. Chem.* **260:**9838-9842.
30. K. Von Figura, V. Gieselmann, and A. Hasilik, 1985, Mannose 6-phosphate-specific receptor is a transmembrane protein with a C-terminal extension oriented towards the cytosol, *Biochem. J.* **225:**543-547.
31. M. R. Natowicz, M. M.-Y. Chi, O. H. Lowry, and W. S. Sly, 1979, Enzymatic identification of mannose 6-phosphate on the recognition marker for receptor-mediated pinocytosis of β-glucuronidase by human fibroblasts, *Proc. Natl. Acad. Sci. (USA)* **76:**4322-4326.
32. K. Von Figura, and U. Klein, 1979, Isolation and characterization of phosphorylated oligosaccharides from α-N-acetylglucosaminidase that are recognized by cell-surface receptors, *Eur. J. Biochem.* **94:**347-354.
33. G. N. Sando, and E. F. Neufeld, 1977, Recognition and receptor-mediated uptake of a lysosomal enzyme, α-*L*-iduronidase, by cultured human fibroblasts, *Cell* **12:**619-627.
34. A. Kaplan, D. T. Achord, and W. S. Sly, 1977, Phosphomannosyl components of a lysosomal enzyme are recognized by pinocytosis receptors on human fibroblasts, *Proc. Natl. Acad. Sci. (USA)* **74:**2026-2030.
35. H. D. Fischer, M. Natowicz, W. S. Sly, and R. K. Bretthauer, 1980, Fibroblast receptor for lysosomal enzymes mediates pinocytosis of multivalent phosphomannan fragment, *J. Cell Biol.* **84:**77-86.
36. R. Myerowitz, and E. F. Neufeld, 1981, Maturation of a -*L*-iduronidase in cultured human fibroblasts, *J. Biol. Chem.* **256:**3044-3048.
37. L. H. Rome, and J. Miller, 1980, Butanedione treatment reduces receptor binding of a lysosomal enzyme to cells and membranes, *Biochem. Biophys. Res. Commun.* **92:**986-993.
38. E. M. Karson, E. F. Neufeld, and G. N. Sando, 1980, *p*-Isothiocyanatphenyl 6-phospho-α-*D*-mannopyranoside coupled to albumin. A model compound recognized by the fibroblast lysosomal enzyme uptake system. 2. Biological properties, *Biochemistry* **19:**3856-3860.
39. J. Distler, V. Hieber, G. Sahagian, R. Schmickel, and G. W. Jourdian, 1979, Identification of mannose 6-phosphate in glycoproteins that inhibit the assimilation of β-galactosidase by fibroblasts, *Proc. Natl. Acad. Sci. (USA)* **76:**4235-4239.
40. I. Tabas, and S. Kornfeld, 1980, Biosynthetic intermediates of β-glucuronidase contain high mannose oligosaccharides with blocked phosphate residues, *J. Biol. Chem.* **255:**6633-6639.
41. A. Hasilik, U. Klein, A. Waheed, G. Strecker, and K. Von Figura, 1980, Phosphorylated oligosaccharides in lysosomal enzymes: Identification of α-N-acetylglucosamine(1)-phospho(6)mannose diester groups, *Proc. Natl. Acad. Sci. (USA)* **77:**7074-7078.
42. A. Varki, and S. Kornfeld, 1980, Structural studies of phosphorylated high mannose-type oligosaccharides, *J. Biol. Chem.* **255:**10847-10858.

43. M. Natowicz, J. U. Baenziger, and W. S. Sly, 1982, Structural studies of the phosphorylated high mannose-type oligosaccharides on human β-glucuronidase, *J. Biol. Chem.* **257:**4412-4420.
44. K. E. Creek, and W. S. Sly, 1982, Adsorptive pinocytosis of phosphorylated oligosaccharides by human fibroblasts, *J. Biol. Chem.* **257:**9931-9937.
45. C. A. Gabel, and S. Kornfeld, 1982, Lysosomal enzyme phosphorylation in mouse lymphoma cell lines with altered asparagine-linked oligosaccharides, *J. Biol. Chem.* **257:**10605-10612.
46. M. Natowicz, D. W. Hallett, C. Frier, M. Chi, P. H. Schlesinger, and J. U. Baenziger, 1983, Recognition and receptor-mediated uptake of phosphorylated high mannose-type oligosaccharides by cultured human fibroblasts, *J. Cell Biol.* **96:**915-919.
47. A. H. Erickson, G. E. Conner, and G. Blobel, 1981, Biosynthesis of a lysosomal enzyme: Partial structure of two transient and functionally distinct NH_2-terminal sequences in cathepsin D, *J. Biol. Chem.* **256:**11224-11231.
48. A. H. Erickson, and G. Blobel, 1979, Early events in the biosynthesis of the lysosomal enzyme cathepsin D., *J. Biol. Chem.* **254:**11771-11774.
49. M. G. Rosenfeld, G. Kreibich, D. Popov, K. Kato, and D. D. Sabatini, 1982, Biosynthesis of lysosomal hydrolases: their synthesis in bound polysomes and the role of co- and post-translational processing in determining their subcellular distribution, *J. Cell Biol.* **93:**135-143.
50. A. H. Erickson, P. Walter, and G. Blobel, 1983, Translocation of a lysosomal enzyme across the microsomal membrane requires signal recognition particle, *Biochem. Biophys. Res. Commun.* **115:**275-280.
51. M. Muller, and H. Muller, 1981, Synthesis and processing of in vitro and in vivo precursors of the vacuolar yeast enzyme carboxypeptidase Y, *J. Biol. Chem.* **256:**11962-11965.
52. R. L. Proia, and E. F. Neufeld, 1982, Synthesis of β-hexosaminidase in cell-free translation and in intact fibroblasts: An insoluble precursor chain in a rare form of Tay-Sachs disease, *Proc. Natl. Acad. Sci. (USA)* **79:**6360-6364.
53. D. E. Goldberg, and S. Kornfeld, 1981, The phosphorylation of β-glucuronidase oligosaccharides in mouse P388D$_1$ cells, *J. Biol. Chem.* **256:**13060-13067.
54. P. Lemansky, G. Volkmar, A. Hasilik, and K. Von Figura, 1984, Cathepsin D and β-hexosaminidase synthesized in the presence of 1-deoxynojirimycin accumulate in the endoplasmic reticulum, *J. Biol. Chem.* **259:**10129-10135.
55. S. S. Krag, and A. R. Robbins, 1982, A Chinese hamster ovary cell mutant deficient in glucosylation of lipid-linked oligosaccharide synthesizes lysosomal enzymes of altered structure and function, *J. Biol. Chem.* **257:**8424-8431.
56. H. F. Lodish, and N. Kong, 1984, Glucose removal from N-linked oligosaccharides is required for efficient maturation of certain secretory glycoproteins from the rough endoplasmic reticulum to the Golgi complex, *J. Cell Biol.* **98:**1720-1729.
57. M. L. Reitman, and S. Kornfeld, 1981, Lysosomal enzyme targeting: N-acetylglucosaminylphosphotransferase selectively phosphorylates native lysosomal enzymes, *J. Biol. Chem.* **256:**11977-11980.
58. A. Waheed, A. Hasilik, and K. Von Figura, 1982, UDP-N-acetylglucosamine: Lysosomal enzyme precursor N-acetylglucosamine-1-phosphotransferase: Partial purification and characterization of the rat liver Golgi enzyme, *J. Biol. Chem.* **257:**12322-12331.
59. A. Waheed, A. Hasilik, and K. Von Figura, 1981, Processing of the phosphorylated recognition marker in lysosomal enzymes: Characterization and partial purification of a microsomal α-N-acetylglucosaminyl phosphodiesterase, *J. Biol. Chem.* **256:**5717-5721.
60. A. Varki, and S. Kornfeld, 1980, Identification of a rat liver α-N-acetylglucosaminyl phosphodiesterase capable of removing "blocking" α-N-acetylglucosamine residues from phosphorylated high mannose oligosaccharides of lysosomal enzymes, *J. Biol. Chem.* **255:**8398-8401.

61. A. Waheed, R. Pohlmann, A. Hasilik, and K. Von Figura, 1981, Subcellular location of two enzymes involved in the synthesis of phosphorylated recognition markers in lysosomal enzymes, *J. Biol. Chem.* **256:**4150-4152.
62. R. Pohlmann, A. Waheed, A. Hasilik, and K. Von Figura, 1982, Synthesis of phosphorylated recognition marker in lysosomal enzymes is located in the cis part of Golgi apparatus, *J. Biol. Chem.* **257:**5323-5325.
63. D. E. Goldberg, and S. Kornfeld, 1983, Evidence for extensive subcellular organization of asparagine-linked oligosaccharide processing and lysosomal enzyme phosphorylation, *J. Biol. Chem.* **258:**3159-3165.
64. W. S. Sly, and P. Stahl, 1978, Receptor-mediated uptake of lysosomal enzymes, in *Transport of Macromolecules in Cellular Systems,* S. C. Silverstein, ed., Dahlem Konferenzen, Berlin, pp. 229-244.
65. A. Hasilik, and E. F. Neufeld, 1980, Biosynthesis of lysosomal enzymes in fibroblasts: Synthesis as precursors of higher molecular weight, *J. Biol. Chem.* **255:**4937-4945.
66. K. Von Figura, and E. Weber, 1978, An alternative hypothesis of cellular transport of lysosomal enzymes in fibroblasts: Effect of inhibitors of lysosomal enzyme endocytosis on intra- and extra-cellular lysosomal enzyme activities, *Biochem. J.* **176:**943.
67. E. F. Neufeld, 1981, Recognition and processing of lysosomal enzymes in cultured fibroblasts, in *Lysosomes and Lysosomal Storage Diseases,* J. W. Callahan and J. A. Lowden, eds., Raven Press, New York, pp. 115-129.
68. J. S. Mort, M. S. Leduc, and A. D. Recklies, 1983, Characterization of a latent cysteine proteinase from ascitic fluid as a high molecular weight form of cathepsin B, *Biochim. Biophys. Acta* **775:**369-375.
69. A. Hasilik, K. Von Figura, E. Conzelmann, H. Nehrkorn, and K. Sandhoff, 1982, Lysosomal enzyme precursors in human fibroblasts: Activation of cathepsin D precursor in vitro and activity of β-hexosaminidase A precursor towards ganglioside GM_2, *Eur. J. Biochem.* **25:**317-321.
70. A. Hasilik, and E. F. Neufeld, 1980, Biosynthesis of lysosomal enzymes in fibroblasts: Phosphorylation of mannose residues, *J. Biol. Chem.* **255:**4946-4950.
71. G. Bach, R. Bargal, and M. Cantz, 1979, I-cell disease: Deficiency of extracellular hydrolase phosphorylation, *Biochem. Biophys. Res. Commun.* **91:**976-981.
72. P. G. Robey, and E. F. Neufeld, 1982, Defective phosphorylation and processing of β-hexosaminidase by intact cultured fibroblasts from patients with mucolipidosis III, *Arch. Biochem. Biophys.* **213:**251-257.
73. A. Hasilik, A. Waheed, and K. Von Figura, 1981, Enzymatic phosphorylation of lysosomal enzymes in the presence of UDP-N-acetylglucosamine. Absence of the activity in I-cell fibroblast, *Biochem. Biophys. Res. Commun.* **98:**761-767.
74. M. L. Reitman, A. Varki, and S. Kornfeld, 1981, Fibroblasts from patients with I-cell disease and pseudo-Hurler polydystrophy are deficient in uridine 5′-diphosphate-N-acetylglucosamine: Glycoprotein N-acetylglucosaminylphosphotransferase activity, *J. Clin. Invest.* **67:**1574-1579.
75. A. P. Varki, M. L. Reitman, and S. Kornfeld, 1981, Identification of a variant of mucolipidosis III (pseudo Hurler polydystrophy): A catalytically active N-acetylglucosaminyl-phosphotransferase that fails to phosphorylate lysosomal enzymes, *Proc. Natl. Acad. Sci. USA* **78:**7773-7777.
76. A. R. Robbins, and R. Myerowitz, 1981, The mannose 6-phosphate receptor of Chinese hamster ovary cells: Compartmentalization of acid hydrolases in mutants with altered receptors, *J. Biol. Chem.* **256:**10623-10627.
77. C. A. Gabel, D. E. Goldberg, and S. Kornfeld, 1983, Identification and characterization of cells deficient in the mannose 6-phosphate receptor: Evidence for an alternate pathway for lysosomal enzyme targeting, *Proc. Natl. Acad. Sci. (USA)* **80:**775-779.

78. A. Hasilik, and E. F. Neufeld, 1980, Biosynthesis of lysosomal enzymes in fibroblasts: Synthesis as precursors of higher molecular weight, *J. Biol. Chem.* **255:**4937-4945.
79. K. Von Figura, V. Gieselmann, and A. Hasilik, 1984, Antibody to mannose 6-phosphate specific receptor induces receptor deficiency in human fibroblasts, *EMBO J.* **3:**1281-1286.
80. C. A. Gabel, and S. Kornfeld, 1984, Targeting of β-glucuronidase to lysosomes in mannose 6-phosphate receptor-deficient MOPC 315 cells, *J. Cell Biol.* **99:**296-305.
81. E. F. Neufeld, T. W. Lim, and L. J. Shapiro, 1975, Inherited disorders of lysosomal metabolism, *Annu. Rev. Biochem.* **44:**357-376.
82. M. Owada, and E. F. Neufeld, 1982, Is there a mechanism for introducing acid hydrolases into liver lysosomes that is independent of mannose 6-phosphate recognition? Evidence from I-cell disease, *Biochem. Biophys. Res. Commun.* **105:**814-820.
83. A. Waheed, R. Pohlmann, A. Hasilik, K. Von Figura, A. Van Elsen, and J. G. Leroy, 1982, Deficiency of UDP-N-acetylglucosamine: Lysosomal enzyme N-acetylglucosamine-1-phosphotransferase in organs of I-cell patients, *Biochem. Biophys. Res. Commun.* **105:**1052-1058.
84. A. L. Miller, B. C. Kress, R. Stein, C. Kinon, H. Kern, J. A. Schneider, and E. Harms, 1981, Properties of N-acetyl-β-*D*-hexosaminidase from isolated normal and I-cell lysosomes, *J. Biol. Chem.* **256:**9352-9362.
85. B. Hoflack, and S. Kornfeld, 1985, Lysosomal enzyme binding to mouse P388D$_1$ macrophage membranes lacking the 215-kDa mannose 6-phosphate receptor: Evidence for the existence of a second mannose 6-phosphate receptor, *Proc. Natl. Acad. Sci. (USA)* **82:**4428-4432.
86. B. Hoflack, and S. Kornfeld, 1985, Purification and characterization of a cation-dependent mannose 6-phosphate receptor from murine P388D$_1$ macrophages and bovine liver, *J. Biol. Chem.* **260:**12006-12014.
87. C. A. Gabel, C. E. Costello, V. N. Reinhold, L. Kurz, and S. Kornfeld, 1984, Identification of methylphosphomannosyl residues as components of the high mannose oligosaccharides of *Dictyostelium discoideum* glycoproteins, *J. Biol. Chem.* **259:**13762-13769.
88. M. C. Willingham, I. H. Pastan, G. G. Sahagian, G. W. Jourdian, and E. F. Neufeld, 1981, Morphologic study of the internalization of a lysosomal enzyme by the mannose 6-phosphate receptor in cultured Chinese hamster ovary cells, *Proc. Natl. Acad. Sci. (USA)* **78:**6967-6971.
89. M. C. Willingham, and I. Pastan, 1980, The receptosome: An intermediate organelle of receptor-mediated endocytosis in cultured fibroblasts, *Cell* **21:**67-77.
90. A. Helenius, I. Mellman, D. Wall, and A. Hubbard, 1983, Endosomes, *Trends Biochem. Sci.* **8:**245-250.
91. H. J. Geuzse, J. W. Slot, G. J. A. M. Strous, H. F. Lodish, and A. L. Schwartz, 1983, Intracellular site of asialoglycoprotein receptor-ligand uncoupling: Double-label immunoelectron microscopy during receptor-mediated endocytosis, *Cell* **32:**277-287.
92. M. Merion, and W. S. Sly, 1983, The role of intermediate vesicles in the adsorptive endocytosis and transport of ligand to lysosomes by human fibroblasts, *J. Cell Biol.* **96:**644-650.
93. R. M. Steinman, I. S. Mellman, W. A. Muller, and Z. A. Cohn, 1983, Endocytosis and the recycling of plasma membrane, *J. Cell Biol.* **96:**1-27.
94. M. Merion, P. Schlesinger, R. M. Brooks, J. M. Moehring, T. J. Moehring, and W. S. Sly, 1983, Defective acidification of endosomes in Chinese hamster ovary cell mutants "cross-resistant" to toxins and viruses, *Proc. Natl. Acad. Sci. (USA)* **80:**5315-5319.
95. A. R. Robbins, C. Oliver, J. L. Bateman, S. S. Krag, C. J. Galloway, and I. Mellman, 1984, A single mutation in Chinese hamster ovary cells impairs both Golgi and endosomal functions, *J. Cell Biol.* **99:**1296-1308.
96. B. Tyoko, and F. R. Maxfield, 1982, Rapid acidification of endocytic vesicles containing a$_2$-macroglobulin, *Cell* **28:**643-651.

97. D. J. Yamashiro, B. Tyoko, S. R. Fluss, and F. R. Maxfield, 1984, Segregation of transferrin to a mildly acidic (pH 6.5) para-Golgi compartment in the recycling pathway, *Cell* **37**:789-800.
98. H. J. Geuze, J. W. Slot, G. J. A. M. Strous, A. Hasilik, and K. Von Figura, 1985, Possible pathways for lysosomal enzyme delivery, *J. Cell Biol.* **101**:2253-2262.
99. G. G. Sahagian, 1984, The mannose 6-phosphate receptor: Function, biosynthesis and translocation, *Biol. Cell* **51**:207-214.
100. M. D. Snider, and O. C. Rogers, 1985, Intracellular movement of cell surface receptors after endocytosis: Resialylation of asialo-transferrin receptor in human erythroleukemia cells, *J. Cell Biol.* **100**:826-834.
101. D. S. Friend, and M. G. Farquhar, 1967, Functions of coated vesicles during protein absorption in the rat vas deferens, *J. Cell Biol.* **35**:357-376.
102. Z. A. Cohn, M. E. Fedorko, and J. G. Hirsch, 1966, The in vitro differentiation of mononuclear phagocytes: Formation of macrophage lysosomes, *J. Exp. Med.* **123**:757-766.
103. P. M. Novikoff, A. B. Novikoff, N. Quintana, and J. J. Hauw, 1971, Golgi apparatus, GERL, and lysosomes of neurones in rat dorsal root ganglia, studied by thick section and thin section cytochemistry, *J. Cell Biol.* **50**:859-886.
104. A. B. Novikoff, 1976, The endoplasmic reticulum: A cytochemist's view (A review), *Proc. Natl. Acad. Sci. (USA)* **73**:2781-2787.
105. C. H. Campbell, and L. H. Rome, 1983, Coated vesicles from rat liver and calf brain contain lysosomal enzymes bound to mannose 6-phosphate receptors, *J. Biol. Chem.* **258**:13347-13352.
106. M. C. Willingham, I. H. Pastan, and G. G. Sahagian, 1983, Ultrastructural immunocytochemical localization of the phosphomannosyl receptor in Chinese hamster ovary (CHO) cells, *J. Histochem. Cytochem.* **31**:1-11.
107. C. A. Gabel, D. E. Goldberg, and S. Kornfeld, 1982, Lysosomal enzyme oligosaccharide phosphorylation in mouse lymphoma cells: Specificity and kinetics of binding to the mannose 6-phosphate receptor in vivo, *J. Cell Biol.* **95**:536-542.
108. V. Gieselmann, R. Pohlmann, A. Hasilik, and K. Von Figura, 1983, Biosynthesis and transport of cathepsin D cultured human fibroblasts, *J. Cell Biol.* **97**:1-5.
109. G. D. Vladutiu, 1983, Effect of the co-existence of galactosyl and phosphomannosyl residues on β-hexosaminidase on the processing and transport of the enzyme in mucolipidosis I fibroblasts, *Biochim. Biophys. Acta* **760**:363-370.
110. J. Roth, and E. G. Berger, 1982, Immunocytochemical localization of galactosyltransferase in HeLa cells: Codistribution with thiamine pyrophosphatase in trans-Golgi cisternae, *J. Cell Biol.* **92**:223-229.
111. H. J. Geuze, J. W. Slot, G. J. A. M. Strous, A. Hasilik, and K. Von Figura, 1984, Ultrastructural localization of the mannose 6-phosphate receptor in rat liver, *J. Cell Biol.* **98**:2047-2054.
112. G. Griffiths, S. Pfeiffer, K. Simons, and K. Matlin, 1985, Exit of newly synthesized membrane proteins from the trans cisterna of the Golgi complex to the plasma membrane, *J. Cell Biol.* **101**:949-964.

4

Receptor Recycling and Ligand Processing Mediated by Hepatic Galactosyl Receptors: A Two-Pathway System

Paul H. Weigel
The University of Texas Medical Branch

Ashwell, Morell, and their co-workers discovered and characterized the first mammalian lectin, a hepatic membrane glycoprotein, that specifically binds glycoproteins containing oligosaccharide chains with terminal galactose or N-acetylgalactosamine groups (1). Many desialylated plasma glycoproteins, such as asialo-orosomucoid, are bound by this asialoglycoprotein (or galactosyl; Gal) receptor, which then mediates their endocytosis and subsequent degradation. The rat Gal receptor is a transmembrane glycoprotein (2, 3) whose native molecular weight is not known. In the presence or absence of detergent, the sizes of the active binding fragment determined by radiation inactivation are, respectively, about 148 kilodaltons (kDa) and 105 kDa (4).

Different laboratories have reported that the affinity-purified receptor contains from one to four different subunits, with molecular weights ranging from 32 to 120 kDa, although a major band is found at about 42–45 kDa in most cases (2). Consistently, two larger minor receptor peptides are found at 49 and 54 kDa (5), but the mole ratio of the three peptides is unusual and variable. The major polypeptide has been sequenced and shown to have 28% homology to the analagous chicken lectin (5), which is a transmembrane phosphorylated glycoprotein (6). The minor peptides have regions of extensive sequence homology to the 42-kDa subunit but are structurally distinct (5), which rules out the possibility that the smaller peptide is generated by proteolysis of the larger proteins(s). The protein chemistry suggested that at least two genes must code for receptor polypeptides, and subsequently two genes have been identified and cloned (7, 8). The significance of the minor bands is not known, and there is presently no accepted model for the subunit composition and stoichiometry of the native Gal receptor. The major subunit is also a transmembrane glycoprotein with the amino terminus facing the cytoplasm (3). Serine on a cytoplasmic domain may be phosphorylated in rat hepatocytes and in human hepatoma cells (9, 10). An interesting possibility is that there are at least two structurally different but related receptors. Our laboratory has gathered evidence, some of which will be summarized here, over the last several years that shows there are two separate receptor pathways for internalization, dissociation, and degradation of ligand. It is tempting to suggest that the basis for these dual pathways would be two subpopulations of receptors.

Binding of ligand requires calcium ions, and the receptor complex rapidly dissociates below pH 6 (1). The carbohydrate binding site(s) on the receptor from rabbits (11, 12) and humans (13) have been mapped by using saccharides of known structure. There may be two separate, relatively small binding sites, approximately 2.5-3.0 nm apart (13), which accommodate terminal galactosyl or N-acetylgalactosaminyl residues with D-pyranosyl structure. The functional receptor may therefore be bivalent, although this is not known. We found that the valency of the purified receptor doubles in the presence of a variety of nonionic detergents such as Triton X-100 (14). Lee and co-workers have proposed a novel model to explain how different ligands manifest different physical parameters of binding to the rabbit Gal receptor (15). They propose that binding sites are clustered on the cell surface to form a lattice that can accommodate a wide range of different oligosaccharide structures and give rise in turn to different binding affinities and different apparent numbers of binding sites. Both biantennary and triantennary glycopeptides of known structure are bound by the receptor, with approximately equal affinities, but only the triantennary structures are endocytosed efficiently (16, 17). This result indicates that after binding ligand, the receptor undergoes a conformational change necessary for subsequent endocytosis. In this case the receptor would not recycle in the absence of ligand. We and others have suggested that the receptor recycles or undergoes a reversible internalization/externalization without ligand (18, 19, 20). The resolution of this point is critical to an understanding of the molecular details of endocytosis.

Internalization of galactosyl ligands by rat hepatocytes is thought to occur via coated pits and coated vesicles (21, 22). However, it was recently shown that another receptor, with a similar sugar specificity, on Kupffer cells mediates uptake via both a coated and a noncoated pathway (23). This raises a question as to whether all Gal receptors in hepatocytes are internalized via the same coated pit pathway. Wall, Hubbard, and others reported that in the absence of ligand, receptors were clustered preferentially in coated pits both in liver (21) and in isolated cells (22). They found no evidence for ligand-induced redistribution of receptors. In contrast, Kolb-Bachofen found that receptors in isolated hepatocytes were not localized in coated pits in the absence of ligand, but rather were induced by ligand to form microaggregates and then larger clusters that were finally found in coated pits (24). This latter sequence of events has also been proposed for other receptors (25). The reason for the disparity in results between the two groups is unclear. The resolution of this conflict is important in the effort to understand how the Gal receptor system functions.

Within minutes after endocytosis, the ligand appears in large smooth-surfaced tubules and vesicles termed endosomes (26). During the next 15 min the ligand moves from the peripheral cytoplasm to the lysosome-Golgi region and is degraded in lysosomes (21). The receptor, however, is reutilized (recycled) both in vivo (27, 28) and in vitro (29, 30). One receptor molecule can effect the uptake and subsequent degradation, in a cyclic fashion, of many asialoglycoproteins without itself being destroyed. Delivery of free ligand to lysosomes is not necessary for receptor recycling to occur (31). As in other systems, degradation stops below 20°C (19, 32) and endocytosis stops below 10°C (33). The intracellular pool of receptors, present in Golgi, lysosomal, and smooth microsomal fractions, represents about 60-90% of the total cell receptor content (16, 27, 30, 34, 35).

Receptor-ligand complexes are internalized (36-38) and then dissociate intracellularly in a biphasic process probably mediated at least in part by low pH (39, 40, 41). Receptor binding to ligand is very pH-sensitive, and low pH has been shown to induce conformational changes in the rat Gal receptor (42). Since one or more early endocytic compartments have been shown to be acidic (e.g., about pH 5.4 in the human hepatoma cell line HepG2; 41), it is assumed that low pH is responsible for the dissociation of receptor-ligand complexes. Some of these studies depend on the pH-sensitivity of a fluorochrome (e.g., fluorescein) attached to the ligand. A major difficulty with such experiments, which complicates their interpretation, is that they measure an average pH value for a large number of intracellular vesicles. At present, for example, there is no information on the pH value of a particular compartment or whether all endocytic vesicles follow the same pathway (i.e., do all receptor-ligand complexes encounter the same cellular compartments). Since dissociation occurs by two kinetically different pathways, two different mechanisms could be involved.

FUNCTION OF THE ASIALOGLYCOPROTEIN RECEPTOR

Although the receptor was discovered in 1968 (1), its physiological function remains unknown. An important function for this receptor seems likely, since it is present in large amounts (0.5-1.0 $\times$ 10^6 binding sites/cell) and is very active in mediating endocytosis. There are at least four possible functions for this receptor:

1. *Clearance from plasma of desialylated proteins.* Numerous investigators in many systems have shown that desialylated glycoproteins, when injected into mammals, are rapidly removed from blood by this receptor in liver (1, 2, 43). However, at least three laboratories have shown that under normal circumstances plasma glycoproteins are not desialylated in vivo and are not cleared from the circulation by this receptor (44, 45, 46).
2. *Intracellular routing of appropriate glycoproteins.* By analogy to the mannose-6-phosphate receptor, which uses this oligosaccharide "handle" to sort, deliver, and package lysosomal enzymes (47), the galactosyl/N-acetylgalactosaminyl recognition marker might allow its receptor to transport or sort specific liver proteins within the hepatocyte (48).
3. *Clearance of antibodies or antibody-antigen complexes from serum (49, 50).* Immunoglobulin G (IgG) molecules have a galactosyl terminal oligosaccharide near the hinge region, yet they are not removed by the liver. Formation of an antigen complex that would alter the conformation in this region to provide a signal for clearance and degradation is an attractive idea (49); however, the initial results have not been verified and extended.
4. *Cell adhesion.* Several reports have shown the involvement of the Gal receptor in the metastasis of tumor cells to the liver in mice (51, 52). We have also shown that the receptor mediates cell attachment (53) and cell spreading to synthetic culture surfaces containing galactose (J. A. Oka, and P. H. Weigel, manuscript submitted). The possible role of this receptor in cell-cell or cell-matrix interactions in vivo must therefore be considered.

RECENT RESULTS FROM THIS LABORATORY

We view endocytosis mediated by the Gal receptor as a process that can be divided into at least nine stages:

1. ligand binding
2. receptor-ligand complex internalization
3. partition of complexes into two different Gal receptor pathways
4. dissociation of complexes by either a fast or a slow process to give free ligand and receptor
5. segregation (spatial separation) of free ligand and receptor into different subcellular compartments

6. ligand delivery to lysosomes or other degradative compartments
7. ligand degradation in lysosomes or other degradative compartments
8. receptor processing and recycling to the cell surface
9. reincorporation of receptor into coated pits; integration of coated pit cycle(s) with the receptor cycle(s).

The overall endocytic process is continuous, but these stages, though arbitrary, are useful in discussing, understanding, and developing experimental strategies to study the molecular details of endocytosis. A primary objective of our research during the last few years has been to develop ways to study each of these stages independently of the others. We have performed studies to examine these stages in more detail, and we have obtained evidence for two receptor pathways.

Measuring Intracellular Receptor Activity and Ligand Dissociation

In order to understand how the receptor system functions, it was necessary to develop a fast, reliable method to determine the surface/internal distribution of receptors in isolated hepatocytes. Previous estimates of surface and total cell receptor content employed combinations of sonication (16), subcellular membrane fractionation (27, 34), antireceptor antibody (35), and solubilization with Triton X-100 (16, 29). For a variety of reasons, we chose to develop a new approach in which cells were permeabilized rather than solubilized. Digitonin was found to be an excellent detergent for this, because it provides a broad concentration "window" over which cells lose no additional protein and the increased amount of receptor activity remains constant (54). The method is also applicable to other membrane-bound receptors (or enzymes), provided that the particular receptor is not solubilized by digitonin. Using this new method, we were able to show that the majority of receptors are intracellular (55). We also showed for the first time that the large intracellular pool of receptors functions during the endocytosis of asialo-orosomucoid by rat hepatocytes (55). This is significant, since the majority of receptors are inside the cell and this pool was previously thought to be nonfunctional (35, 56). The important question of whether all cellular Gal receptors are functionally equivalent remains to be answered. The answer to this question should have a large impact on morphologic studies employing various electron microscopic techniques, since the presence of inactive or functionally distinct receptor populations would dramatically affect the interpretations of such studies.

Metabolic Energy is Required for Receptor Recycling but Not for Endocytosis

In an effort to find inhibitors of the different stages of endocytosis and recycling, we have examined the effect of metabolic poisons on the Gal receptor system (57). In these studies, we met several important criteria: (a) We measured adenosine triphosphate (ATP) levels by means of a luciferase/luciferin assay. Measurement of ATP levels is very important, since many cell types, especially hepatocytes, are relatively insensitive to individual metabolic inhibitors and require combinations of inhibitors to deplete ATP pools quantitatively. (b) Final measurements of ATP and intracellular ligand content were made only on viable cells isolated from Percoll density gradients. (c) ATP depletion was complete *prior* to the initiation of endocytosis. We found that a single round of endocytosis occurred even if the cellular ATP levels were depleted by >98%, although the extent was decreased compared with that in controls (Fig. 4-1). The rate of continuous internalization, however, is inhibited dramatically (>90%) when ATP levels decrease by only 25%. The normal ATP content of cells equilibrated at 37°C in complete medium is 21 nmol/mg protein. The continuously recycling receptor system senses a threshold value of ATP at a critical concentration of about 17 nmol/mg cell protein and virtually shuts down at lower ATP concentrations.

Our results suggest that the early stages of receptor-mediated endocytosis, including internalization, either do not require metabolic energy (as reflected by cell ATP levels) or are extremely efficient in the utilization of ATP (i.e., can operate even in the micromolar ATP range). This is consistent with our current view of endocytosis (presented below), in which the generation of vesicles from coated pits is an energy-independent process driven by the favorable thermodynamics associated with the self-assembly of clathrin baskets to form a coated vesicle. In any case, there is a very large difference in sensitivity to cellular ATP content between one round of endocytosis and multiple rounds that require receptor recycling.

Evidence for Two Functionally Distinct Populations of Cell Surface Gal Receptors

We recently showed that increasing or decreasing the temperature in the absence of ligand reversibly increases or decreases the steady-state amount of receptor activity expressed on the hepatocyte surface (20, 58). This modulation of receptor activity by temperature may account for most of the variability in receptor number/cell reported by different laboratories. There are two populations of surface receptor activity with respect to modulation by temperature: insensitive and sensitive. We have designated these, State 1 and State 2 receptors, respectively (59). Freshly isolated hepatocytes, kept on ice after preparation by a collagenase perfusion technique, express almost entirely State 1 receptors on their surface. When these cells are incubated at 37°C, the surface receptor content increases two- to three fold and then remains constant with time (58). This new receptor activity on the cell surface does not arise from preexisting active intracellular receptors (54). Prior to most experiments, we routinely treat cell suspensions at 37°C under defined conditions in order to attain a stable surface receptor content and external/internal distribution of receptors.

Surface receptor activity can also be decreased in the absence of ligand by metabolic energy poisons (60) such as cyanide or azide (Fig. 4-2) and by microtubule depolymerizing agents such as colchicine (Fig. 4-3), colcemide, podophyllotoxin, and vinblastine sulfate, but not lumicolchicine. The receptor activity decrease due to ATP depletion is rapidly reversible even in the presence of cycloheximide, which inhibits protein synthesis by 99%. Neither

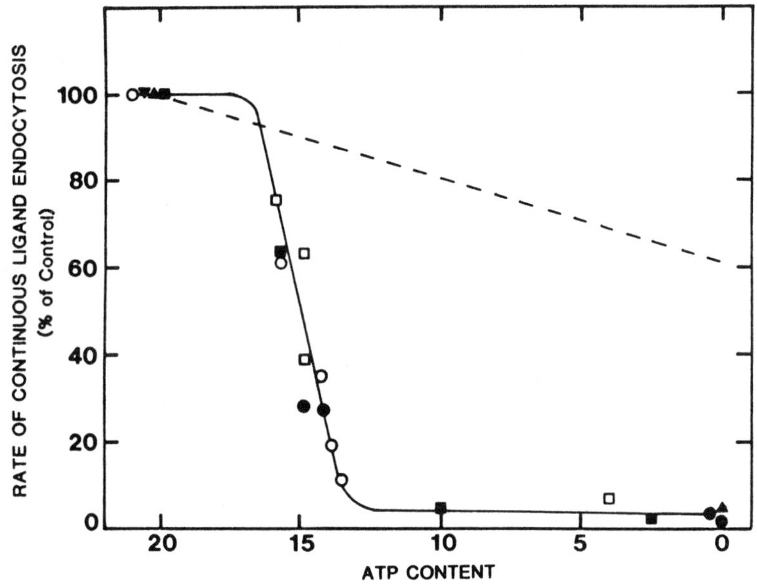

Figure 4-1. Effect of ATP depletion on endocytosis. Cells were incubated at 37°C for 10 min with sodium azide and sodium fluoride (0–10 mM) to achieve different extents of ATP depletion. Each symbol is a separate experiment. ^{125}I-asialo-orosomucoid was then added and the linear steady-state rate of internalization was determined. The dashed line is a similar experiment in which the extent of endocytosis of pre-bound ^{125}I-asialo-orosomucoid was determined.

ATP depletion nor microtubule drugs cause a loss of all cell surface receptor activity. Again, there are two populations of receptors, one whose activity is modulated and one that is unaffected. Both classes of poisons prevent the characteristic increase in receptor activity when freshly isolated cells are incubated at 37°C. For example, colchicine at 3×10^{-7} M causes a 50% inhibition in the receptor activity increase. The expression of State 2 Gal receptor activity on hepatocytes is sensitive to cellular energy depletion, low temperature, and microtubule drugs. State 1 receptor activity is not perturbed by these treatments. Experiments with multiple or successive inhibitor treatments support this conclusion. Monensin also appears to modulate the activity of surface Gal receptors (61).

The distributions of State 1 and State 2 surface receptors at 35°C are roughly 30% and 70%, respectively, of the total. Both State 1 and State 2 receptors have identical affinities for asialo-orosomucoid (58). Whether these two receptor populations can be differentiated by their specificity toward various oligosaccharide structures remains to be determined. An interesting question is whether the two receptor pools correlate with reports

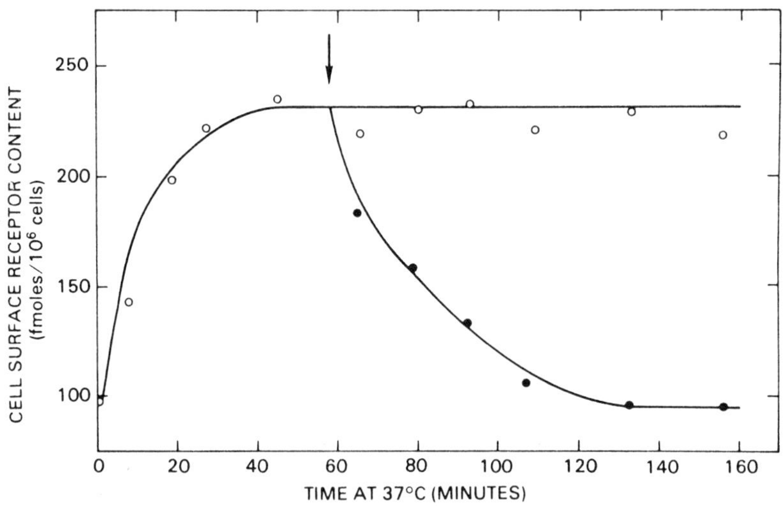

Figure 4-2. Effect of sodium azide on cell surface receptor activity. Freshly isolated cells were incubated at 37°C. At 58 min sodium azide (●) in PBS was added to a final concentration of 10 mM. The controls received phosphate buffered saline (PBS) (○). Samples were removed, chilled to 4°C, and washed, and receptor content was determined by ^{125}I-asialo-orosomucoid binding. Note that the abscissa does not begin at zero and that the final level of surface receptor activity is identical to that at the start of the experiment.

that about 60% of the active surface receptors are localized in coated pits in the absence of ligand (22, 62, 63).

The modulation of State 2 Gal surface receptor activity could be explained by the removal of active receptors from the surface to an inaccessible compartment (e.g., the cell interior), the inactivation of receptor molecules, which remain at the cell surface, or a combination of both. The loss of cell surface receptor activity after exposure of an agent in the absence of ligand is interpreted by most investigators as evidence for continuous receptor recycling. The observed results would be explained if the externalization phase of State 2 recycling receptors was slowed to a greater extent by the inhibitors or low temperature than the internalization phase. The experiment discussed in Figure 4-8 shows that both receptor internalization and inactivation may occur in the absence of ligand when cells are depleted of ATP.

In initial experiments to determine whether unoccupied receptors normally recycle in the absence of ligand, we have used a goat antireceptor antibody and ^{125}I-protein A to determine whether the apparent number of receptor molecules on the cell surface increases when receptor activity

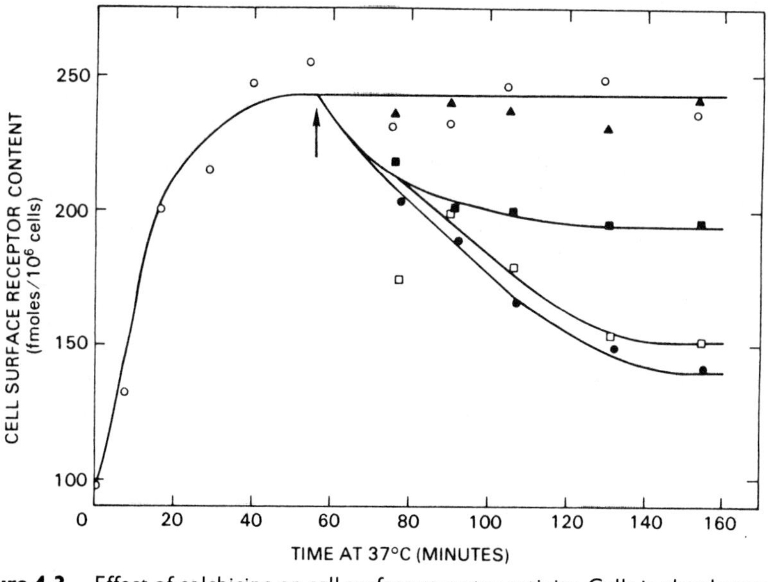

Figure 4-3. Effect of colchicine on cell surface receptor activity. Cells in the absence of ligand were incubated at 37°C. After 58 min, colchicine was added to a final concentration of 0 (○), 10^{-7} (▲), 10^{-6} (■), 10^{-5} (□), or 10^{-4} (●) M. Surface receptor content was assessed at 4°C as indicated in Figure 4-2.

increases after cells have been exposed to higher temperatures. Our preliminary results suggest that the increase in State 2 receptor activity on freshly isolated cells treated at 37°C is not accompanied by a large change in the ratio of bound ^{125}I-asialo-orosomucoid to ^{125}I-protein A (which should occur if inactive receptors on the surface, capable of binding antibody, were activated). We favor the interpretation that the new receptors that appear at 37°C, the State 2 receptors, are inserted into the plasma membrane. However, this is contrary to the conclusion that inactive receptors were still on the cell surface after modulation by ionophores (61). Possibly, the inactive surface receptors are "buried" and inaccessible to antibody. We intend to continue these studies to assess the mechanism of receptor modulation.

Evidence for Two Pathways for the Dissociation of Internalized Receptor-Ligand Complexes

We have exploited the digitonin technique described previously as an easy way to differentiate between internal ligand that is free (i.e., dissociated) and that which is receptor-bound. Only the later ligand remains associated with the digitonin-permeabilized cell pellet; the free ligand leaves the cell and is in the medium. The kinetic process(es) by which endocytosed receptor-ligand complexes dissociate is not as simple as initially assumed. The rate of dissociation is not a single, uniform first-order process with one rate constant, but rather is biphasic (40). Dissociation occurs in two kinetically (and presumably morphologically) different compartments, dramatically different in the rate at which they release free ligand. The apparent first-order rate constants for the fast and slow dissociation processes are 0.28 and 0.014 min^{-1}, respectively (40). In other experiments, we were able to show that the fast dissociation process is mediated by State 2 receptors and the slow dissociation process is mediated by State 1 receptors. For the uptake of surface-bound ligand under non-steady-state conditions, the partition of complexes into State 1/State 2 compartments is about 40/60. This partition is not affected by the percent receptor occupancy (40). A similar slow dissociation compartment was also observed by Harford and co-workers (39) by using detergent solubilization and a filter binding assay to quantitate free intracellular ligand.

Three pieces of evidence support the idea that the slow dissociation compartment is physiologically significant and important for the overall receptor system to function: (a) The slow dissociation process is completely stopped below 18°C, whereas the fast dissociation process still continues (40). At 18°C, however, receptor recycling still occurs, which suggests that all recycling receptors are not obligated to pass through the slow dissociation compartment. These could be two functionally nonequivalent receptor pools. (b) Depletion of ATP levels by about 50% completely stops the slow, but not the fast, dissociation process (64). More extensive depletion of ATP can inhibit dissociation by both pathways. (c) The number of receptor-ligand complexes in the slow dissociation compartment becomes saturated under continuous endocytosis conditions when a steady state is reached (40).

We have shown that internalized State 1 receptor-ligand complexes committed to the slow dissociation compartment can return to the cell surface before they dissociate (65). The return of internalized ligand to the surface has also been reported by others (66–68) and has been termed diacytosis (66). The internalized receptor-asialo-orosomucoid complexes in the slow dissociation compartment lose the capability to return to the cell surface with increasing time after internalization. This loss suggests that complexes in this compartment are either processed or translocated farther along the endocytic pathway and, in a time-dependent manner, pass a point in the pathway after which they cannot return to the surface (see Fig. 2A in ref. 65). We discovered that at 18°C (at which temperature the slow dissociation process ceases), there is no time-dependent loss of the ability of complexes to return to the surface (Fig. 4-4). This finding supports the idea that there *is* a processing or translocating step in the endocytic pathway after commit-

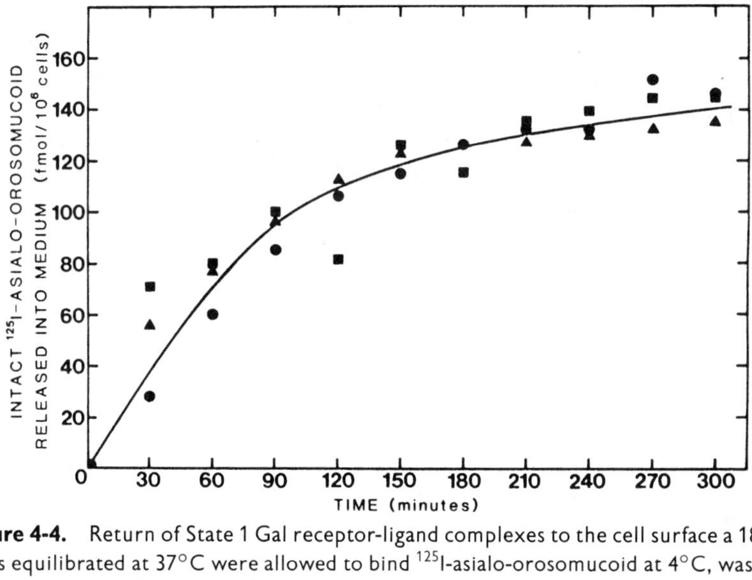

Figure 4-4. Return of State 1 Gal receptor-ligand complexes to the cell surface a 18°C. Cells equilibrated at 37°C were allowed to bind ^{125}I-asialo-orosomucoid at 4°C, washed, and incubated for 7 min at 37°C. The cells, containing endocytosed ligand, were then chilled on ice, washed with EGTA, and incubated at 18°C. EGTA was added to the cell suspensions at 0 (●), 60 (■), or 120 (▲) minutes and samples were taken at different times to determine the amount of intact ligand released into the medium.

ment of complexes to the slowly dissociating compartment but before these complexes can dissociate. This step does not occur at or below 18°C.

Evidence for Two Pathways for Ligand Degradation Mediated by Gal Receptors

Many laboratories, including ours, have studied the continuous internalization and degradation of asialoglycoproteins mediated by the Gal receptor in vivo, in perfused liver, and in isolated hepatocytes. Under conditions of continuous exposure to ligand, the kinetics of appearance of acid-soluble breakdown products consistently shows an apparent lag of 15-20 min and then a rapid increase to a steady-state rate (19). However, when the degradation of a single wave of surface-bound ligand is measured, we can follow the fate of ligand internalized by State 1 receptors alone or by State 1 and State 2 together (Fig. 4-5; 59). In the latter case, one observes similar kinetics to those seen under continuous uptake conditions; the kinetics are biphasic. However, during the apparent lag phase there is a genuinely slow rate of degradation that begins immediately. This usually has not been noticed under continuous uptake conditions because investigators have not been interested in collecting early kinetic data; the specific activity of the ligand has been too low; and the final steady-state rate overwhelms the brief early degradation rate. Cells that express only State 1 receptors at the start of the experiment show only the slower immediate degradation pathway. Cells that contain State 2 receptors also show the rapid degradation following a 20-min lag. Surface-bound asialo-orosomucoid bound to either State 1 or State 2 receptors is essentially completely degraded over about 6 hr at 37°C. The rates of ligand degradation are very similar ($\sim 0.006\ min^{-1}$) when expressed relative to the amount of ligand entering that pathway.

The immediate appearance of acid-soluble, low-molecular-weight products in the State 1 receptor pathway is a surprising result and suggests that degradation in this pathway may be nonlysosomal. Since there is no significant lag before degradation occurs, the primary endosomes containing receptor-ligand complexes must either already contain membrane-bound proteases (e.g., leucine aminopeptidase or dipeptidyl peptidase) or must fuse with a class of lysosomes or other protease-containing vesicles found at the cell periphery. The kinetics of ligand degradation by the State 1 pathway suggest that ligand cannot be delivered to lysosomes within the cell interior. By several other nonkinetic criteria, however, the State 1 degradation pathway has characteristics similar to those of the lysosomal pathway (21). Ligand degradation mediated by both State 1 and 2 Gal receptors is inhibited with almost identical sensitivity by chloroquine, leupeptin, and monensin (59). Microtubule drugs, discussed below, show a differential effect. Under conditions of continuous ligand uptake and degradation at 37°C, the State 1 pathway accounts for only 10–20% of the total ligand processing.

The State 2 Gal Receptor Pathway Uses Microtubules to Deliver Ligand to Lysosomes

Endocytosed molecules are transferred through a series of intracellular membrane compartments from the peripheral plasma membrane to the interior Golgi-lysosome region. This translocation process might involve the

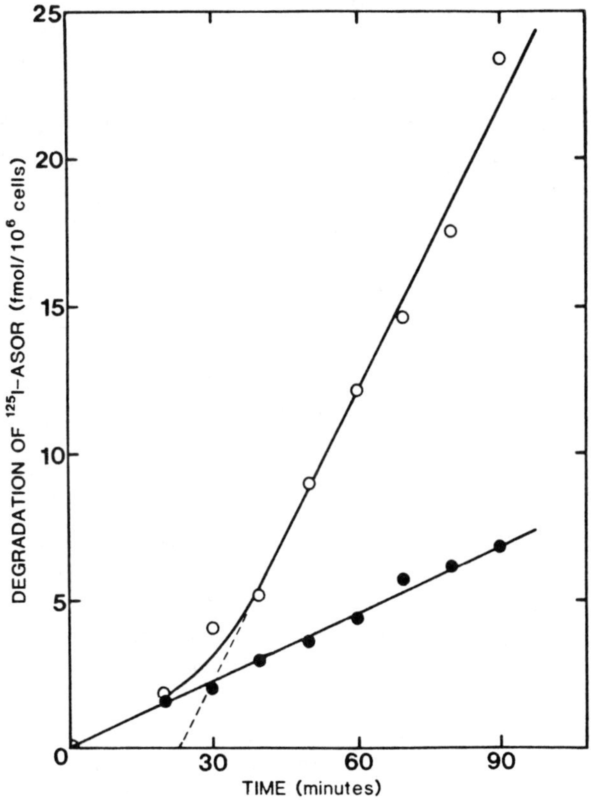

Figure 4-5. Degradation of ligand pre-bound to State 1 or to State 1 and State 2 surface Gal receptors. Cells containing only State 1 receptor activity (●) or 37°C equilibrated cells with both State 1 and State 2 receptor activity (○) were allowed to bind ^{125}I-asialo-orosomucoid at 4°C. The cells were washed and incubated at 37°C, and kinetics of appearance of acid-soluble radioactivity in the medium were determined.

movement of membrane vesicles and require the participation of cytoskeletal elements such as microfilaments or microtubules. Concentrating on the latter structures, we investigated the effects of microtubule depolymerizing drugs on the various stages of endocytosis (69). Microtubule drugs did not inhibit (a) the binding of asialo-orosomucoid to Gal receptors, (b) the endocytosis of a single round of surface-bound protein, (c) the biphasic dissociation of internalized receptor-ligand complexes (discussed above) or (d) the degradation of already internalized protein that had passed a certain point in the process by which ligand is delivered to lysosomes. However, these drugs very effectively inhibit degradation (by about 80-90%) if they are added before or with asialo-orosomucoid. The two receptor pathways discussed previously show a very different sensitivity to colchicine. The major pathway mediated by State 2 receptors is more sensitive to the drug by several orders of magnitude (59). We conclude that either the segregation of dissociated asialo-orosomucoid and State 2 receptors or the intracellular routing of internalized asialo-orosomucoid to lysosomes in the State 2 receptor pathway is inhibited by microtubule drugs. We believe that the State 2 Gal receptor pathway is the one that has been characterized by others investigating the Gal receptor system (e.g., 9, 21, 37, 39, 41).

Effects of Monensin on Endocytosis, Dissociation, and Receptor Recycling

Differential inhibition of the overall endocytic pathway by various inhibitors or low temperature provides a way to isolate and study its "separate" stages and to synchronize the system—to accumulate ligand or receptor at the point of inhibition. Such blockades should help define and correlate subcellular compartments at the electron microscope and the biochemical levels. We have therefore screened compounds to find potential inhibitors of the fast and slow dissociation processes, of receptor/ligand segregation, and of recycling. In addition to the results described above, we and others (70) find that monensin blocks the fast and slow dissociation processes. Harford and co-workers assessed the effect of short-term exposure to monensin but did not study the kinetics of receptor-ligand dissociation. When we examined the kinetics of dissociation in the presence of monensin at different concentrations, we observed that the inhibition was only temporary (Fig. 4-6). The apparently complete inhibition by monensin is actually transient; after about 20 min dissociation resumes. For example, even in the presence of 50 μM monensin, 40% of the internalized receptor-ligand complexes dissociate by 80 min. Further addition of monensin after 20 min has no affect on this dissociation, suggesting that it is not mediated by low pH.

At high concentrations (50 μM), monensin completely blocks continuous internalization, receptor recycling, and degradation, although a single round of endocytosis is not affected. Our results and the work of others indicate that monensin may exert various effects, depending on the concentration used. If monensin does block the transfer or processing of endosomes to the appropriate compartment(s) where dissociation occurs, then this could be used as a tool to synchronize the system and to define the biochemistry and morphology of those compartments. In contrast, we find that 1 mM chloroquine has no effect on the endocytosis of surface-bound ^{125}I-asialo-orosomucoid but completely prevents dissociation (Fig. 4-7). Unlike that for monensin, the effect of chloroquine is not temporary.

A transient, rather than complete, inhibition of dissociation greatly changes the interpretation of the mechanism of monensin inhibition. Monensin may not simply raise the pH of an acidic endosomal compartment in which dissociation occurs (41, 70), but rather it may delay the processing or

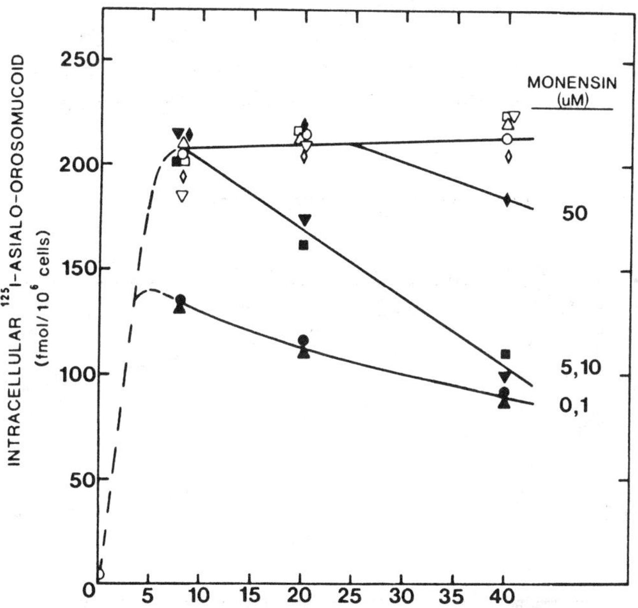

Figure 4-6. Effect of monensin on the dissociation of internalized receptor-ligand complexes. Cells containing surface-bound ^{125}I-asialo-orosomucoid were incubated for 10 min at 4°C in the presence of the indicated concentrations of monensin and then transferred to 37°C. Samples were removed and washed with EGTA, and the amounts of total (open symbols) and receptor-bound (filled symbols) intracellular ^{125}I-asialo-orosomucoid were determined by the digitonin technique.

translocation of nascent endocytic vesicles to the compartment(s) in which dissociation takes place. Acidification may not be necessary for the dissociation of all complexes. We believe that the fast dissociation process (involving State 2 receptors) is mediated by low pH, whereas the slow dissociation process (involving State 1 receptors) is mediated by a different mechanism and occurs, after a lag, even in the presence of monensin.

Modulation of Total Cell Gal Receptor Activity: Evidence for an Inactivation/Activation Cycle

The decrease in surface receptor activity in the presence of 10 mM sodium azide is about 50% as shown in Figure 4-2. The intracellular content of active receptors also decreases by about 50%. This decrease is not due to receptor degradation or loss, since recovery of receptor activity occurs as soon as the inhibitor is removed, even in the presence of cycloheximide (Fig. 4-8). The results strongly suggest that State 2 Gal receptors can be inactivated and then reactivated, even in the absence of ligand, in an energy-dependent process. In addition, several other results suggest the existence of an activation/inactivation cycle for the receptor: (a) The increased content of active

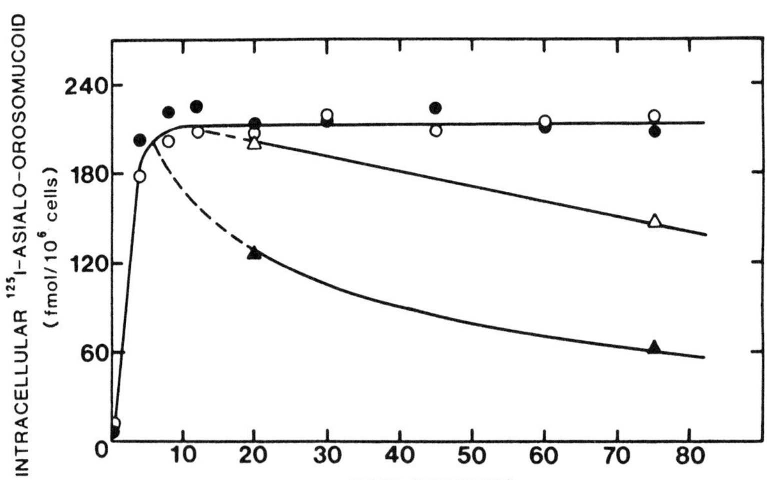

Figure 4-7. Effect of chloroquine on endocytosis and dissociation. Cells containing surface-bound ^{125}I-asialo-orosomucoid were incubated at 4°C with (○, ●) or without (▲, △) 1 mM chloroquine for 10 min and then put at 37°C. Total (○, △) and receptor-bound (▲, ●) ligand were determined as in Figure 4-6. Cell-associated radioactivity decreased in the control due to protein degradation.

surface receptors when freshly isolated cells are put at 37°C does not come from an existing internal pool of active receptors, but rather comes from an inactive pool (Fig. 1 in ref. 54). (b) When metabolic energy poisons, microtubule drugs, or low temperatures cause a decrease in surface receptor activity, there is *not* a corresponding increase in intracellular active receptor content (as assessed with digitonin). Active surface receptors do not simply move into the cell.

These data support the conclusion that the activity of State 2 Gal receptors can be modulated, although such regulation, whether covalent or noncovalent, could be at the cell surface (61) and/or inside the cell. One possible explanation for these results is that State 2 receptors undergo a covalent modification, such as phosphorylation, which regulates the activity of the protein. Studies in human hepatoma cells (9) and rat hepatocytes (10) showed that the Gal receptor is a phosphoprotein but failed to find evidence

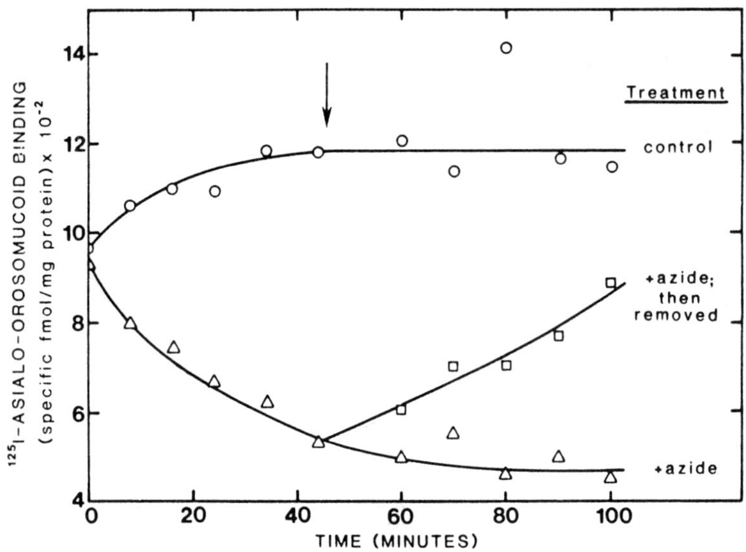

Figure 4-8. Effect of sodium azide on total cellular receptor activity. Freshly isolated cells were incubated at 0°C for 15 min with 2×10^{-5} M cycloheximide in the presence (□, △) or absence (○) of 10 mM sodium azide (NaN_3) and then incubated at 37°C. After 45 min (arrow) all the cell suspensions were washed once at 37°C by centrifugation and resuspended in fresh medium containing cycloheximide with (△) or without (□) azide. The incubation was continued at 37°C. Samples were removed and washed, and the total cell binding of ^{125}I-asialo-orosomucoid was determined at 4°C in the presence of digitonin. Nonspecific binding, measured by the inclusion of a 50-fold excess of nonradioactive protein, was 7% of the total.

for a functional role of phosphorylation/dephosphorylation in the receptor cycle. We believe this is still an exciting possibility, and it is a major focus of our current research.

A WORKING MODEL OF THE GALACTOSYL RECEPTOR SYSTEM

Our present model of how the Gal receptor system functions is shown in Figure 4-9. This model is based on various compartments containing ligand or receptor-ligand complexes, which have been kinetically defined (Table 4-1). The points at which the various inhibitors or treatments block the system are indicated. The nine stages in the endocytic process are designated by their corresponding numbers, as enumerated previously. We cannot explain our data with a model in which all cellular receptors are functionally

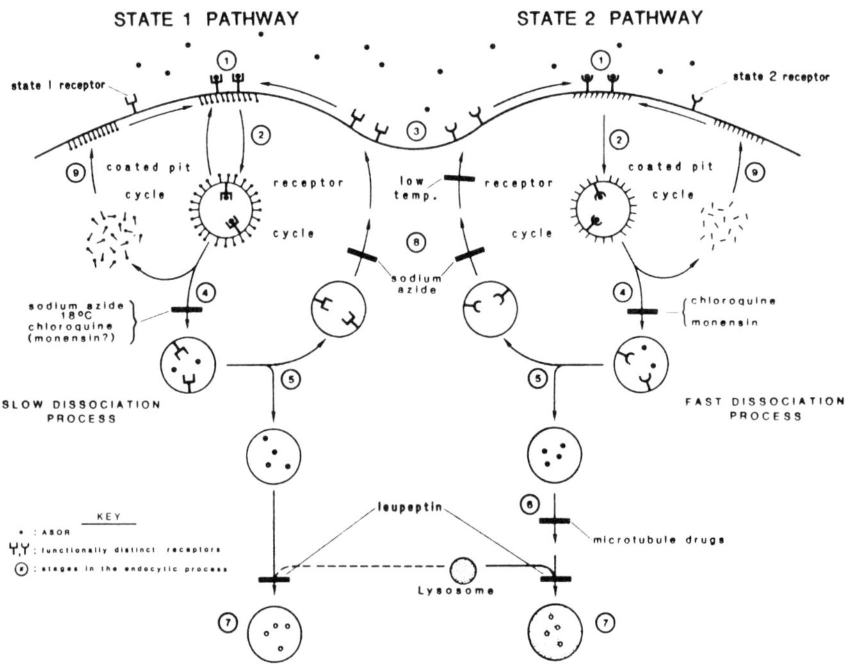

Figure 4-9. Functional model of the hepatic Gal receptor system. The two parallel pathways for internalization and ligand dissociation and degradation mediated by State 1 and State 2 Gal receptors are discussed in the text.

Table 4-1. Kinetic Rate Constants of the Hepatic Galactosyl Receptor System (37°C)[a]

Binding: $k_{on} = 3.2 \times 10^5\ M^{-1}\ s^{-1}$
Binding: k_{on} (4°C) $= 1.8 \times 10^4\ M^{-1}\ s^{-1}$
Binding: k_{off} (4°C) $= 0.9 \times 10^{-5}\ s^{-1}$
$K_a = 2 \times 10^9\ M^{-1}$
Endocytosis = 0.27 min^{-1}
Fast dissociation; State 2 pathway = 0.28 min^{-1}
Slow dissociation; State 1 pathway = 0.014 min^{-1}
Return of State 1 receptor-ASOR complexes to the cell surface = 0.07 min^{-1}
ASOR degradation; State 1 plus State 2 = 0.005-0.008 min^{-1}

[a]The rate constants for various stages of the endocytic process in isolated rat hepatocytes have been determined in this laboratory. In all cases the ligand used was asialo-orosomucoid (ASOR).

equivalent and are internalized via one pathway. Our model differs from others previously discussed (37, 71) in that it proposes two functionally different surface receptor populations and two different endocytic, dissociation, and degradation pathways that operate simultaneously. This unusual model explains our recent data and the results from other laboratories. In this model, Stage 3 does not occur as a separate partition event after internalization. Partition of receptor-ligand complexes into either the slow or fast dissociation pathway is inherent in the system, because surface receptors are functionally distinct and/or in different types of coated pits.

In the model, one pathway is mediated by State 2 receptors and is characterized by modulatable receptor activity, fast dissociation, and a 20-min lag prior to the onset of ligand degradation. State 2 receptors recycle in the absence of ligand. The other pathway is mediated by State 1 receptors, which are not modulatable and do not recycle in the absence of ligand. The State 1 receptor pathway is characterized by slow dissociation and the immediate onset of degradation. Degradation in the State 1 pathway may involve lysosomes (dashed line in Fig. 4-9), but this is not yet known.

The two functionally distinct surface receptor populations could reflect structural differences (5) in the receptors (represented by the two different symbols), which determine whether or not they are in coated pits. In this case the model would propose a coated and a noncoated endocytic pathway (28); the State 2 recycling receptors would follow the coated pit endocytic pathway and correspond to the rapidly dissociating pool. Alternatively, as is shown in Figure 4-9, both endocytic pathways could be mediated by different types of coated pits. In this case the receptors might be structurally different, and thereby be directed into the proper type of coated pit or they could be identical and randomly partitioned into coated pits, which are functionally different. There are already precedents for considering different functional roles for subpopulations of coated vesicles within the cell (72-74). As is the case for tubulin, a family of different clathrin proteins or different clathrin-associated proteins could modify the function of clathrin within the cell. It is not difficult to envision subpopulations of distinct coated membrane regions or vesicles with different functions and intracellular fates. Such a notion provides and attractive mechanism for regulating intracellular traffic of various types of molecules. At the electron microscopic level, such subpopulations of coated vesicles could easily be missed if the morphological differences were minor.

In addition to the results presented previously, several other recent experiments using low temperature to dissect the system provide compelling support for proposing two parallel endocytic pathways. At 18°C both the fast dissociation and receptor recycling processes occur (40). Asialoorosomucoid uptake is six times the surface (or two times the total) receptor number/cell within 5 hr. At 18°C the number of internalized complexes committed to the slowly dissociating compartment saturates, but these complexes do not dissociate. The slow dissociation process does not occur at 18°C, due to the inhibition of a temperature-sensitive translocation or processing step necessary for dissociation. At 37°C the amount of ligand that can return to the cell surface after internalization decreases with time (65), but at 18°C this decrease does not occur (Fig. 4-4). A stable percentage of surface receptors (State 2 receptors) does not bring ligand into the slow dissociation compartment, but rather enters the rapidly dissociating compartment exclusively and recycles. The State 1 receptors mediating uptake via the slow dissociation process are trapped at 18°C and cannot dissociate. This result suggests that the receptors which can be committed to the rapidly or to the slowly dissociating compartments are not functionally the same. Internalized receptor-ligand complexes, committed to the slow dissociation compartment, can return to the cell surface at 18°C (65), yet further processing and dissociation of these complexes does not occur. These complexes, once having returned to the surface, do not reenter the cell via the State 2 pathway; they cannot follow the route of the rapidly dissociating complexes. We know the fast dissociation pathway can occur because additional asialoorosomucoid added to these cells is processed this way. This again demonstrates that the surface Gal receptors, capable of bringing ligand into the fast or slow dissociation pathways, are somehow different.

If no trigger (or push button) is needed to initiate endocytosis of receptors, then the process could be likened to an escalator, rather than an elevator. The escalator could accommodate a variety of different receptors. We believe that the self-assembly and budding into the cell of clathrin-coated vesicles from coated pits provides the main driving force for endocytosis (75). A later energy-requiring step (such as clathrin uncoating of vesicles; 76) would satisfy the laws of thermodynamics (57) and keep the escalator "cycling." This view of the process requires two other interactions. One is between clathrin or an associated protein and the membrane to anchor the lipid bilayer to the dynamically growing clathrin basket. The favorable energetics of forming this structure can supply the energy to overcome the surface tension of the bilayer to form a droplet (vesicle). The other necessary interaction is to target and anchor receptors to the coated pit so that they are able to ride the escalator. This latter interaction would presumably occur between a cytoplasmic domain of the receptor and a specific component of the coated pit. Since the low-density lipoprotein (LDL; 77), epidermal growth factor (EGF; 78), chicken glycoprotein (79), and rat Gal receptors (5) do not contain a common cytoplasmic primary sequence, a different signal must be involved.

We have observed that each of the latter three receptors contains a Ser or Thr residue that can be phosphorylated (6, 9, 10, 80) and one to three Arginine (Arg) or lysine (Lys) residues within one to several amino acids on both the amino and carboxyl sides (Table 4-2). The transferrin receptor (81) and the V-erb-B oncogene protein (80) also have similar sequences. Introduction of a negatively charged phosphoryl group into a sequence with a positive charge could induce a dramatic conformational change. These receptors may form a class of molecules that are activated by phosphorylation and inactivated by dephosphorylation. The action of specific kinases and phosphates could regulate both receptor activity and targeting to coated pits and satisfy the thermodynamic requirement for energy to sustain a true cycle. The receptor phosphorylated at a cytoplasmic domain containing Arg or Lys could be recognized by a coated pit component and then be able to enter the cell, whether or not it is bound to the appropriate ligand. After internalization the receptor would be dephosphorylated and subsequently inactivated by an as yet unknown transmembrane mechanism. Transmembrane regulation of receptor activity (i.e., the ability of a cytoplasmic modification of the protein to affect the function of a domain on the other side of the membrane) has been postulated for the Gal receptor (61). Receptor inactivation would make dissociation irreversible and would greatly increase the efficiency of the next step (Stage 5), segregation of the ligand and the receptor.

As investigators continue to closely examine the LDL, EGF, transferrin,

Table 4-2. Phosphorylated Cytoplasmic Sequences of the EGF, Chicken GlcNAc, and Rat Gal Receptors[a]

EGF receptor[b]:

645 *** 662
-*ARG*-*ARG*-*ARG*-HIS-ILE-VAL-*ARG*-*LYS*-*ARG*-THR-LEU-*ARG*-*ARG*-LEU-LEU-GLN-GLU-*ARG*-

Gal receptor:

26 *** 39
-ALA-PRO-*ARG*-LEU-LEU-GLN-*ARG*-LEU-CYS-SER-GLY-PHE-*ARG*-LEU-

GlcNAc receptor:

3 *** 15
-GLU-GLU-*ARG*-LEU-SER-ASP-ASN-VAL-*ARG*-LEU-TYR-*LYS*-GLY-

Transferrin receptor[c]:

51 +++ 64
-ASN-THR-*LYS*-ALA-ASN-VAL-THR-*LYS*-PRO-*LYS*-*ARG*-CYS-SER-GLY-

[a]The phosphorylated serine or threonine residues are indicated by three asterisks, and the positively charged lysine and arginine residues are underlined.
[b]The V-erb-B oncogene product has the same sequence from amino acids 89-110 as shown for the EGF receptor (80). There are also two other similar cytoplasmic sequences in EGF relatively close to the transmembrane spanning region; 681-692 and 721-733 (78).
[c]The transferrin receptor has a similar sequence (81). The potentially phosphorylated threonine is indicated by three + signs.

immunoglobulin Fc, and other receptor systems, they will probably recognize that the two-pathway model presented here for the Gal receptor system is generally applicable to these systems as well.

ACKNOWLEDGMENTS

This research was supported by National Institutes of Health Grant GM 30218. I thank Betty Jackson for her help in preparing the manuscript and Janet Oka for her help with the experiments and figures.

REFERENCES

1. G. Ashwell, and A. G. Morell, 1974, The role of surface carbohydrates in the hepatic recognition and transport of circulating glycoproteins, *Adv. Enzymol.* **41**:99-128.
2. G. Ashwell, and J. Harford, 1982, Carbohydrate-specific receptors of the liver, *Annu. Rev. Biochem.* **51**:531-554.
3. K. B. Chiacchia, and K. Drickamer, 1984, Direct evidence for the transmembrane orientation of the hepatic glycoprotein receptors, *J. Biol. Chem.* **259**:15440-15446.
4. C. J. Steer, E. S. Kempner, and G. Ashwell, 1981, Molecular size of the hepatic receptor for asialoglycoproteins determined in situ by radiation inactivation, *J. Biol. Chem.* **256**:5851-5856.
5. K. Drickamer, J. F. Mamon, G. Binns, and J. O. Leung, 1984, Primary structure of the rat liver asialoglycoprotein receptor. Structural evidence for multiple polypeptide species, *J. Biol. Chem.* **259**:770-778.
6. K. Drickamer, and J. F. Mamon, 1982, Phosphorylation of the membrane receptor for glycoproteins. Possible transmembrane orientation of the chicken hepatic lectin, *J. Biol. Chem.* **257**:15156-15161.
7. E. C. Holland, J. O. Leung, and K. Drickamer, 1984, Rat liver asialoglycoprotein receptor lacks a cleavable NH_2-terminal signal sequence, *Proc. Natl. Acad. Sci. (USA)* **81**:7338-7342.
8. M. Spiess, and H. F. Lodish, 1985, Sequence of a second human asialoglycoprotein receptor: Conservation of two receptor genes during evolution, *Proc. Natl. Acad. Sci. (USA)* **82**:6465-6469.
9. A. L. Schwartz, 1984, Phosphorylation of the human asialoglycoprotein receptor, *Biochem. J.* **223**:481-486.
10. T. Takahashi, H. Nakada, T. Okumura, T. Sawamura, and Y. Tashiro, 1985, Phosphorylation of the rat hepatocyte asialoglycoprotein receptor, *Biochem. Biophys. Res. Commun.* **126**:1054-1060.
11. M. Sarkar, J. Liao, E. A. Kabat, T. Tanabe, and G. Ashwell, 1979, The binding site of rabbit hepatic lectin, *J. Biol. Chem.* **254**:3170-3174.
12. R. T. Lee, 1982, Binding site of the rabbit liver lectin specific for galactose/N-acetyl galactosamine, *Biochemistry* **21**:1045-1050.
13. J. U. Baenziger, and Y. Maynard, 1980, Human hepatic lectin, physiochemical properties and specificity, *J. Biol. Chem.* **255**:4607-4613.
14. D. A. Ray, J. A. Oka, and P. H. Weigel, 1984, The stoichiometry of ligand binding by the hepatic galactosyl receptor is increased by detergents, *Fed. Proc.* **43**:2043.
15. M. R. Hardy, R. R. Townsend, S. M. Parkhurst, and Y. C. Lee, 1985, Different modes of ligand binding to the hepatic galactose/N-acetylgalactosamine lectin on the surface of rabbit hepatocytes, *Biochemistry* **24**:22-28.
16. J. U. Baenziger, and D. Fiete, 1980, Galactose and N-acetyl-galactosamine-specific endocytosis of glycopeptides by isolated rat hepatocytes, *Cell* **22**:611-620.

17. Y. C. Lee, R. R. Townsend, M. R. Hardy, J. Lönngren, J. Arnarp, M. Haraldsson, and H. Lönn, 1983, Binding of synthetic oligosaccharides to the hepatic Gal/Gal NAc lectin. Dependence on fine structural features, *J. Biol. Chem.* **258:**199-202.
18. H. Tolleshaug, and T. Berg, 1979, Chloroquine reduces the number of asialoglycoprotein receptors in the hepatocyte plasma membrane, *Biochem. Pharmacol.* **28:**2919-2922.
19. P. H. Weigel, and J. A. Oka, 1982, Endocytosis and degradation mediated by the asialoglycoprotein receptor in isolated rat hepatocytes, *J. Biol. Chem.* **257:**1201-1207.
20. P. H. Weigel, and J. A. Oka, 1983, The surface content of asialoglycoprotein receptors on isolated hepatocytes is reversibly modulated by changes in temperature, *J. Biol. Chem.* **258:**5089-5094.
21. D. A. Wall, G. Wilson, and A. L. Hubbard, 1980, The galactose-specific recognition system of mammalian liver: The route of ligand internalization in rat hepatocytes, *Cell* **21:** 79-93.
22. D. A. Wall, and A. L. Hubbard, 1981, Galactose-specific recognition system of mammalian liver: Receptor distribution on the hepatocyte cell surface, *J. Cell Biol.* **90:**687-696.
23. V. Kolb-Bachofen, J. Schlepper-Schafer, and H. Kolb, 1983, Receptor-mediated particle uptake by liver macrophages. The galactose-particle receptor mediates uptake by a coated and also non-coated structures, *Exp. Cell Res.* **148:**173-182.
24. V. Kolb-Bachofen, 1981, Hepatic receptor for asialoglycoprotein: Ultrastructural demonstration of ligand-induced microaggregation of receptors, *Biochim. Biophys. Acta* **645:**293-299.
25. I. Pastan, and M. C. Willingham, 1983, Receptor-mediated endocytosis: Coated pits and receptosomes in the Golgi, *Trends Biochem. Sci.* **506:**250-254.
26. A. Helenius, I. Mellman, D. Wall, and A. Hubbard, 1983, Endosomes, *Trends Biochem. Sci.* **506:**245-254.
27. T. Tanabe, W. E. Pricer, Jr., and G. Ashwell, 1979, Subcellular membrane topology and turnover of a rat hepatic binding protein specific for asialoglycoproteins, *J. Biol. Chem.* **254:**1038-1043.
28. E. Regoeczi, M. T. Debanne, M. W. C. Hatton, and A. Koj, 1978, *Biochim. Biophys. Acta* **541:**372-384.
29. C. J. Steer, and G. Ashwell, 1980, Studies on a mammalian hepatic binding protein specific for asialoglycoproteins: Evidence for receptor recycling in isolated hepatocytes, *J. Biol. Chem.* **255:**3008-3013.
30. R. Warren, and D. Doyle, 1981, Turnover of the surface proteins and the receptor for serum asialoglycoproteins in primary cultures of rat hepatocytes, *J. Biol. Chem.* **256:**1346-1355.
31. J. U. Baenziger, and D. Fiete, 1982, Recycling of the hepatocyte asialoglycoprotein receptor does not require delivery of ligand to lysosomes, *J. Biol. Chem.* **257:**6007-6009.
32. W. A. Dunn, A. L. Hubbard, and N. N. Aronson, Jr., 1980, Low temperature selectively inhibits fusion between pinocytic vesicles and lysosomes during heterophagy of ^{125}I-asialofetuin by their perfused rat liver, *J. Biol. Chem.* **255:**5971-5978.
33. P. H. Weigel, and J. A. Oka, 1981, Temperature dependence of endocytosis mediated by the asialoglycoprotein receptor in isolated rat hepocytes: Evidence for two potentially rate limiting steps, *J. Biol. Chem.* **256:**2615-2617.
34. W. E. Pricer, Jr., and G. Ashwell, 1976, Subcellular distribution of a mammalian hepatic binding protein specific for asialoglycoproteins, *J. Biol. Chem.* **251:**7539-7544.
35. R. J. Stockert, U. Gartner, A. G. Morell, and A. W. Wolkoff, 1980, Effects of receptor-specific antibody on the uptake of desialylated glycoproteins in the isolated perfused rat liver, *J. Biol. Chem.* **255:**3830-3831.
36. P. H. Weigel, 1981, Evidence that the hepatic asialoglycoprotein receptor is internalized during endocytosis and that receptor recycling can be uncoupled from endocytosis at low temperature, *Biochem. Biophys. Res. Commun.* **101:**1419-1425.
37. K. Bridges, J. Harford, G. Ashwell, and R. D. Klausner, 1982, Fate of receptor and ligand

during endocytosis of asialoglycoproteins by isolated hepatocytes, *Proc. Natl. Acad. Sci. (USA)* **79:**350-354.

38. M. C. Herzig, and P. H. Weigel, 1986, Preparation and characterization of chemical affinity probes for the hepatic galactosyl receptor, *Fed. Proc.* (in press).
39. J. Harford, K. Bridges, G. Ashwell, and R. D. Klausner, 1983, Intracellular dissociation of receptor-bound asialoglycoproteins in cultured hepatocytes: A pH-mediated nonlysosomal event, *J. Biol. Chem.* **258:**3191-3197.
40. J. A. Oka, and P. H. Weigel, 1983, Recycling of the asialoglycoprotein receptor in isolated rat hepatocytes: Dissociation of internalized ligand from receptor occurs in two kinetically and thermally distinguishable compartments, *J. Biol. Chem.* **258:**10253-10262.
41. B. Tycko, C. H. Keith, and F. R. Maxfield, 1983, Rapid acidification of endocytic vesicles containing asialoglycoprotein in cells of a human hepatoma line, *J. Cell Biol.* **97:**1762-1776.
42. M. DiPaola, and F. R. Maxfield, 1984, Conformational changes in the receptors for epidermal growth factor and asialoglycoproteins induced by the mildly acidic pH found in endocytic vesicles, *J. Biol. Chem.* **259:**9163-9171.
43. J. Harford, and G. Ashwell, 1982, The hepatic receptor for asialoglycoproteins, *The Glycoconjugates* **4:**27-55.
44. K. -L. Wong, P. A. Charlwood, M. W. C. Hatton, and E. Regoeczi, 1974, Studies of the metabolism of asialo-transferrin: Evidence that transferrin does not undergo desialylation in vivo *Clin. Sci. Mol. Med.* **46:**763-774.
45. R. Clarenburg, 1983, Asialoglycoprotein receptor is uninvolved in clearing intact glycoproteins from rat blood, *Am. Phys. Soc.* **244:**G247-G253.
46. M. J. Kuranda, and N. N. Aronson, Jr., 1983, Tissue locations for the turnover of radioactively labeled rat orosomucoid in vivo, *Arch. Biochem. Biophys.* **224:**526-533.
47. W. S. Sly, and H. D. Fischer, 1982, The phosphylmannosyl recognition system for intracellular and intercellular transport of lysosomal enzymes, *J. Cell. Biochem.* **18:**67-85.
48. K. Olden, J. B. Parent, S. L. White, 1982, Carbohydrate moieties of glycoproteins: A reevaluation of their function, *Biochim. Biophys. Acta* **650:**209-232.
49. R. W. Thornburg, J. F. Day, J. W. Baynes, and S. R. Thorpe, 1980, Carbohydrate-mediated clearance of immune complexes from the circulation: A role for galactose residues in the hepatic uptake of IgG-antigen complexes, *J. Biol. Chem.* **255:**6820-6825.
50. R. J. Stockert, M. S. Kressner, J. C. Collins, I. Sternlieb, and A. G. Morell, 1982, IgA interaction with the asialoglycoprotein receptor, *Proc. Natl. Acad. Sci. (USA)* **79:**6229-6231.
51. G. F. Springer, R. Cheingsong-Popov, V. Schirrmacher, P. R. Desai, and H. Tegtmeyer, 1983, Proposed molecular basis of murine tumor cell-hepatocyte interaction, *J. Biol. Chem.* **258:**5702-5706.
52. R. Cheingsong-Popov, P. Robinson, P. Altevogt, and V. Schirrmacher, 1983, A mouse hepatocyte carbohydrate-specific receptor and its interaction with liver-metastasizing tumor cells, *Int. J. Cancer* **32:**359-366.
53. P. H. Weigel, 1980, Rat hepatocytes bind to synthetic galactoside surfaces via a patch of asialoglycoprotein receptors, *J. Cell Biol.* **87:**855-861.
54. P. H. Weigel, D. A. Ray, and J. A. Oka, 1983, Quantitation of intracellular membrane-bound enzymes and receptors in digitonin-permeabilized cells, *Anal. Biochem.* **133:**437-449.
55. P. H. Weigel, and J. A. Oka, 1983, The large intracellular pool of asialoglycoprotein receptors functions during the endocytosis of asialoglycoproteins by isolated rat hepatocytes, *J. Biol. Chem.* **258:**5095-5102.
56. R. J. Stockert, D. J. Howard, A. G. Morell, and I. H. Scheinberg, 1980, Functional segregation of hepatic receptors for asialoglycoproteins during endocytosis, *J. Biol. Chem.* **255:**9028-9029.
57. B. L. Clarke, and P. H. Weigel, 1985, Recycling of the asialoglycoprotein receptor in isolated

rat hepatocytes: ATP depletion blocks receptor recycling but not a single round of endocytosis, *J. Biol. Chem.* **260:**128-133.
58. P. H. Weigel, 1980, Characterization of the asialoglycoprotein receptor on isolated rat hepatocytes, *J. Biol. Chem.* **255:**6111-6120.
59. P. H. Weigel, B. L. Clarke, and J. A. Oka, 1986, Endocytosis and ligand dissociation and degradation mediated by the hepatic galactosyl receptor occur via two different pathways, *Fed. Proc.* (in press).
60. P. H. Weigel, B. L. Clarke, and J. A. Oka, 1984, Surface and intracellular galactosyl receptor activity is reversibly decreased by metabolic energy poisons, *J. Cell Biol.* **99:**327a.
61. D. Fiete, M. D. Brownell, and J. U. Baenziger, 1983, Evidence for transmembrane modulation of the ligand-binding site of the hepatocyte galactose/N-acetyl galactosamine-specific receptor, *J. Biol. Chem.* **258:**817-823.
62. P. L. Zeitlin, and A. L. Hubbard, 1982, Cell surface distribution and intracellular fate of asialoglycoproteins: A morphological and biochemical study of isolated rat hepatocytes and monolayer cultures, *J. Cell Biol.* **92:**634-647.
63. S. Matsura, H. Nakada, T. Sawamura, and Y. Tashiro, 1982, Distribution of asialoglycoprotein receptor on rat hepatocyte cell surface, *J. Cell Biol.* **95:**864-875.
64. J. A. Oka, B. L. Clarke, and P. H. Weigel, 1983, The internal pool of slowly dissociating receptor-asialoglycoprotein complexes returns to the cell surface prior to dissociation, *J. Cell Biol.* **97:**428a.
65. P. H. Weigel, and J. A. Oka, 1984, Recycling of the hepatic asialoglycoprotein receptor in isolated rat hepatocytes: Receptor-ligand complexes in an intracellular slowly dissociating pool return to the cell surface prior to dissociation, *J. Biol Chem.* **259:**1150-1154.
66. H. Tolleshaug, P. A. Chindemi, and E. Regoeczi, 1981, Diacytosis of human asialo-transferrin type III by isolated rat hepatocytes, *J. Biol. Chem.* **256:**6526-6528.
67. D. T. Connolly, R. R. Townsend, K. Kawaguchi, W. R. Bell, and Y. C. Lee, 1982, Binding and endocytosis of cluster glycosides by rabbit hepatocytes: Evidence for a short-circuit pathway that does not lead to degradation, *J. Biol. Chem.* **257:**939-945.
68. E. Regoeczi, P. A. Chindemi, M. T. Debanne, and M. W. C. Hatton, 1982, Dual nature of the hepatic lectin pathway for human asialo-transferrin type III in the rat, *J. Biol. Chem.* **257:**5431-5436.
69. J. A. Oka, and P. H. Weigel, 1983, Microtubule-depolymerizing agents inhibit asialo-orosomucoid delivery to lysosomes but not its endocytosis or degradation in isolated rat hepatocytes, *Biochim. Biophys. Acta* **763:**368-376.
70. J. Harford, A. W. Wolkoff, G. Ashwell, and R. D. Klausner, 1983, Monensin inhibits intracellular dissociation of asialoglycoproteins from their receptor, *J. Cell Biol.* **96:**1824-1828.
71. A. L. Schwartz, S. E. Fridovich, and H. F. Lodish, 1982, Kinetics of internalization and recycling of the asialoglycoprotein receptor in a hepatoma cell line, *J. Biol. Chem.* **257:**4230-4237.
72. J. L. Salisbury, J. S. Condeelis, and P. Satir, 1983, Receptor-mediated endocytosis: Machinery and regulation of the clathrin-coated vesicle pathway, *Int. Rev. Exp. Pathol.* **24:**1-62.
73. C. J. Steer, and R. D. Klausner, 1983, Clathrin-coated pits and coated vesicles: Functional and structural studies, *Hepatology* **3:**437-454.
74. S. R. Pfeffer, and R. B. Kelly, 1985, The subpopulation of brain coated vesicles that carries synaptic vesicle proteins contains two unique polypeptides. *Cell* **40:**949-957.
75. M. H. Fagan, and T. G. Dewey, 1984, Statistical mechanical model of the energetics of coated vesicle formation, *Biophys. J.* **45:**383-387.
76. S. L. Schmid, and J. E. Rothman, 1985, Enzymatic dissociation of clathrin cages in a two stage process, *J. Biol. Chem.* **260:**10044-10049.
77. D. W. Russell, W. J. Schneider, T. Yamamoto, K. L. Luskey, M. S. Brown, and J. L. Goldstein,

1984, Domain map of the LDL receptor: Sequence homology with the epidermal growth factor precursor, *Cell* **35:**577-585.

78. A. Ulrich, L. Coussens, J. S. Hayflick, T. J. Dull, A. Gray, A. W. Tam, J. Lee, Y. Yarden, T. A. Libermann, J. Schlessinger, J. Downward, E. L. V. Mayes, N. Whittle, M. D. Waterfield, and P. H. Seeburg, 1984, Human epidermal growth factor receptor cDNA sequence and aberrant expression of the amplified gene in A431 epidermoid carcinoma cells, *Nature* **309:**418-425.
79. K. Drickamer, 1981, Complete amino acid sequence of a membrane receptor for glycoproteins: Sequence of the chicken hepatic lectin, *J. Biol. Chem.* **256:**5827-5839.
80. T. Hunter, N. Ling, and J. A. Cooper, 1984, Protein kinase C phosphorylation of the EGF receptor at a threonine residue close to the cytoplasmic face of the plasma membrane, *Nature* **311:**480-483.
81. C. Schneider, M. J. Owen, D. Danville, and J. G. Williams, 1984, Primary structure of human transferrin receptor deduced from the mRNA sequence, *Nature* **311:**675-678.

5

Binding Protein Specific for Mannose and N-acetylglucosamine (Mannan-Binding Protein, MBP)

Toshisuke Kawasaki, Kazutoshi Mori, Shogo Oka, Nobuko Kawasaki, Yasunori Kozutsumi, and Ikuo Yamashina
Kyoto University

The isolation and characterization of binding proteins specific for galactose- or N-acetylglucosamine-terminated glycoproteins from respective livers of mammalian (1-4) and avian species (5, 6) implied the presence of another class of binding protein specific for mannose residues. Utilizing the yeast *(Saccharomyces cerevisiae)* mannan as a ligand, we could isolate binding proteins with dual specificity for mannose and N-acetylglucosamine from mammalian and avian tissues and sera, and they were referred to tentatively as *mannan-binding protein* (MBP). In this chapter, properties of MBP isolated from various sources will be described together with the possible functions of these proteins.

ISOLATION AND CHARACTERIZATION OF MBP

MBP of Mammalian Liver

The finding that ^{125}I-labeled mannan binds to the particulate fraction, in the presence of Ca^{2+}, of rat liver homogenates was exploited for the purpose of isolating the binding protein. The Triton X-100 extract of the acetone powder of rat liver was mixed with a slurry of affinity resin of Sepharose 4B to which yeast mannan had been covalently attached. The binding protein bound to the resin in the presence of Ca^{2+}, and was then eluted with a buffer

Table 5-1. Properties of MBP Isolated from Liver and Serum

MBP	*Mr intact protein*	*Mr subunit*	*Scatchard plot of mannan binding*	*Kd for mannan (M)*	*Glycine content (residues/ 100 residues)*	*Yield*[a] *(mg)*
Rat liver	194,000	32,000	c[b]	1.6×10^{-8} [d]	11.7	0.3
Rabbit liver	250,000	31,000	c[b]	1.6×10^{-9} [d]	13.9	1-2
Chicken liver	280,000	26,000	c[b]	7.1×10^{-9} [d]	14.9	0.2
	~740,000	31,000				
Rabbit serum	>500,000	29,000	s[c]	2.3×10^{-9}	12.5	2-3
Human serum	650,000	31,000	s[c]	2.2×10^{-9}	14.8	0.1

[a]The value means the milligrams of the purified MBP obtained from 100 g of liver or 100 ml or serum.
[b]c = curvelinear line.
[c]s = straight line.
[d]The value represents the Kd of a higher affinity site.

containing ethylene diaminetetra-acetic acid (EDTA) but no Ca^{2+}. The eluate was rerun over the column two more times to achieve complete removal of the binding protein. Properties of MBP isolated by these procedures are summarized in Table 5-1. Thus, the isolated binding protein behaved as a homogeneous protein with an estimated molecular weight of 190,000 by gel filtration and polyacrylamide gel electrophoresis (PAGE) in the presence of Triton X-100. On SDS (sodium dodecyl sulfate)-PAGE under reduced conditions, the binding protein migrated as a single band with an estimated molecular weight of 32,000, while under nonreduced conditions, it migrated as two bands corresponding to a dimer (61,000) and a trimer (89,000) of the 32,000 subunit, respectively. Thus, the intact binding protein appears to be a hexamer of the single subunit. An interesting chemical characteristic of the binding protein is its high content of glycine, which constitutes more than 10% (mol/mol) of the amino acid residues (7).

A similar, and probably identical, binding protein was isolated from rat liver by Townsend and Stahl (8), Maynard and Baenziger (9), and Wild and co-workers (10) by using a modification of the procedures described above. Analogous binding proteins were also obtained from rabbit (11, 12) and human (10) livers.

The binding reaction of MBP with mannan was quick and complete, occurring within 5 min at 25°C, and was essentially independent of the amount of ligand added. The reaction was reversible, with the bound ligand being displaced within 10 min upon the addition of excess free ligand (7).

Ca^{2+} is required for binding, as no binding occurred either in its absence or in the presence of EDTA. The concentration of Ca^{2+} giving the half maximum binding activity was 0.2 mM. Ca^{2+} protected MBP from inactivation by proteases, suggesting that the Ca^{2+} directly interacts with MBP, presumably to provide adequate protein conformation to bind. The fact that ligands showed a significant protective effect in the presence of Ca^{2+}, but not in its absence, is in accordance with this presumption (13).

The stability of the binding complex of MBP and mannan depends upon the pH. After incubation of the binding complex for 30 min in the pH range from 7 to 10, no dissociation was observed. Under acidic conditions, on the other hand, the complex became progressively more unstable and dissociation was 50% at pH 5.5 and complete at pH 5.0 (7, 12).

A Scatchard plot of the binding data between rat or rabbit liver MBP and mannan showed a curvilinear profile (see Table 5-1), suggesting the presence of more than one binding site (7, 12).

Yeast mannan is a glycoprotein consisting of an oligosaccharide portion, the structure of which is similar to high-mannose-type oligosaccharides of mammalian glycoproteins (14), and a peptide portion; however, the oligosaccharide portion was responsible for the specific binding (11).

Of the various sugars tested as haptenic inhibitors for mannan binding, mannose was, as expected, a potent inhibitor; N-acetylglucosamine, a minor component of mannan, was a more potent inhibitor, and N-acetylmannosamine, which is not normally present in mannan, was the most potent inhibitor. On the other hand, mannosamine, glucosamine, galactose, N-acetylgalactosamine, mannose 6-phosphate, glucose 6-phosphate, and N-acetylneuraminic acid were very weak inhibitors or not inhibitory at all (7, 8, 11). Studies using oligomannosides prepared from yeast mannan demonstrated that α-1, 6-linked mannose was a more potent inhibitor than α 1, 2- or α 1, 3-linked mannose when the sizes of the oligosaccharides were comparable (11). The effect of anomeric configuration, tested by using p-nitrophenyl glycosides, indicated that α-mannoside and α-N-acetylglucoaminide were better inhibitors than the corresponding β-isomers (15).

The dual specificity of binding toward mannose and N-acetylglucosamine might indicate two distinct binding sites on the binding protein. To help solve the problem, neoglycoproteins, $GlcNAc_{32}$-BSA and Man_{38}-BSA (BSA = bovine serum albumin), were used as ligands. These neoglycoproteins have the advantage of being simple not only in their carbohydrate structure but also in their binding kinetics, giving a straight line on a Scatchard plot analysis over a broader range of ligand concentrations. In hapten inhibition studies of neoglycoprotein binding, mannose and N-acetylglucosamine inhibited the binding almost indistinguishably from each other irrespective of the ligand used. Furthermore, the affinity of Man-BSA for the MBP is as high as that of GlcNAc-BSA, and inhibited the binding of ^{125}I-GlcNAc-BSA competitively when added together (7). These results suggest the involvement of

Table 5-2. Affinity of Various Glycoproteins to MBP

Glycoprotein	*Ki (M)*[a]	*Ratio*
OR	4.7×10^{-4}	29,000
ASOR	2.5×10^{-5}	1,600
AGOR	9.8×10^{-7} [b]	61
Cathepsin D (porcine spleen)	5.4×10^{-6} [c]	340
Ovalbumin	3.8×10^{-6}	240
Gal/GalNAc-binding protein (rabbit liver)	3.6×10^{-7}	23
α-Mannosidase (porcine kidney)	3.1×10^{-8} [c]	1.9
Thyroglobulin (bovine)	2.1×10^{-8}	1.3
Mannan	1.6×10^{-8} [b]	1.0
Man_{33}-BSA	4.8×10^{-9} [b]	0.30
$GlcNAc_{43}$-BSA	1.8×10^{-9} [b]	0.11

[a]The inhibition constant (Ki) of glycoproteins was determined by a Scatchard plot analysis of the binding data between rat liver MBP and ^{125}I-Man_{33}-BSA in the presence or absence of the glycoproteins tested.
[b]The value denotes Kd determined by a Scatchard plot analysis.
[c]The value is for rabbit liver MBP.

a common binding site recognizing both mannose and N-acetylglucosamine. However, the unlikely possibility that MBP has two binding sites, each specific for N-acetylglucosamine and mannose, which are located close enough to be blocked nonspecifically by the ligand on the other site, is not completely eliminated.

The affinity of various glycoproteins for the binding protein is shown in Table 5-2. Since kinetic analysis revealed that all the glycoproteins listed in this table inhibited the binding of Man-BSA competitively, the affinity was expressed as the inhibition constant (Ki). Despite the fact that these glycoproteins share a common inner core structure consisting of mannose and N-acetylglucosamine, Man α1, 6[Man α1, 3]Man β1, 4GlcNAc β 1, 4GlcNAc β, Asn, marked variations were found in their affinities. Orosomucoid (OR) and its derivatives are complex type glycoproteins (shown at the top). Of these, Agalacto-orosomucoid (AGOR), which has terminal N-acetylglucosamine residues, showed a high affinity, while asialo-orosomucoid (ASOR) and OR, in which these N-acetylglucosamine residues are covered with galactose and N-acetylneuraminylgalactose, respectively, showed a very low or negligible affinity. In contrast, glycoproteins with high-mannose-type oligosaccharides (shown in the middle), which have *per se* terminal mannose residues, bound to the binding protein with high affinity. Gal/GalNAc binding protein (3, 4), thyroglobulin, and α-mannosidase, which are glycoproteins with multi-high-mannose-type oligosaccharides, showed a much higher affinity than ovalbumin, which has single oligosaccharide. The former two are known to bear complex-type oligosaccharides as well, indicating that the coexistence of complex-type oligosaccharides with high-mannose-type oligosaccharides does not seem to significantly affect the binding affinity.

Evidently, all the information required for recognition and binding resides in the carbohydrate moiety of these glycoproteins. Thus, upon selective destruction of the carbohydrate moiety by $NaIO_4$ oxidation, α-mannosidase and Gal/GalNAc binding protein completely lost their inhibitory activity, while the enzymatic and the binding activities of respective molecules were retained fully (11, 12). However, the glycopeptides and oligosaccharides prepared from glycoproteins showed a much lower affinity (Ki = on the order of 10^{-4} M) than the respective parental molecules. This was true even for a glycoprotein with a single oligosaccharide chain like ovalbumin. Thus, in contrast to the considerable high affinity of ovalbumin (Ki = 3.8×10^{-6} M), Ki of a glycopeptide (Fr. IV-D) and an oligosaccharide $(Man)_6$GlcNAcol, obtained therefrom was 4.6×10^{-4} M and 5.4×10^{-4} M, respectively (13). Similar results were also reported by Maynard and Baenziger (9). These results, together with the earlier observation that reduced and alkylated ribonuclease B was approximately 60-fold more potent than ribonuclease B in inhibiting binding (9), suggest a role for the polypeptide portion in determining their affinity. Although the nature of the role of the polypeptide portion is not clear, it may involve hydrophobic interactions. The finding that *p*-nitrophenyl glycosides, mannosides, and N-acetylglucosaminides were more potent inhibitors than the corresponding monosaccharides (15) is consistent with this view.

The specificity of MBP deduced from the above findings may be summarized as follows: MBP recognizes and binds primarily mannose and/or N-acetylglucosamine residues exposed at the nonreducing termini of glycoproteins. As might be expected, glycoproteins with multi-high-mannose-type oligosaccharides are the most potent natural ligands, although a role of the polypeptide portion in determining the affinity of a molecule cannot be dismissed. Maynard and Baenziger concluded that the binding protein binds to the core sugars in asparagine-linked oligosaccharides Man α1, 6[Man α1, 3]Man β1, 4GlcNAc β1, 4 GlcNAc β, Asn and the effect of peripheral sugars, galactose and N-acetylneuraminic acid, is not significant, as is the case with concanavalin A (9). However, this is not consistent with the results shown in Table 5-2 or with the findings that ASOR is over 100-fold less effective than AGOR (9). In addition, the dissociation constant (Kd) for AGOR shown in Table 5-1 is over 450-fold larger than that reported by Maynard and Baenziger (1.8-2.1×10^{-9} M).

MBP of Mammalian Serum

The serum binding protein specific for mannose and N-acetylglucosamine was first discovered in the rabbit (16), followed by the isolation of an analogous binding protein from the rat (15) and humans (17). From 100 ml of rabbit serum, 2-3 mg of the purified binding protein was obtained, indicating that the plasma is a major source of the binding protein as well as the liver. The binding protein isolated from serum is almost indistinguishable from the rabbit liver MBP in the profiles of double immunodiffusion, antibody neutralization of the binding activity, the molecular size of the subunit, and amino acid composition. However, the serum binding protein is clearly differentiated from the liver protein in its large molecular size, sensitivity to monosaccharides as haptenic inhibitors of mannan binding, and the presence of a single binding site (12, 16).

An analogous binding protein was isolated from human serum. Extensive characterization of serum components from humans permitted identification of MBP as a new protein distinct from C-reactive protein (CRP) and amyloid P-component, both of which bind polysaccharides in the presence of Ca^{2+}. Besides the Ca^{2+}-dependent binding activity, a Ca^{2+}-insensitive mannan binding activity was found in human sera. This is characteristic of human sera and is associated with immunoglobulin G (IgG). These IgG molecules may represent naturally produced heterophile antibody as described by Sela et al. (18) and Snyder and Fleissner (19).

MBP of Chicken Liver

Because of the similarity of the sugar specificity between the chicken liver binding lectin specific for N-acetylglucosamine-terminated glycoprotein (5, 6, 20), and mammalian liver MBP, a possibility that GlcNAc binding protein might be MBP in chicken liver or vice versa was sought.

A Ca^{2+}-dependent mannan activity was found in the homogenates of chicken liver. The purified GlcNAc binding protein did not bind mannan at all, suggesting the presence of MBP distinct from GlcNAc binding protein. Sepharose 4B-mannan affinity chromatography of the Triton X-100 extract of chicken liver, followed by chromatography on diethylaminoethyl (DEAE)-Sepharose CL-6B, resulted in the isolation of a binding protein with properties similar to those of mammalian liver MBP. Binding of GlcNAc binding protein and MBP with glycoproteins is sometimes complementary; that is, GlcNAc binding protein bound to AGOR but not to mannan, while MBP bound to mannan but not to AGOR. This can be explained by the fact that inhibition of GlcNAc binding protein by mannose is less than one-eightieth of that by N-acetylglucosamine (20), while MBP is inhibited almost equally by mannose or N-acetylglucosamine (21). Also, liver MBP binds preferentially to high-mannose-type glycoproteins as discussed above.

CELLULAR AND SUBCELLULAR DISTRIBUTION OF MBP AND ITS ENDOGENOUS INHIBITORS IN RAT LIVER

Cellular Distribution of MBP

Liver MBP was originally thought to be a receptor that mediates the uptake of glycoproteins with terminal mannose and/or N-acetylglucosamine into hepatic sinusoidal cells (22-26). In fact, isolated sinusoidal cells of rat liver showed a potent ability to incorporate mannan but not ASOR. In contrast, hepatocytes had basically no activity to incorporate mannan, but had a marked ability to incorporate ASOR (27). However, when the mannan binding activity of the cells was measured in Triton X-100-solubilized cell extracts of hepatocytes, an approximately 26-fold higher activity was found relative to that of sinusoidal cells, indicating that MBP is found primarily in hepatocytes rather than in sinusoidal cells. Furthermore, the lack of surface binding activity of hepatocytes, corresponding to less than 1% of the total cell activity, indicated that MBP is located intracellularly and thereby ruled out the proposed cell surface receptor function (27). In contrast, approximately one-fourth of the total cell binding activity was found on the surface of sinusoidal cells. The properties of this receptor will be described below.

Maynard and Baenziger initially proposed that liver MBP is a hepatocyte protein (9) and this was proved by our quantitative data (27).

Subcellular Distribution and Topology of MBP

When rat liver homogenate was fractionated into five subcellular fractions by conventional differential centrifugation, approximately 75% of the total activity was detected in the microsomal fraction, 20% in the nuclear fraction, less than 3% in the light mitochondrial fraction, 1.4% in the soluble fraction, and less than 1% in the heavy mitochondrial fractions. When microsomes were separated into rough and smooth microsomes, approximately 76% of the activity was recovered in rough microsomes, with a fivefold increase in the specific activity over the homogenate. Of the constituents of smooth microsomes, plasma membranes had very low activity, in agreement with the fact that hepatocytes neither take up nor bind mannan. Golgi apparatus, on the other hand, showed high binding activity over the subfractions, GF1, GF2, and GF3, isolated from ethanol-intoxicated rats by the method of Ehrenreich et al. (28). Purified lysosomes prepared from Triton WR-1339-treated rats were almost completely inactive (see Fig. 5-1, *A* and *B*). These results indicate that MBP is located predominantly in the rough endoplasmic reticulum, but is also distributed rather widely in the Golgi apparatus from trans-Golgi vacuoles to cis-Golgi vesicles and probably also in the smooth endoplasmic reticulum.

Next, we investigated the topology of MBP in microsomes by utilizing selective protease digestion of proteins on the outer surface of the microsomal vesicles. Upon incubation of freshly prepared microsomes with various concentrations of subtilisin, MBP in the microsomal vesicles retained its activity as completely as nucleoside diphophatase did, which is known to be located inside of microsomal vesicles (29). In contrast, under the same conditions, nicotinamide adenine dinucleotide phosphate (NADPH)-cytochrome c reductase, an integral membrane protein whose active site is exposed on the outer or cytosolic surface of microsomal vesicles (30), was completely inactivated. However, when the same experiments were performed in the presence of Triton X-100 at a low concentration, which makes microsomal vesicles leaky, MBP became as susceptible to subtilisin as NADPH-cytochrome c reductase. Additionally, the mode of association of MBP with membranes was investigated by utilizing the differential solubilization of microsomal proteins with detergents. Upon incubation of microsomal vesicles with Triton X-100 at increasing concentrations, MBP was readily solubilized similar to nucleoside diphosphatase. The detergent concentration required to obtain the maximum release of MBP was over tenfold lower than that required for the solubilization of the bulk of the microsomal proteins, which suggests that MBP is located within the lumen or the cisternal space of microsomal vesicles as a peripheral membrane protein (31).

Subcellular Distribution of Endogenous Inhibitors

The initial observation of the augmentation of binding activity resulting from the purification of MBP was explained by the detection of competitive inhibitors in the Triton X-100 extract of the liver, which were separated from MBP by affinity chromatography on Sepharose 4B-mannan (7). Since competitive inhibitors presumably represent endogenous or natural ligands of MBP, their distribution among subcellular fractions was investigated (31). The overall profiles of the distribution of inhibitors are similar to those of MBP as shown in Figure 5-1, *C* and *D*. Thus, more than 60% of the total inhibitory activity of the homogenate was recovered in microsomes, more

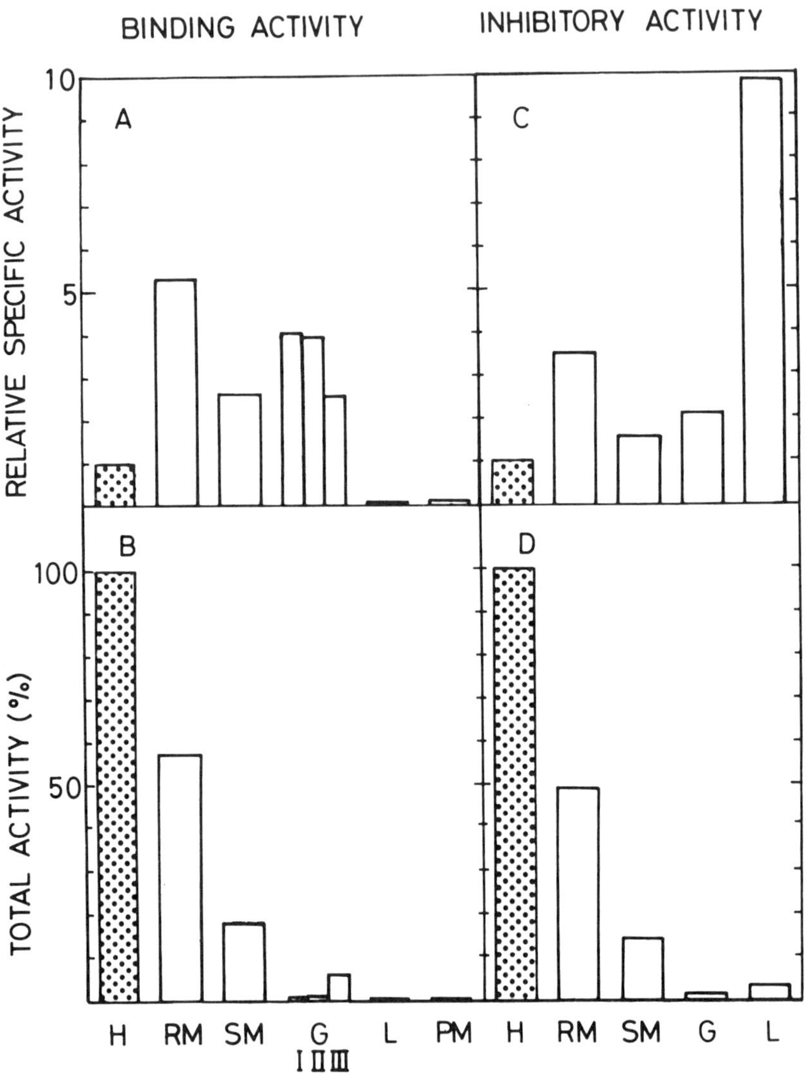

Figure 5-1. Subcellular distribution of MBP and its endogenous inhibitors. *(A)* and *(C)* show the specific activity of mannan binding and its inhibition of subcellular fractions of rat liver, respectively. The value of the homogenate is taken as 1.0. *(B)* and *(D)* show the distribution of MBP and its endogenous inhibitors among subcellular fractions, respectively. The amount of the homogenate is taken as 100%. The symbols are as follows: H, homogenate; RM, rough microsomes; SM, smooth microsomes; G, Golgi apparatus; G I, GF1; G II, GF2; G III, GF3; L, lysosomes; PM, plasma membranes.

than 75% of which was distributed in rough microsomes, with a fourfold increase in the specific activity over the homogenate. In contrast, little activity was detected in the soluble (4.8%) and heavy mitochondrial (2.6%) fractions. The most potent inhibitory activity, almost tenfold higher than that of the homogenate, was detected in the purified lysosomes, where no MBP activity was detected (Fig. 5-1, *A* and *C*). This fact is in accordance with the observations that purified lysosomal enzymes such as α-mannosidase (12, 32), β-glucuronidase (11), and cathepsin D (33), all of which bear high-mannose-type oligosaccharides, bound to MBP with high affinity (Ki = 10^{-6}-10^{-9} M). Due to the presence of these inhibitors, binding activity of subcellular organelles was masked to various extents, which basically correlated with their inhibitor content. Thus, in rough microsomes, more than 86% of the binding activity could not be detected until inhibitors were separated from MBP, 36% for smooth microsomes and 35% for common Golgi. Therefore, binding activity of subcellular organelles described in the previous section was estimated after purification of MBP by a Sepharose 4B-mannan column.

Function of Liver MBP

In the cisternal space of the endoplasmic reticulum, where MBP is predominantly localized, high-mannose-type oligosaccharides are present as the constituents of newly synthesized glycoproteins. These glycoproteins are subsequently transferred to Golgi apparatus, where multiple processing reactions occur to produce more differentiated oligosaccharides, that is, complex-type and hybrid-type oligosaccharides, followed by targeting to respective destinations, that is, lysosomes, various membranes, and extracellular milieu (34). At the various stages in these pathways, the involvement of carrier proteins or sorting mechanisms has been assumed (35, 36). Although the function of liver MBP is currently unknown, it is interesting to speculate that it may serve as a carrier protein for the transport of biosynthetic intermediates of glycoproteins with high-mannose-type oligosaccharides from the rough endoplasmic reticulum to the Golgi apparatus. The Ca^{2+}-dependency of binding and acid ability of the preformed complex substantiate the proposed function of MBP, since they provide a means to control the association-dissociation reaction between MBP and endogenous ligands within organelles.

MBP resembles the receptor for phosphomannosyl recognition in its intracellular localization, association with endogenous ligands and its high affinity for lysosomal enzymes (37). The phosphomannosyl receptor has been shown to be involved in the delivery of enzymes to lysosomes in cultured fibroblasts (38-40), while in the liver, regardless of its presence (41), the functional significance of the receptor is still in doubt (42). Hence, MBP may be associated with the process of packaging of acid hydrolases into lysosomal compartments in this liver. The lack of MBP in lysosomes appears to be incompatible with the possibility; however, the situation is similar to that of the phosphomannosyl receptor in fibroblasts, in which there exists a mechanism that immediately recycles the receptors after enzyme delivery to lysosomes.

Another possible role of MBP may be to anchor glycoproteins within the cisternal space of the endoplasmic reticulum, preventing them from moving to the Golgi apparatus. The presence of microsomal enzymes that bear high-mannose-type oligosaccharides support this hypothesis (43-45).

Alternatively, MBP may be a regulatory protein that controls glycosidases or glycosyltransferases involved in the biosynthetic processing reactions as has been implied for a lactose-binding protein (46).

Finally, despite all the discussions described above, it is clear that the metabolic and chemical properties of the endogenous ligands should be explored in more detail prior to defining the physiological roles of MBP in the liver.

ENDOCYTOSIS RECEPTOR SPECIFIC FOR MANNOSE/L-FUCOSE/ N-ACETYLGLUCOSAMINE (RECEPTOR-MBP)

Receptor-MBP of Liver Sinusoidal Cells

As described above, the sinusoidal cells isolated from rat liver, which include Kupffer cells and endothelial cells, incorporated mannan markedly when incubated at 37°C. The number of the binding sites on the surface of the sinusoidal cells was 3,500 per cell, as estimated by incubating the cells at 4°C. The value corresponds to approximately one-fourth of that of the total binding protein present in the whole cells and one-fiftieth of the amount of mannan accumulated within the cells after incubation at 37°C for 60 min (27). Those binding sites responsible for the mannan uptake have many properties in common with liver MBP. Thus, K_{uptake}, the concentration of ligand required to produce half maximum uptake, was 7.4×10^{-8} M (3 μg/ml) for mannan as estimated by Lineweaver-Burk regression analysis; the value was similar to the Kd of liver MBP (1.6×10^{-8} M). It should be noted that the

K_{uptake} for mannan is also similar to those reported previously for several glycoproteins; 4.3×10^{-8} M for β-hexosaminidase A, 3.6×10^{-7} M for ahexosamino-orosomucoid (AHOR) (47), and 1.1 to 6.4×10^{-7} M for AGOR, AHOR, ribonuclease B, and ovalbumin glycopeptides (48). Ca^{2+} is required for the uptake of mannan into the cells as well as for the binding on the cell surface. Despite these similarities, there are definite differences between the two proteins. First, IgG or its Fab fragment prepared from rabbit antiserum raised against rat liver MBP did not inhibit the mannan uptake by the sinusoidal cells at all, whereas they completely inhibited the binding activity of rat liver MBP. Second, liver MBP was quantitatively isolated from hepatocytes by affinity chromatography on a column of Sepharose 4B-mannan, whereas by the same procedures only 11% of the mannan binding activity detected in the cells was recovered from the sinusoidal cells. In addition, the recovered activity from the latter cells was not inhibited significantly (less than 20%) by the treatment with anti-liver MBP IgG. Third, although these two proteins have similar sugar specificity, which is primarily directed toward equatorial 4-OH supplemented with axial 2-OH, there are differences in the effects caused by the substitution at 2-OH by acetamido group between them. Thus, the acetamido substitution increases the affinity of the sugars to liver MBP, while it sometimes decreases the affinity to receptor MBP; N-acetylmanosamine and N-acetylglucosamine were far better ligands than mannosamine and glucosamine for liver MBP, and mannose or fucose was a more potent inhibitor than N-acetylmanosamine or N-acetylglucosamine for receptor MBP. Differences are also noted in the specificity spectrum for glycoproteins. For example, liver MBPs have a rather narrow specificity to glycoproteins with multi-high-mannose-type oligosaccharides (see Table 5-2), while the specificity of that of sinusoidal cells appears to be broader. Thus, the glycoprotein uptakes by sinusoidal cells, expressed as a ratio to mannan uptake, were 0.02, 0.49, 0.46, 2.92, and 6.56, for ASOR, AGOR, AHOR, Man_{33}-BSA, and $GlcNAc_{43}$-BSA, respectively. These figures suggest that the receptor does not require a highly ordered structure of oligosaccharides as long as they have terminal mannose or N-acetylglucosamine (27). A similar study has been done by Maynard and Baenziger (48), with the conclusion that the minimum structure required for binding and endocytosis is GlcNAc β1,6 or Manα1, 6Manα1, 6[Man 1,3]Manβ1, 4GlcNAc β1, 4GlcNAc β, Asn. Hybrid-type oligosaccharides are not endocytosed even though they have the required core structure, indicating that the β1-4-linked GlcNAc located on the β-mannose inhibits binding (48). Thus, these authors proposed that the specificity of the sinusoidal cell receptor is more narrow and restricted than that of the liver MBP. It seems hard to reconcile this conclusion with ours.

Receptor MBP of Other Tissues

Following intravenous injection of glycoproteins terminated with mannose and/or N-acetylglucosamine, a significant amount of these glycoproteins is specifically incorporated into macrophages of various tissues other than liver sinusoidal cells (26, 49, 50). In accordance with these lines of evidence obtained in vivo, the homogenate of lymphoid tissues such as mesenteric lymph nodes, lung, and spleen showed mannan binding activity; the specific activities were much higher than that of the liver (50). Furthermore, adherent cells from lymph nodes and spleen, in which macrophages are known to be concentrated, had specific binding activities approximately fivefold higher than those of the unfractionated cells.

An attempt to isolate receptor MBP from mesenteric lymph nodes, the organs with the highest specific binding activity, by a modification of the procedures described above resulted in the isolation of a binding protein. The isolated binding protein can be classified as MBP on the grounds that binding to mannan was dependent upon the presence of Ca^{2+} and was inhibited most effectively by mannose or N-acetylglucosamine, among the sugars tested. However, lymph node MBP, which presumably represents receptor MBP, was differentiated from liver MBP by a couple of parameters. Thus, in contrast to liver MBP, which has a single subunit, lymph node MBP showed one major (molecular weight, Mr = 58,000) and one minor (Mr = 32,000) band upon SDS-PAGE under reduced conditions. The explanation of this observation must await the final assessment of the homogeneity of the binding protein, however. In addition, mannosamine inhibited the binding of lymph node MBP as potently as its N-acetyl derivative, while the sugar has been shown to be a very weak inhibitor of liver MBP. Another parameter of differentiation was provided by an immunochemical test; the anti-IgG raised against liver MBP inhibited lymph node MBP only slightly (25%) (50).

Similar receptors were found on alveolar macrophages, peritoneal macrophages, macrophages derived from bone marrow in culture, and also their hybrids (51-53). Stahl and his associates have studied mechanisms of receptor recycling associated with the ligand internalization on alveolar macrophages (54); that subject is, however, outside the scope of this chapter.

Finally, immunochemical distinctions from liver MBP and a higher affinity to L-fucose than to N-acetylglucosamine appear to be the properties that are commonly shared by receptor MBP of tissue macrophages of various origins, including alveolar (8), lymph node, and peritoneal macrophages (15, 50). However, a question as to whether these similarities mean the occurrence of an identical MBP on the macrophages of different tissues has not been answered yet.

ACKNOWLEDGMENT

We wish to thank Dr. Y. C. Lee for providing us with neoglycoproteins and Dr. K. Sugahara for careful reading of the manuscript.

REFERENCES

1. R. L. Hudgin, W. E. Pricer, Jr., G. Ashwell, R. J. Stockert, and A. G. Morell, 1974, The isolation and properties of a rabbit liver binding protein specific for asialoglycoproteins, *J. Biol. Chem.* **249:**5536-5543.
2. T. Kawasaki, and G. Ashwell, 1976, Chemical and physical properties of an hepatic membrane protein that specifically binds asialoglycoproteins, *J. Biol. Chem.* **251:**1296-1302.
3. T. Kawasaki, and G. Ashwell, 1976, Carbohydrate structure of glycoproteins isolated from an hepatic membrane-binding protein specific for asialoglycoproteins, *J. Biol. Chem.* **251:**5292-5299.
4. M. Lowe, B. Nilsson, 1983, The structure of the complex type oligosaccharide from rabbit hepatic binding protein. A re-examination, *J. Biol. Chem.* **258:**1885-1887.
5. J. Lunney, and G. Ashwell, 1976, A hepatic receptor of avian origin capable of binding specifically modified glycoproteins, *Proc. Natl. Acad. Sci. (USA)* **73:**341-343.
6. T. Kawasaki, and G. Ashwell, 1977, Isolation and characterization of an avian hepatic binding protein specific for N-acetylglucosamine-terminated glycoproteins, *J. Biol. Chem.* **252:**6536-6543.
7. Y. Mizuno, Y. Kozutsumi, T. Kawasaki, and I. Yamashina, 1981, Isolation and characterization of a mannan-binding protein from rat liver, *J. Biol. Chem.* **256:**4247-4252.
8. R. Townsend, and P. Stahl, 1981, Isolation and characterization of a mannose/N-acetylglucosamine/fucose binding protein from rat liver, *Biochem. J.* **194:**209-214.
9. Y. Maynard, and J. U. Baenziger, 1982, Characterization of a mannose and N-acetylglucosamine-specific lectin present in rat hepatocytes, *J. Biol. Chem.* **257:**3788-3794.
10. J. Wild, D. Robinson, and B. Winchester, 1983, Isolation of mannose-binding proteins from human and rat liver, *Biochem. J.* **210:**167-174.
11. T. Kawasaki, R. Etoh, and I. Yamashina, 1978, Isolation and characterization of a mannan-binding protein from rabbit liver, *Biochem. Biophys. Res. Commun.* **81:**1018-1024.
12. Y. Kozutsumi, T. Kawasaki, and I. Yamashina, 1981, Kinetical properties of the serum mannan-binding protein from rabbit, *J. Biochem.* **90:**1799-1807.
13. S. Oka, K. Mori, T. Kawasaki, and I. Yamashina, 1983, unpublished results.
14. C. E. Ballou, 1974, Some aspects of the structure, immunochemistry, and genetic control of yeast mannans, *Adv. Enzymol.* **40:**239-270.
15. T. Kawasaki, 1982, Studies on "mannan-binding proteins." Mammalian lectin specific for mannose and N-acetylglucosamine residues, *Seikagaku* **54:**1019-1032.
16. Y. Kozutsumi, T. Kawasaki, and I. Yamashina, 1980, Isolation and characterization of a mannan-binding protein from rabbit serum, *Biochem. Biophys. Res. Commun.* **95:**658-664.
17. N. Kawasaki, T. Kawasaki, and I. Yamashina, 1983, Isolation and characterization of a mannan-binding protein from human serum, *J. Biochem.* **94:**937-947.
18. B. A. Sela, J. L. Wang, and G. M. Edelman, 1975, *Proc. Natl. Acad. Sci. (USA)* **72:**1127-1131.
19. H. W. Snyder, Jr., and E. Flessner, 1980, Specificity of human antibodies to oncovirus glycoproteins: Recognition of antigen by natural antibodies directing against carbohydrate structures, *Proc. Natl. Acad. Sci. (USA)* **77:**1622-1626.

20. S. K. Sikder, E. A. Kabat, C. J. Steer, and G. Ashwell, 1983, The binding site of chicken hepatic lectin, *J. Biol. Chem.* **258:**12520-12525.
21. S. Oka, T. Kawasaki, and I. Yamashina, 1985, Isolation and characterization of mannan-binding proteins from chicken liver, *Arch. Biochem. Biophys.* **241:**95-105.
22. R. J. Stockert, A. G. Morell, and I. H. Scheinberg, 1976, The existence of a second route for the transfer of certain glycoproteins from the circulation into the liver, *Biochem. Biophys. Res. Commun.* **68:**988-993.
23. J. W. Baynes, and F. Wold, 1976, Effect of glycosylation on the in vivo circulating half-life of RNase, *J. Biol. Chem.* **251:**6016-6024.
24. P. Stahl, P. H. Schelsinger, J. S. Rodman, and T. Doebber, 1976, Recognition of lysosomal glycosidases in vivo inhibited by modified glycoproteins, *Nature* **264:**86-88.
25. P. H. Schlesinger, T. W. Doebber, B. F. Mandell, R. White, C. DeSchryver, J. S. Rodman, M. J. Miller, and P. Stahl, 1978, *Biochem. J.* **176:**103-109.
26. D. T. Achord, F. E. Brot, C. E. Bell, and W. S. Sly, 1978, Human β-glucuronidase: In vivo clearance and in vitro uptake by a glycoprotein recognition system on reticuloendothelial cells, *Cell* **15:**269-278.
27. K. Mori, T. Kawasaki, and I. Yamashina, 1983, Identification of the mannan-binding protein from rat livers as a hepatocycte protein distinct from the mannan receptor on sinusoidal cells, *Arch. Biochem. Biophys.* **222:**542-552.
28. J. H. Ehrenreich, J. J. M. Bergeron, P. Siekevitz, and G. E. Palade, 1973, *J. Cell Biol.* **59:**45-72.
29. Y. Kuriyama, 1972, Studies on microsomal nucleotide diphosphatase of rat hepatocytes. Its purification, intramembranous localization, and turnover, *J. Biol. Chem.* **247:**2979-2988.
30. A. Ito, and R. Sato, 1969, Proteolytic microdissection of smooth-surfaced vesicles of liver microsomes, *J. Cell Biol.* **40:**179-189.
31. K. Mori, T. Kawasaki, and I. Yamashina, 1984, Subcellular distribution of the MBP and its endogenous inhibitors in rat liver, *Arch. Biochem. Biophys.* **232:**223-233.
32. T. Kawasaki, Y. Kozutsumi, Y. Nakao, T. Teramura, and I. Yamashina, 1981, Mannan-binding protein and carbohydrate structure of lysosomal enzymes, in *Proceedings, 6th International Symposium on Glycoconjugates,* T. Yamakawa, T. Osawa, and S. Handa, eds. Japan Scientific Societies Press, Tokyo, pp. 6-7.
33. Y. Nakao, Y. Kozutsumi, T. Kawasaki, I. Yamashina, H. van Halbeek, J. F. G. Vliegenthart, 1984, Oligosaccharides on cathepsin D from porcine spleen, *Arch. Biochem. Biophys.* **229:**43-54.
34. S. C. Hubbard, and R. J. Ivatt, 1981, Synthesis and processing of asparagine-linked oligosaccharides, *Ann. Rev. Biochem.* **50:**555-583.
35. J. E. Rothman, and E. F. Richard, 1980, Coated vesicles transport newly synthesized membrane glycoproteins from endoplasmic reticulum to plasma membrane in two successive stages, *Proc. Natl. Acad. Sci. (USA)* **77:**780-784.
36. M. F. Pearse, and M. S. Bretscher, 1981, Membrane recycling by coated vesicles, *Ann. Rev. Biochem.* **50:**85-101.
37. H. D. Fischer, A. Gonzalez-Noriega, W. S. Sly, and D. J. Moore, 1980, Phosphomannosyl enzyme receptors in rat liver: Subcellular distribution and role in intracellular transport of lysosomal enzymes, *J. Biol. Chem.* **255:**9608-9615.
38. A. Hasilik, and E. F. Neufeld, 1980, Biosynthesis of lysosomal enzymes in fibroblasts. Phosphorylation of mannose residues, *J. Biol. Chem.* **255:**4946-4950.
39. A. Hasilik, 1980, Biosynthesis of lysosomal enzymes, *Trends Biochem. Sci.* **5:**237-240.
40. D. E. Goldberg, and S. Kornfeld, 1983, Evidence for extensive subcellular organization of asparagine-linked oligosaccharides processing and lysosomal enzyme phosphorylation, *J. Biol. Chem.* **258:**3159-3165.

41. G. G. Sahagian, J. Distler, and G. W. Jourdian, 1981, Characterization of a membrane-associated receptor from bovine liver that binds phosphomannosyl residues of bovine testicular β-galactosidase, *Proc. Natl. Acad. Sci. (USA)* **78:**4289-4293.
42. M. Owada, and E. F. Neufeld, 1982, Is there a mechanism for introducing acid hydrolases into liver lysosomes that is independent of mannose 6-phosphate recognition, *Biochem. Biophys. Res. Commun.* **105:**814-820.
43. T. Mizuochi, Y. Nishimura, K. Kato, and A. Kobata, 1981, Comparative studies of aspargine-linked oligosaccharide structures of rat liver microsomal and lysosomal β-glucuroidase, *Arch. Biochem. Biophys.* **209:**298-303.
44. M. Moriyasu, A. Ito, and T. Omura, 1982, Purification and properties of arylsulfatase C from rat liver microsomes, *J. Biochem.* **92:**1189-1195.
45. L. Liscum, R. D. Cummings, R. G. W. Anderson, G. N. DeMartino, J. L. Goldstein, and M. S. Brown, 1983, 3-Hydroxy-3-methylglutaryl-CoA reductase: A transmembrane glycoprotein of the endoplasmic reticulum with N-linked "high-mannose" oligosaccharide, *Proc. Natl. Acad. Sci. (USA)* **80:** 7165-7169.
46. P. Scudder, R. A. Childs, T. Feizi, D. H. Joziasse, W. E. C. M. Schiphorst, and D. H. Van den Eijnden, 1982, Bovine heart lectin stimulates β-D-galactoside α 2→6 sialyltransferase of bovine colostrum, *Biochem. Biophys. Res. Commun.* **104:**272-279.
47. C. J. Steer, J. W. Kusiak, R. O. Brady, and E. A. Jones, 1979, Selective hepatic uptake of human β-hexosaminidase A by a specific glycoprotein recognition system on sinusoidal cells, *Proc. Natl. Acad. Sci. (USA)* **76:**2774-2778.
48. Y. Maynard, and J. U. Baenziger, 1981, Oligosaccharide specific endocytosis by isolated rat hepatic reticuloendothelial cells, *J. Biol. Chem.* **256:**8063-8068.
49. P. H. Schlesinger, J. S. Rodman, T. W. Doebber, P. D. Stahl, Y. C. Lee, C. P. Stowell, and T. B. Kuhlenschmidt, 1980, The role of extra-hepatic tissues in the receptor mediated plasma clearance of glycoproteins terminated by mannose and N-acetylglucosamine, *Biochem. J.* **192:**597-606.
50. T. Kawasaki, Y. Mizuno, T. Masuda, and I. Yamashina, 1980, Mannan-binding protein in lymphoid tissues of rats, *J. Biochem.* **88:**1891-1894.
51. R. A. B. Ezekowitz, J. Austyn, P. D. Stahl, and S. Gordon, 1981, Surface properties of bacillus Calmette-Guerin-activated mouse macrophages. Reduced expression of mannose-specific endocytosis, Fc receptors, and antigen F4/80 accompanies induction of Ia, *J. Exp. Med.* **154:**60-76.
52. P. Stahl, and S. Gordon, 1982, Expression of a mannosyl-fucosyl receptor for endocytosis on cultured primary macrophages and their hybrids, *J. Cell Biol.* **93:**49-56.
53. V. L. Shepherd, Y. C. Lee, P. H. Schlesinger, and P. D. Stahl, 1981, L-Fucose-terminated glycoconjugates are recognized by pinocytosis receptors on macrophages, *Proc. Natl. Acad. Sci. USA* **78:**1019-1022.
54. C. Tietze, P. Schlesinger, and P. Stahl, 1982, Mannose-specific endocytosis receptor of alveolar macrophages. Demonstration of two functionally distinct intracellular pools of receptor and their roles in receptor recycling, *J. Cell Biol.* **92:**417-424.

6

Receptors for Glycoproteins with Oligosaccharides Containing α-Glucose-1-Phosphate

Richard B. Marchase
Duke University Medical Center

Both histochemical studies and fractionations of the various cellular organelles demonstrate a remarkable specificity in the localization of most proteins to one or a few subcellular compartments. The sorting of newly synthesized proteins to produce these specializations and the mechanisms necessary for maintaining them are major problems facing cell biologists today.

The receptor-mediated trafficking to lysosomes of acid hydrolases with oligosaccharides containing phosphorylated mannose is the best understood pathway by which proteins translated into the endoplasmic reticulum are segregated and delivered to their ultimate destinations (for reviews, see refs. 1 and 2 and chap. 3 of this volume). Our understanding of this mechanism arose from work demonstrating the receptor-mediated uptake of exogenous lysosomal hydrolases by fibroblasts from their culture medium (3, 4). Mannose-6-phosphate (Man-6-P) was found to be a potent inhibitor of this reuptake (5, 6) and soon thereafter was identified in the hydrolases' oligosaccharides (7, 8). The Man-6-P present in the oligosaccharides was shown to be crucial to the lysosomal localization mechanism, and a receptor for Man-6-P, primarily intracellular, was described (see Sahagian, chap. 3 of this volume). The biosynthetic pathway of the Man-6-P recognition marker has also been elucidated (9-12). This synthesis includes an enzyme that possesses a substrate specificity that would seem to provide the first step in the discrimination by which lysosomal-destined enzymes are segregated from the bulk of newly synthesized proteins (10).

While covalently bound carbohydrates were detected on vertebrate proteins near the turn of the century, it was only in the last five years with the work described above that sugar phosphates were established as components of vertebrate glycoproteins. This initial work with lysosomal hydrolases was followed by the report of a distinct class of phosphoglycoproteins whose oligosaccharides, unlike those of lysosomal hydrolases, contained phosphodiester-linked glucose (13). A protein with a binding site for such a phospho-oligosaccharide has been isolated from membranes of chick neural retina (13, 14, 15) and several other tissues (16, 17). This protein, called ligatin, seems to utilize this binding site to secure the phosphoglucose-containing glycoproteins to the membranes of cells. Neither the complete subcellular localization nor movements of α-glucose-1-phosphate (α-Glc-1-P)-binding proteins has as yet been elucidated. Nonetheless, the presence of such a binding protein and the distinct substrate specificity of the enzyme responsible for the synthesis of the phosphodiester-linked glucose, compared with that responsible for the lysosomal recognition signal (18), suggest that these proteins may participate in a subcellular localization mechanism. This chapter will review the work pertinent to this possibility.

ELECTRON MICROSCOPIC OBSERVATIONS OF THE ILEUM OF SUCKLING RATS

In suckling neonates, epithelial cells of the ileum are characterized by the endocytic complex, a dense array of membrane vesicles located beneath the microvilli near the apical borders of the cells but not near their basolateral surfaces. In electron microscopic studies (19, 20), the lumenal surfaces of these membranes were found to be partially covered by two-dimensional arrays of 7.5-nm particles (Figs. 6-1 and 6-2). Knutton et al. (21) suggested that these arrays were composed of aligned rows of the particles, since during the preparation of these membranes, individual filaments with a periodic unidimensional structure were dissociated from the membrane surface.

Jakoi et al. (16) pursued this observation and found that extraction of these filaments was enhanced by treatment of the membranes with 10-mM $CaCl_2$ (Fig. 6-3). In addition they showed that dialysis against a calcium chelator resulted in depolymerization of the filaments. Molecular sieve chromatography separated the constituent proteins into two populations. The excluded peak contained β-N-acetylhexosaminadase activity and was in part morphologically similar to the 7.5-nm particles that appeared as rows on the intact ileal membranes. Eluting later from the sizing column with an apparent molecular weight of 20,000 daltons or less was a second peak of protein that, upon addition of 20 mM $CaCl_2$, formed 3-nm filaments of varying lengths (Fig. 6-4). The authors suggested that these filaments, formed from the calcium-dependent polymerization of the smaller protein, acted as attachment sites for the 7.5-nm particles to produce the decorated filaments seen on the membranes. This model provided an explanation for the regular

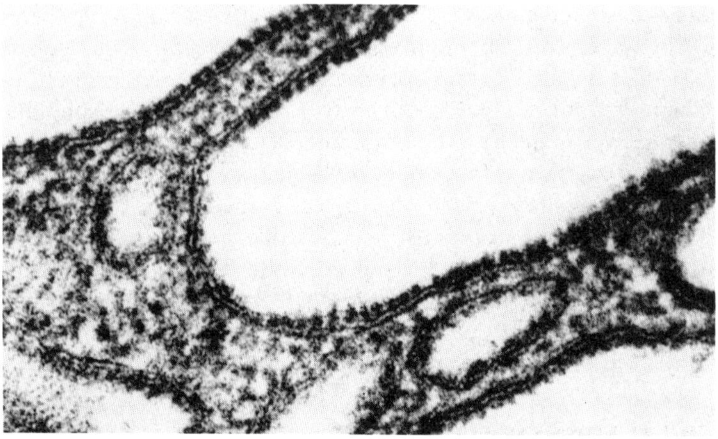

Figure 6-1. Electron micrograph of a thin section through the endocytic complex of the epithelial cells in the suckling rat ileum. This dense array vesicles located at the base of the microvilli is covered in part by 7.5–nm particles with a center-to-center spacing of 14 nm (see arrow). This micrograph shows that the particles are located on the lumenal surface of the membrane and above the unit membrane. [× 177,500] (micrograph courtesy of J. D. Robertson).

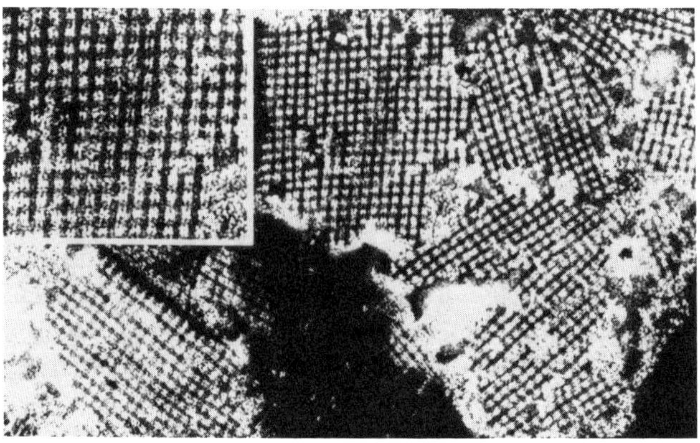

Figure 6-2. Negatively stained (1% phosphotungstic acid) membrane vesicle preparation from the suckling rat ileum. The membranes, viewed en face, are seen to be covered with a regular two-dimensional array of the 7.5-nm particles. [Main micrograph × 85,200; inset × 131,350] (21).

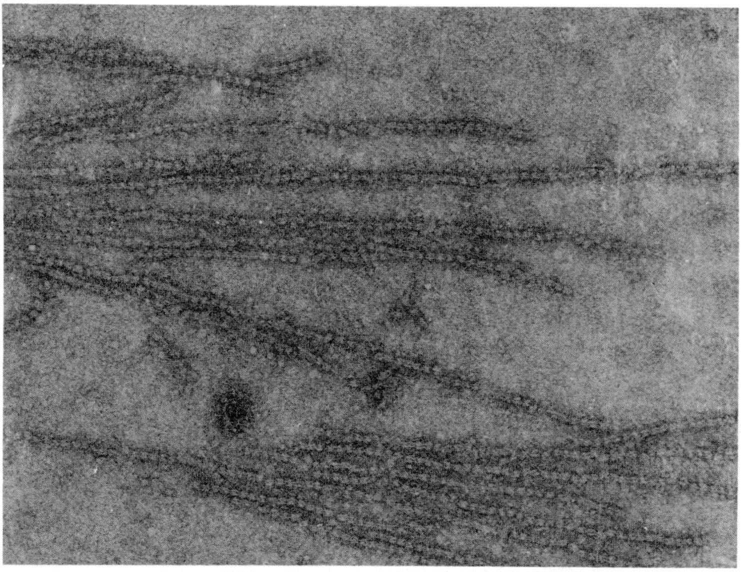

Figure 6-3. Filaments of the 7.5-nm particles visualized by negative stain following treatments of ileal membranes with 10-mM $CaCl_2$. [× 120,000] (16).

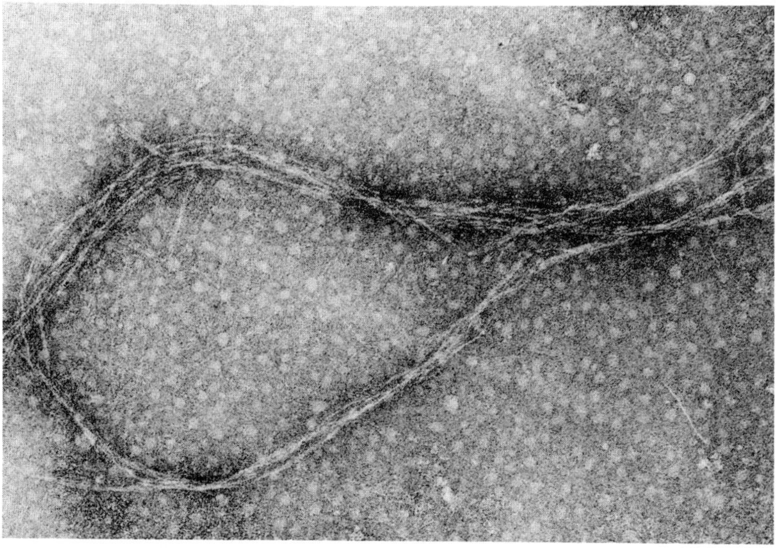

Figure 6-4. Negatively stained micrograph demonstrating the repolymerization of isolated ligatin by $CaCl_2$ after its separation from the 7.5-nm particles by molecular sieve chromatography. The ligatin monomer by itself forms filaments 3 nm in diameter. [× 150,000] (16).

arrays of these particles seen in electron micrographs. It was also consistent with the morphological observation that the particles seemed to be only peripherally associated with the lumenal surface of the unit membrane (Fig. 6-1). A schematic of this model is provided in Figure 6-5. The filament-forming protein was named ligatin, from the Latin *ligare*, which translates as "to bind together."

LIGATIN FROM EMBRYONIC CHICK NEURAL RETINA

Merrell et al. (22) reported the extraction of a proteinaceous component from neural retina plasma membranes of embryonic chick. When added to dissociated retinal cells, this extract inhibited the reaggregation of the cells. The fraction showing inhibitory activity had an apparent molecular weight of approximately 10,000 daltons. Physical similarities between this preparation and ileal ligatin led Jakoi and Marchase (14) to attempt a calcium extraction of chick neural retina membranes similar to that which had successfully isolated ligatin from the membranes of the suckling rat ileum (16). Following treatment with 20 mM $CaCl_2$, electron microscopic observations of retinal membranes revealed individual filaments 4.5 nm in diameter

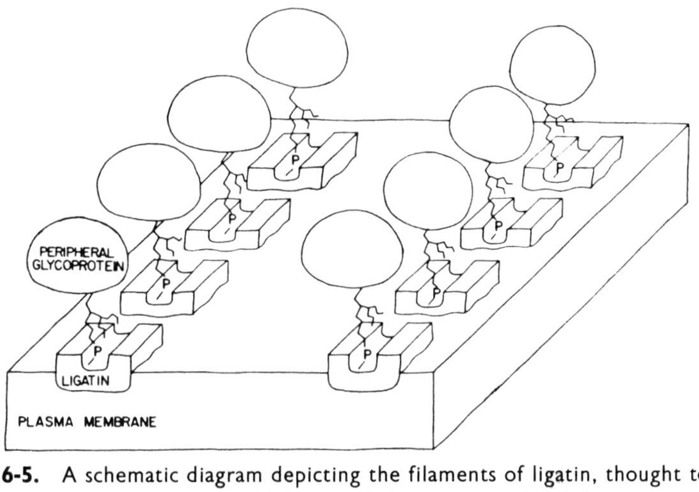

Figure 6-5. A schematic diagram depicting the filaments of ligatin, thought to be at least partially embedded in the lipid bilayer, and the more peripheral glycoproteins bound by ligatin. In ileum, the peripheral glycoproteins are the 7.5-nm particles seen in electron micrographs, Figures 6-1 through 6-3. As is discussed in the section entitled "Affinity Chromatography to Immobilized Ligatin," the interaction between ligatin and the peripheral glycoproteins is mediated by a phospho-oligosaccharide on the bound glycoprotein.

(14). After dialysis against EGTA, the filaments could no longer be seen in the preparation. The release of filaments from membrane surfaces by treatment with calcium and the subsequent disappearance of these filaments after dialysis against EGTA were identical to results obtained with the filaments isolated from ileum. However, the filaments observed from neural retina preparations were only 4.5 nm in diameter and differed in morphology from the 7.5-nm filaments seen in ileal preparations.

Subsequent molecular sieve chromatography of the retinal extracts yielded two protein peaks, a heterogeneous peak eluting first from the sizing column and a second protein chromatographing with an apparent molecular weight of approximately 10,000 daltons (14). Unlike the preparations from ileum, the retinal proteins eluting early from the sizing column were found to have minimal β-N-acetylhexosaminidase activity. They also bore no morphological resemblance to the 7.5 nM ileal particles. In contrast to these results, the smaller protein seemed very similar to its analog from ileum. After dialysis against 5 mM $CaCl_2$, negative stain electron microscopy of the protein that previously exhibited a molecular weight of 10,000 daltons revealed polymerization of the protein into filaments approximately 3 nm in diameter. These filaments were identical in morphology to the repolymerized filaments of ileal ligatin. Sodium dodecyl sulfate (SDS)-polyacrylamide gel electrophoresis (SDS-PAGE) confirmed an apparent molecular weight of approximately 10,000 daltons. An amino acid composition test performed on retinal ligatin (14) revealed a relatively polar profile (Table 6-1), and was similar to the composition subsequently obtained for the ileal protein (23).

Jakoi and Marchase (14) proposed that the retinal protein also be considered ligatin. Because of the electron microscopic findings in ileum, ligatin in neural retina was suggested to serve as a baseplate on membranes of the cells for the attachment of other, more peripheral glycoproteins. The finding that retinal ligatin, following its separation by molecular sieve chromatography, repolymerized into 3-nm filaments, but that native filaments from membranes of retina had diameters of 4.5 nm, suggested that the membrane filaments were decorated with additional proteins. Both electron microscopic and biochemical observations, however, suggested that the glycoproteins bound to the ligatin filaments in retina differed from the 7.5-nm particles found decorating ileal ligatin.

As was mentioned earlier, a primary stimulus for seeking ligatin in neural retina was the similarity between ileal ligatin and the proteinaceous inhibitor of retinal cell reaggregation described by Merrell et al. (22). Since ileal ligatin seemed to function as a baseplate for the attachment of more peripheral membrane proteins, it seemed possible that an analogous function by a similar protein in retina could be responsible for the inhibitory activity: If cell-surface proteins that mediated intercellular adhesion contained ex-

Table 6-1. Amino Acid Composition of Retinal Legatin

Amino acid	*Mol per 100 Mol of amino acid*
Asx	10.6
Thr	4.4
Ser	13.1
Glx	22.2
Pro	2.6
Gly	14.1
Ala	7.6
Val	3.4
Met	0.6
Ile	2.4
Leu	4.7
Tyr	1.8
Phe	2.9
His	2.0
Lys	5.1
Arg	2.7
Trp	0.0
Cys	Not determined

posed binding sites recognized by ligatin, then exogenous soluble ligatin might bind to this material and thus inhibit adhesion rates in much the same way that Fab fragments to cell adhesion molecules do (24).

In order to ascertain whether retinal ligatin possessed biological activity similar to the preparation of Merrell et al. (22), single cell reassociation studies were carried out in the absence and presence of retinal ligatin (15). In a typical experiment, 70% of the single cells present at time zero became associated with other cells by 30 min of incubation in the absence of ligatin. When ligatin at 20 μg/ml was added to the assay, only 30% of the single cells reassociated. This represented about a 50% inhibition of aggregation.

AFFINITY CHROMATOGRAPHY TO IMMOBILIZED LIGATIN

Electron microscopic observations (16) led to the proposal that ligatin functions as a membrane baseplate for the attachment of other, more peripheral glycoproteins to these membranes. The association that occurs between ligatin and these more peripheral glycoproteins has since been investigated utilizing ligatin covalently attached to inert matrices (13, 17, 23). In experiments with neural retina, ^{32}p was used to metabolically label the glycoproteins that cosolubilized with ligatin; their binding to a ligatin affinity column was then monitored radioactively (13).

Table 6-2. Elution of ^{32}P-Proteins from Retinal Ligatin Affinity Column

Possible eluant	*Concentration (mM)*	*Percent Bound* ^{32}P Eluted
Glucose	100	0
Mannose	100	0
Galactose	100	0
Fructose	100	0
GlcNAc	100	0
GalNAc	100	0
Fucose	100	0
α-Glc-1-P	10	87
β-Glc-1-P	10	3
α-Gal-1-P	10	11
α-Man-1-P	10	6
α-Fru-1-P	10	18
α-GlcNAc-1-P	10	12
Glc6P	10	3
Gal6P	10	6
Fru6P	10	9
Man6P	10	22
Man6P	40	38
Glc-1, 6diP	10	11
Fru-1, 6diP	10	8
Pi	100	11
NaCl	750	100

When proteins that had cosolubilized with ligatin from neural retina were passed over an Affi-gel 10 column covalently coupled with 700 μg of ligatin, approximately 25% of the ^{32}P applied to the column was retained (13). As is shown in Table 6-2, the bound radioactivity was resistant to removal by neutral sugars but was almost completely eluted with 10 mM α-Glc-1-P. α-Glucose-1-phosphate was by far the most effective among the hexoses phosphorylated at the 1-carbon, while β-Glc-1-P was ineffective. It is significant that α-GlcNAc-1-P, the phosphodiester-linked residue of lysosomal hydrolases (25), was relatively ineffective. Among the sugars phosphorylated at carbon 6, Man-6-P was most effective, although it required higher concentrations than α-Glc-1-P for comparable levels of elution. The efficacy of α-Glc-1-P in eluting retained proteins from the ligatin affinity column is consistent with the presence of oligosaccharides on the bound proteins that contain glucose residues attached via phosphodiester bonds through their 1-carbons and that mediate the attachment of the glycoproteins to ligatin. The efficacy of Man-6-P suggests that the phosphodiester bonds could link the terminal glucose residues to the 6-carbons of penultimate mannose residues.

Data supporting this hypothesis were obtained when ^{32}P-containing oligosaccharide chains were isolated from the glycoproteins following treatment with endoglycosidase H (endo H). The substrate specificity of this enzyme (26) suggests that the oligosaccharide chains mediating this binding were of the high-mannose type. This procedure resulted in deglycosylated proteins that would no longer bind to the affinity column. However, the isolated ^{32}P-labeled oligosaccharides displayed affinity for the column (13).

PARTIAL CHARACTERIZATION OF THE ^{32}P-OLIGOSACCHARIDES

The ^{32}P-oligosaccharides cleaved from ligatin-associated glycoproteins by endo H were fractionated by the cation exchange resin QAE Sephadex (13). A portion of the endo H-sensitive ^{32}P was eluted from the resin with a salt concentration typical of the presence of a single phosphodiester bond (25). The presence of a phosphodiester was further suggested by experiments in which alkaline phosphatase was shown to be ineffective in removing the ^{32}P from these oligosaccharides unless they had been previously treated with mild acid. This hydrolysis (0.1 N HCl, 30 min, 100°C) is insufficient to effectively degrade glycosidic bonds but is capable of removing groups distal to a phosphodiester. It resulted in the selective elimination from the oligosaccharide of glucose, which was detected by both linked enzymatic assays and by $NaB[^3H_4]$ reduction. Mannose-6-phosphate was also detected in complete acid hydrolysates of the oligosaccharide using linked enzymatic assays, and was found to be in approximately a one-to-one ratio to glucose. The identification of glucose as the sugar distal to the phosphate has been confirmed directly by the appearance of glucose-1-phosphate following alkaline hydrolysis of these oligosaccharides (27). The alkaline hydrolysis seems to take place through the formation of glucose-1, 2-cyclic phosphated (28), and supports the suggestion from affinity column studies that glucose is present in the alpha configuration.

SYNTHESIS OF THE PHOSPHO-OLIGOSACCHARIDES FROM NEURAL RETINA

The synthetic origin of the phosphate present in the ligatin-associated oligosaccharide has been investigated by Koro and Marchase (18). Similar to the results found by Reitman and Kornfeld (9, 10) and Hasilik et al. (11) in their examinations of the phosphorylated oligosaccharides on lysosomal enzymes, the presence of phosphate in the ligatin-associated oligosaccharides seems to occur due to the intact transfer of a sugar-1-phosphate by a phosphoglycosyltransferase.

The reaction observed by Koro and Marchase (18) for the uridine diphosphate (UDP)-Glc glycoprotein glucose-1-phosphotransferase (GlcPTase) in the retinal system is:

$$\text{UDP-Glc} + \text{Oligosaccharide} \rightarrow \text{Oligosaccharide-P-Glc} + \text{UMP.}$$

The activity was detected in homogenates from 10-day embryonic chick neural retina using doubly-labeled Ud^{32}P-^{3}H-glucose. ^{32}P was incorporated solely into endo H-sensitive oligosaccharides, and was not affected by inclusion in the assays of excesses of unlabeled breakdown products or potential contaminants. Excess unlabeled UDP-Glc, however, completely abolished the ^{32}P incorporation. The reaction had a pH optimum of 6.2 and was stimulated by manganese, with 5 mM being the optimal concentration. The latter characteristics are similar to those displayed by the GlcNAcPTase active in lysosomal localization (29). However, several results suggest that the GlcPTase is distinct from the GlcNAcPtase. First, unlabeled UDP-GlcNAc was not an effective competitor of UD^{32}P-^{3}H-Glc in the phosphoglucosyltransferase reaction (18), although it was completely effective in abolishing incorporation from UD^{32}P-GlcNAc (29). Also, SDS-PAGE autoradiographs of endogenous acceptor glycoproteins in embryonic chick neural retina homogenates labeled by the presence of (β^{32}P)UDP-Glc were shown to be distinct from those labeled by (β^{32}P)UDP-GlcNAc, indicating that the two enzymatic activities recognize different populations of endogenous glycoproteins (30). To further probe the acceptor specificities of these enzymes, three glycoproteins known to be exogenous acceptors for GlcNAcPTase were included in assays for GlcPTase from retinal homogenates. Cathepsin D and β-N-acetylhexosaminidase had no significant effects on phosphoglucose incorporation. Uteroferrin, an acid phosphatase, had a pronounced inhibitory effect on incorporation from UDP-Glc, and subsequent experiments suggested that phosphorylation of the GlcPTase or another protein may be necessary for maximal activity to be seen. Last, I-cells, which have previously been shown to possess no GlcNAcPTase activity, and control human fibroblasts were assayed for both GlcPTase and GlcNAcPTase. GlcNAcPtase activity was observed only in control cells, whereas GlcPTase was observed in both I-cells and controls at similar specific activities (30).

A POSSIBLE ROLE FOR THE GLCPTASE IN INTRACELLULAR TRAFFICKING

The phosphodiester linkage described for ligatin-associated retinal cell-surface proteins does not seem to undergo cleavage in vivo to expose Man-6-P, since phosphodiester-linked glucose is present in cell-surface preparations (13) and seems to be required for binding to ligatin. The inefficiency with which the Man-6-P receptor binds to phosphodiester-containing oligosaccharides (31, 32) raises the possibility that the phosphodiester-linked glucose could be a distinct subcellular trafficking signal. A diagrammatic summary of this proposal is presented below (Fig. 6-6).

Three classes of polypeptides manufactured on membrane-associated ribosomes and then released into the lumen of the endoplasmic reticulum are depicted: △, those destined for secretion; ○, those destined for lysosomes; and ⬡, those destined for association with ligatin. These classes of proteins are shown to bear "core" oligosaccharides produced in the endoplasmic reticulum (ER) and the *cis* regions of the Golgi. In the absence of a diverting signal, the first class of polypeptides traverse the Golgi and are localized to secretory vesicles. The second and third classes are marked via the acquisition of phosphorylated sugars for segregation to other vesicles, and for eventual localization to specific subcellular destinations. Most of the 30 to 40 glycoproteins destined for localization to lysosomes have been suggested to possess a common characteristic in their tertiary structures (10) that codes for their recognition by the GlcNAcPTase. This enzyme adds phosphodiester-linked GlcNAc to the high-mannose-type oligosaccharides on these proteins. The derivatized proteins are the further processed by the

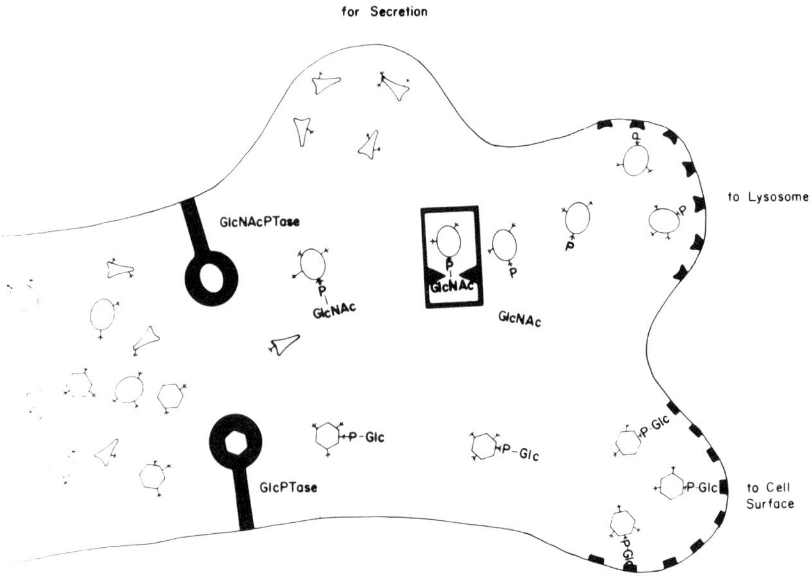

Figure 6-6. A model suggesting roles for the GlcNAcPTase and the GlcPTase in intracellular trafficking. See text for details.

enzymatic removal of the terminal GlCNAc residues, exposing the Man-6-Ps that are recognized by the Man-6-P receptor. This recognition ultimately leads to the segregation of these proteins in vesicles destined to be primary lysosomes.

We propose a parallel mechanism that could segregate another class of glycoproteins from those just mentioned and from those destined for secretion. We suggest that these glycoproteins could bear a characteristic in their tertiary structures that would code for their recognition by the GlcPTase. The phosphodiester-linked glucose would then be recognized by an intracellular receptor distinct from that utilized in the lysosomal localization scheme. This receptor could be ligatin or another protein of similar binding specificity. This recognition would ultimately lead to the segregation of these proteins in vesicles destined for a distinct subcellular destination, such as a limited region of the cell surface.

PROTEINS CAPABLE OF BINDING α-Glc-1-P-6-MAN

In investigations designed to purify any cellular proteins that might act as receptors for the phospho-oligosaccharide synthesized by the GlcPTase, we (33) have utilized an artificial ligand that is analogous to this structure. The ligand, α-glucose-1-P-6-α-mannose-phenylamine (34), was coupled to cyanogen bromide-activated Sepharose and used in affinity chromatography studies. Two results merit special mention. First, crude membrane preparations from either suckling pig ileum or embryonic chick brain were extracted with the chaotropic agent lithium diiodosalicylate (LIS). The resulting supernatant was passed over the affinity column described above. Greater than 95% of the protein ran directly through the column. However, elution with NaCl at high concentrations released proteins that had been retained. Polyacrylamide gel electrophoresis of these proteins from either source revealed only three bands, ranging in size from 10,000 daltons to 17,000 daltons. Second, membrane fractions from these same sources were extracted with $CaCl_2$ as described in sections A and B. The protein peak that eluted late from molecular sizing columns and has been identified with ligatin was then applied to the α-Glc-1-P column. Nearly 100% of the protein bound to the column, and after its release, electrophoresed identically to the retained proteins from the LIS extract. Other purified proteins, such as bovine serum albumin, showed no affinity for the column.

The finding of three low-molecular-weight peptides capable of binding to this phospho-oligosaccharide is intriguing. The manner in which the forms of the binding protein relate to one another and the subcellular distributions of each are currently under investigation.

LIGATIN IN OTHER TISSUES AND RELATED PROTEINS

The term ligatin has been used to describe proteins possessing the following properties:

1. They are dissociable in filamentous heteropolymers from plasma membranes by treatment with $CaCl_2$ at high levels.
2. Upon dialysis against EGTA, proteins with apparent molecular weights of 10,000–20,000 daltons, the ligatin subunits, can be separated from other constituents.
3. The return of $CaCl_2$ to isolated subunits causes polymerization into 3-mm filaments.
4. They possess amino acid compositions characterized by relatively high levels of acidic residues and low levels of hydrophobic residues.
5. In affinity studies, they bind proteins that cosolubilize with them and then release these proteins upon treatment with specific phosphorylated sugars.

Proteins from suckling rat ileum (16, 17), chick neural retina (13, 14), and newborn rat brain (23) have been isolated that meet all of these criteria. In affinity column studies with ligatin from neural retina (13), brain (17), and a similar preparation from macrophage α-Glc-1-P was overall the most effective of the sugar phosphates in eluting bound proteins, while Man-6-P was the most effective of the sugars derivatized at the 6-carbon. In ileum (17) Man-6-P was the most effective of the sugar phosphates tested.

Proteins solubilized from membrane preparations by treatment with $CaCl_2$ and shown to possess at least some of the characteristics that contribute to the operational definition of ligatin have been obtained from mouse peritoneal macrophages (35), rabbit sperm (36), and sea urchin embryos (37). Proteins of similar molecular weight and amino acid composition have been obtained from red blood cells (38), and skeletal muscle (39).

These findings suggest that ligatin or similar proteins are present in the membranes of a wide variety of animals and tissues. Both biochemical and electron microscopic comparisons are consistent with the proposal that ligatin may serve as a membrane receptor for peripheral phosphoglycoproteins in these tissues, but that the identities of the bound proteins are different in the various preparations. We suggest that ligatin's ability to secure these diverse proteins to the cell surface is not directly a function of their primary amino acid sequence, but rather is due to the presence on all of the proteins of a common oligosaccharide terminating in phosphodiester-linked glucose.

In speculating as to possible evolutionary ties of ligatin to other proteins, consideration must be given to a group of small (10,000–20,000 daltons), soluble acidic proteins capable of binding Ca^{++} (40, 41). Of special interest may be two serum proteins found following acute inflammation, called C-reactive protein and serum amyloid protein (42).

These proteins not only form complexes in the presence of Ca^{++} but also bind phosphate-containing organics such as phosphorylcholine (43). Colley and Baenziger (44) suggest a relation between these proteins and a small, soluble Ca^{++}-dependent lectin they are studying that is capable of binding mannose. In experiments with mannan affinity columns, serum amyloid protein bound to the column in the presence of Ca^{++} and was elutable with Man-6-P and α-glycerolphosphate. The two serum proteins, the mannose-binding protein of Baenziger and ligatin (14, 23), are all similar in size and amino acid composition. That they also all seem to have Ca^{++}-dependent affinities for related ligands certainly suggests possible relationships among them.

SUMMARY AND PROSPECTS

At present, only two proteins have been described that bind oligosaccharides containing phosphate. However, it was only in 1979 that phosphate was established as a component of the oligosaccharides of vertebrate glycoproteins. It would seem likely that future investigations might well discover distinct phospho-oligosaccharides and proteins capable of binding them. That such binding proteins might play roles in subcellular localization mechanisms, as is the case with the Man-6-P receptor and perhaps with ligatin as well, remains an interesting possibility.

ACKNOWLEDGMENTS

I thank Lillian Koro, Ann Hiller, and Ronald Myatich for helpful comments, and Bernard Eden and Ursula Chesney for their assistance. I also thank Rachel Fink for the illustrations, and I am deeply indebted to Barbara McPartland for her patience and excellent secretarial support. Work from this laboratory is supported by NIH EY 04480, NS 06233, and GM 31381, and by a Nanaline H. Duke Scholarship.

REFERENCES

1. W. S. Sly, and H. D. Fischer, 1982, The phosphomannosyl recognition system for intracellular and intercellular transport of lysosomal enzymes, *J. Cell Biochem.* **18**:531-549.
2. R. B. Marchase, L. A. Koro, and A. M. Hiller, 1984, Receptors for glycoproteins with phosphorylated oligosaccharides, in *The Receptors,* vol. 1, P. M. Conn, ed., Academic Press, New York, pp. 261-313.
3. S. Hickman, and E. F. Neufeld, 1972, A hypothesis for I-cell disease: Defective hydrolases that do not enter lysosomes, *Biochem. Biophys. Res. Commun.* **49**:992-999.

4. S. Hickman, L. J. Shapiro, and E. F. Neufeld, 1974, A recognition marker for uptake of a lysosomal enzyme by cultured fibroblasts, *Biochem. Biophys. Res. Commun.* **57:**55-61.
5. A. Kaplan, D. T. Achord, and W. S. Sly, 1977, Phosphohexosyl components of a lysosomal enzyme are recognized by pinocytosis receptors on human fibroblasts, *Proc. Natl. Acad. Sci. (USA)* **74:**2026-2030.
6. A. Kaplan, D. Fischer, D. Achord, and W. Sly, 1977, Phosphohexosyl recognition is a general characteristic of pinocytosis of lysosomal glycosidases by human fibroblasts, *J. Clin. Invest.* **60:**1088-1093.
7. M. R. Natowicz, M. M. Chi, O. H. Lowry, and W. S. Sly, 1979, Enzymatic identification for receptor-mediated pinocytosis of β-glucuronidase by human fibroblasts, *Proc. Natl. Acad. Sci. (USA)* **76:**4322-4326.
8. J. Distler, U. Hieber, G. Sahagian, R. Schmickel, and G. W. Jourdian, 1979, Identification of mannose-6-phosphate in glycoproteins that inhibit the assimilation of β-galactosidase by fibroblasts, *Proc. Natl. Acad. Sci. (USA)***76:**4235-4239.
9. M. L. Reitman, and S. Kornfeld, 1981, UDP-N-Acetylglucosamine: glycoprotein N-acetylglucosamine-1-phosphotransferase. Proposed enzyme for the phosphorylation of the high-mannose oligosaccharide units of lysosomal enzymes. *J. Biol. Chem.* **256:**4257-4281.
10. M. L. Reitman, and S. Kornfeld, 1981, Lysosomal enzyme targeting, *J. Biol. Chem.* **256:**11977-11980.
11. A. Hasilik, U. Klein, A. Waheed, G. Strecker, and K. von Figura, 1980, Phosphorylated oligosaccharides in lysosomal enzymes: Identification of α-N-acetylglucosamine(1)-phospho(6)mannose diester groups. *Proc. Natl. Acad. Sci. (USA)* **77:**7074-7078.
12. A. Varki, and S. Kornfeld, 1980, Structural studies of phosphorylated high mannose-type oligosaccharides, *J. Biol. Chem.* **255:**10847-10858.
13. R. B. Marchase, L. A. Koro, C. M. Kelly, and D. R. McClay, 1982, Retinal ligatin recognizes glycoproteins bearing oligosaccharides terminating in phosphodiester-linked glucose, *Cell* **28:**813-820.
14. E. R. Jakoi, and R. B. Marchase, 1979, Ligatin from embryonic chick neural retina, *J. Cell Biol.* **80:**642-650.
15. R. B. Marchase, P. Harges, and E. R. Jakoi, 1981, Ligatin from embryonic chick neural retina inhibits retinal cell adhesion, *Developmental Biol.* **86:**250-255.
16. E. R. Jakoi, G. Zamphghi, and J. D. Robertson, 1976, Regular structures in unit membranes: II. Morphologial and biochemical characterization of two water-soluble membrane proteins isolated from the suckling rat ileum, *J. Cell Biol.* **70:**97-111.
17. E. R. Jakoi, K. Kemp, and S. M. Gaston, 1981, Ligatin binds phosphohexose residues on acid hydrolases, *J. Supramolec. Struct. and Cell. Biochem.* **16:**139-153.
18. L. A. Koro, and R. B. Marchase, 1982, A UDP-glucose: Glycoprotein glucose-1-phosphotransferase in embryonic chicken neural retina, *Cell* **31:**739-748.
19. K. R. Porter, K. Kenyon, and S. Badenhauser, 1967, Specializations of the unit membrane, *Protoplasma* **63:**262-274.
20. S. L. Wissig, and D. O. Graney, 1968, Membrane modifications in the apical endocytic complex of ileal cells, *J. Cell Biol.* **39:**564-579.
21. S. Knutton, A. R. Limbrick, and J. D. Robertson, 1974, Regular structures in membranes. I. Membranes in the endocytic complex of ileal epithelial cells, *J. Cell. Biol.* **62:**679-694.
22. R. Merrell, D. I. Gottleib, and L. Glaser, 1975, Embryonal cell surface recognition. Extraction of an active plasma membrane component, *J. Biol. Chem.* **250:**5655-5659.
23. S. M. Gaston, R. B. Marchase, and E. R. Jakoi, 1982, Brain ligatin: A membrane lectin that binds acetylcholinesterase, *J. Cell. Biochem.* **18:**447-459.
24. G. M. Edelman, 1983, Cell adhesion molecules, *Science* **219:**450-457.

25. I. Tabas, and S. Kornfeld, 1980, Biosynthetic intermediates of β-glucuronidase contain high-mannose oligosaccharides with blocked phosphate residues, *J. Biol. Chem.* **255:**6633-6639.
26. A. Tarentino, and F. Maley, 1974, Purification and properties of an endo-β-N-acetylglucosaminidase from *streptomyces griseus, J. Biol. Chem.* **249:**811-817.
27. R. B. Marchase, J. M. Cook, R. Myatich, and N. J. Pollack, Alkaline hydrolysis of glycoproteins containing phosphodiester-linked glucose, submitted for publication.
28. A. C. Paladini, and L. T. Leloir, 1952, Studies on uridine-diphosphate glucose, *Biochem. J.* **51:**426-430.
29. A. Waheed, A. Hasilik, and K. von Figura, 1982, UDP-N-acetylglucosamine: Lysosomal enzyme precursor N-acetylglucosamine-1-phosphotransferase, *J. Biol. Chem.* **257:**12322-12331.
30. A. M. Hiller, L. A. Koro, and R. B. Marchase, 1986, Glucose-1-phosphotransferase and GlcNac-1-phosphotransferase have distinct acceptor specificities. *J. Biol. Chem.*, in press.
31. H. D. Fischer, K. E. Creek, and W. S. Sly, 1982, Binding of phosphorylated oligosaccharides to immobilized phosphomannosyl receptors, *J. Biol. Chem.* **257:**9938-9943.
32. C. A. Gabel, D. E. Goldberg, and S. Kornfeld, 1982, Lysosomal enzyme oligosaccharide phosphorylation in mouse lymphoma cells: Specificity and kinetics of binding to the mannose-6-phosphate receptor in vivo, *J. Cell Biol.* **95:**536-542.
33. A. M. Hiller, O. Hindsgaul, and R. B. Marchase, 1984, α-Glucose-1-phosphate binding proteins isolated by affinity chromatography, *J. Cell Biol.* **99:**178a.
34. O. P. Srivastara, and O. Hindsgaul, 1985, Synthesis of glycosides of 6-O-(α-D-glucopyranosyl phosphate)-α-D-mannopyranose: The putative ligatin receptor, *Carbohydrate Res.* **143:**77-84.
35. E. R. Jakoi, and R. B. Corley, 1979, Ligatin: An external cell surface receptor specific for hexosaminidase A isolated from mouse peritoneal macrophages, *J. Cell Biol.* **83:**65 (abst.).
36. R. B. Marchase, K. A. Lakosi, M. R. Truschel, and M. G. O'Rand, 1982, A sperm cell-surface receptor for a seminal fluid protein, *J. Cell Biol.* **95:**165a.
37. R. B. Marchase, and D. R. McClay, 1979, A plasma membrane protein from embryonic sea urchins that resembles vertebrate ligatin, *J. Cell Biol.* **83:**65 (abst.).
38. E. Gaetjens, 1976, Isolation of a ^{32}P-labeled polypeptide of low molecular weight from phosphorylated human erythrocyte membranes, *Biochemistry* **15:**40-45.
39. M. Barany, K. Barany, E. Gaetjens, and A. Steinschneider, 1977, Isolation of phosphorylated acid chloroform/methanol-soluble proteins from live frog muscle, *Biochem. Biophys. Acta* **491:**387-397.
40. R. H. Kretsinger, 1976, Calcium-binding proteins, *Annu. Rev. Biochem.* **45:**239-284.
41. R. H. Kretsinger, 1980, Structure and evolution of calcium-modulated proteins, *CRC Crit. Rev. Biochem.* **8:**119-223.
42. I. Kushner, J. E. Volankis, and H. Gewurz, 1982, *Annals N. Y. Acad. Sci.* **389:**1-479.
43. E. C. Gotschlich, T.-Y. Liv, and E. Oliveira, 1982, Binding of C-reactive protein to C-carbohydrate and PC-substituted protein. *Annals N. Y. Acad. Sci.* **389:**163-171.
44. K. J. Colley, and J. U. Baenziger, 1983, The man/GlcNAc-specific lectin of hepatocytes may be related to serum amyloid component and C-reactive protein, *Abst. Soc. Complex Carbohyd.* **41.**

7

Asialotransferrins: Lectin Modulation of a Physiological Cell-Ligand Interaction

Erwin Regoeczi
McMaster University

It is probably true that no ligand has created so much confusion since the advent of receptor-mediated endocytosis as asialotransferrin, and this seems to hold true up to the present time. Confusion has arisen from the interpretation of certain observations rather than because of erroneous experimental designs. The laboratory of this writer has had its share in the comedy of errors that evolved around asialotransferrin, as may be concluded from the brief historical retrospect given below.

In the first comprehensive study on asialoglycoproteins, in 1971, Morell and his co-workers (1) observed that desialylation of human transferrin, in contradistinction to that of many other glycoproteins, failed to induce a rapid elimination of the protein by the galactose-*N*-acetylgalactosamine (Gal/GalNAc)-specific hepatin lectin. That study was performed in rats. When, in 1974, we studied human asialotransferrin in rabbits, quite a different picture was obtained: The modified preparation disappeared from the plasma much faster than the simultaneously injected native transferrin (2). This discrepancy gave rise to the first confusion, which ended in 1980 with the discovery of a recognition system for biantennary asialoglycans in the bone marrow of rabbits that is apparently poorly expressed in rats (3).

If human asialotransferrin is injected in a rat together with a plasma volume marker, such as transferrin or albumin, it is possible to show that a substantial fraction of the dose (>50%) disappears from the circulation during the first 5 min. (There is no further decrease in plasma radioactivity during the ensuing half an hour, and, unless the dual-isotope technique is taken advantage of, the initial drop is easily overlooked; this is a probable explanation of how asialotransferrin initially became classified as an exception.) In 1978, we traced the 'vanished' asialotransferrin to the liver, from which it was readily recoverable (i.e., brought back to the plasma) by the injection of an excess of asialo-orosomucoid (4). Furthermore, in experiments with small doses (<5 μg/100 g), catabolism of the radioligand was conspicuously absent. Since the uptake and subsequent release, in an intact form, of an asialoglycoprotein by the hepatocyte were inconceivable in those days, we ventured to speculate that human asialotransferrin, in trace amounts, only bound to the plasmalemma without actually getting inside the cell, and that endocytosis of this ligand—leading to its prompt catabolism—required the synergistic action of several asialotransferrin molecules. This turned out to be the second confusing interpretation, which lasted until 1981; that is, the demonstration with suspended hepatocytes of the diacytic movement of asialotransferrin (5). (The term *diacytosis* denotes a sequence of endo- and exocytosis with no ligand degradation in between.)

Y. C. Lee found some months later that certain synthetic cluster glycosides (Tris derivatized with one to three galactose or lactose residues and conjugated with ^{125}I-tyrosine) were handled by isolated hepatocytes in a manner similar to that for asialotransferrin (6). This was an important observation, because it demonstrated that lectin-mediated diacytosis was not unique to asialotransferrin. By the time of the 1982 Gordon Conference on Lysosomes, research on the receptor-mediated recycling of diferric transferrin was in full swing in many laboratories, extending over a variety of cells. The subcellular transferrin pathway, as reviewed elsewhere (7), turned out to share some striking similarities with the route taken by asialotransferrin. Not surprisingly, therefore, the question lingering on many investigators' minds was whether the behavior of asialotransferrin at the subcellular level was determined by the Gal/GalNAc-specific lectin or, perhaps, by transferrin receptors.

Young and his co-workers (8) conducted pertinent studies in 1983 on suspended rat hepatocytes with homologous transferrin and asialotransferrin. They came to the conclusion that rat asialotransferrin may be taken up by the hepatocyte either through the Gal/GalNAc-specific lectin or the transferrin receptor; in the former case, asialotransferrin is handled like any other asialoglycoprotein, that is, it is destroyed; in the latter case, however, the protein is returned to the medium undegraded. For reasons explained in the section on the mechanism of diacytosis, this interpretation became a third area of confusion. Its end came in 1985 with the study of Dekker et al. (9), showing that rat asialotransferrin is in fact not degraded by rat hepatocytes when taken up by—what was perceived to be—the Gal/GalNAc-specific lectin. Regrettably, however, these workers did not fully appreciate the glycan differences between human and rat transferrins, nor did they realize the fact that human transferrin possesses a significantly higher affinity for the transferrin receptor of the rat than the homologous protein. This situation prompted the fourth confusing interpretation; namely, the postulation of a "hitherto unknown recognition system for species specificity" that protects rat asialotransferrin from degradation. The thought seemed attractive and we considered it ourselves in 1974 (2), but abandoned it very soon because of the finding that the behavior of bovine, canine, and porcine asialotransferrins was determined by their carbohydrate contents rather than by their deployment in a homologous or heterologous setting (10).

It is difficult to appreciate the biological behavior of asialotransferrins without some knowledge of their glycans. This information is outlined in the following section.

HUMAN TRANSFERRIN GLYCANS

Human transferrin phenotype C, as shown by Jamieson and co-workers (11), contains two *N*-glycosidically linked complex glycans that were originally thought to be uniformly of the biantennary type. However, the protein fraction prepared by these investigators for carbohydrate analyses encompassed only one (the major) of the plasma transferrin components, whereby the inherent glycan heterogeneity of transferrin has remained unnoticed. The first evidence for such a heterogeneity was furnished by Spik and her colleagues in 1975 by the demonstration that some of the glycopeptides obtained from the C-terminal portion of the molecule contain an additional sialyl-*N*-acetyllactosamine sequence (12). By 1977, it became clear that immunologically pure transferrin obtained from healthy individual donors can always be resolved into three electrophoretic components in alkaline polyacrylamide gel, provided the iron distribution is rendered uniform beforehand (13). Transferrin subfractions that differ by 1, 2, 3, and so on sialyl residues per molecule, band in 7.5% polyacrylamide gels at regular intervals like the rungs of a ladder, and isolation of these subfractions can be accomplished by chromatography on diethylaminoethyl (DEAE)-cellulose (14).

From DEAE chromatograms, the proportion of transferrin molecules bearing an additional sialyl-*N*-acetyllactosamine sequence was calculated to be approximately 33%, but, in sharp contrast, only 16–17% of desialylated transferrin was firmly bound by the Gal/GalNAc-specific lectin immobilized on Sepharose 6B (15). The reason for this discrepancy was resolved by the finding of Hatton et al. (16) that the interaction of triantennary asialoglycopeptides from transferrin with the immobilized Gal/GalNAc-specific lectin is not uniform: Some of these peptides are bound by the lectin column, while

Table 7-1. Human Asialotransferrin Types: Proposed Nomenclature and Properties

Abbreviated name	*Glycans*	K_A *(M^{-1}, 22°C)*[a]	*Preparation*[b]	*Lectin binding*[c]
HAsTf-1	2 biantennary	1.5×10^6	DEAE, sialidase	–
HAsTf-2	1 bi-, 1 triantennary	1.4×10^7	DEAE, sialidase, lectin chromatography	–
HAsTf-3	1 bi-, 1 triantennary	1.1×10^8	DEAE, sialidase, lectin chromatography	+
HAsTf-4	2 triantennary	?	Con-A, sialidase	?

[a]Association constants determined by using purified rat liver plasma membranes (18).
[b]Steps are performed in the order listed. Lectin chromatography refers to immobilized Gal/GalNAc-specific lectin.
[c]Binding refers to immobilized Gal/GalNAc-specific lectin. "Minus" signifies no binding or slight retardation, and "plus" denotes firm binding that requires elutin by galactopyranoside.

others are only slightly retarded. The same study also showed that a triantennary glycan may be present at either glycosylation site in transferrin.

The developments outlined above ultimately gave rise to the isolation of three human asialotransferrin types in 1979 (17). Their properties are summarized in Table 7-1, which also lists an additional form, designated HAsTf-4. This component, recently identified by Spik and her colleagues (19), contains two triantennary glycans. In normal serum it is present merely in traces, but its proportion may rise to 4-5% in liver disease. The characteristics of this component as an asialoglycoprotein await exploration in the future; according to Dekker et al., it donates iron to suspended hepatocytes rapidly (9).

It is noticeable in Table 7-1 that HAsTf-2 and HAsTf-3, both possessing one bi- and one triantennary glycan, bind to purified plasma membranes with affinities that differ by almost an order of magnitude. This raises the question as to the nature of the difference between these two molecules. The two most likely answers are schematically illustrated in Figure 1. One of these envisages glycan topoisomerism; that is, an interchange of the bi- and triantennary glycans between the two sites of glycosylation. For topoisomerism to hold true as an explanation, all the triantennary glycans of transferrin must be of the same structure and they must be present at either of the attachment sites. Indeed, as has already been pointed out, triantennary glycans are available in transferrin in association with two different sequons (16), and März et al. (20) found only a single kind of triantennary glycan in transferrin; namely, one in which the third antenna is attached to the $\alpha(1\rightarrow3)$-linked mannose. However, there is now mounting evidence suggesting that human transferrin, in addition to the structure identified by März and his colleagues, also contains a triantennary glycan in which the third antenna is located on the $\alpha(1\rightarrow6)$-linked mannose (19, 21). (This combined occurrence of two different triantennary glycans makes human transferrin, for the time being, a unique plasma glycoprotein.)

The following considerations strongly favor the view that the unequal affinities of HAsTf-2 and HAsTf-3 for the Gal/GalNAc-specific lectin are due to two structurally different triantennary glycans, rather than to the topoisomerism of one type of triantennary glycan: (1) Triantennary asialoglycopeptides from human transferrin exhibit a marked disparity with respect to binding to the immobilized Gal/GalNAc-specific lectin (16). (2) The two kinds of triantennary glycan are available in human transferrin in approximately equal amounts (19, 21, 22); that is, in the same proportion as HAsTf-2 and HAsTf-3 (15).

Two aspects of the human transferrin glycans remain to be solved. One is a practical one: Which type of triantennary glycan imparts the strong binding properties on HAsTf-3?[1] The other one is more relevant to glycoprotein

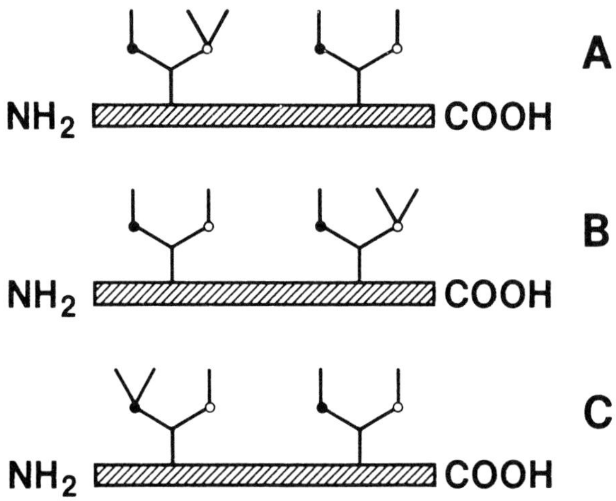

Figure 7-1. Glycan arrangements that may underlie the difference between HAsTf-2 and HAsTf-3. The transferrin molecule is depicted schematically as a shaded rod with its N and C terminals marked as such. Closed circle denotes the mannose that is linked $\alpha1\rightarrow3$ to the core (i.e., β-linked) mannose, and open circle the one that is $\alpha1\rightarrow6$ linked. Both HAsTf-2 and HAsTf-3 are assumed to possess one bi- and one triantennary glycan. The asialotransferrin types may differ either because of glycan topoisomerism (A vs. B), or because of the biological variation in the site of attachment of the third branch in the triantennary glycan (A vs. C). For further explanations see the text.

[1]In all likeihood, the active triantennary glycan is the one with the third antenna on the $\alpha(1\rightarrow3)$-linked mannose. This may be predicted from the facts that the desialylated triantennary glycan of the purified lectin from rabbit liver is a potent inhibitor of asialo-orosomucoid binding (23), and that the third antenna of the glycan is attached to the $\alpha(1\rightarrow3)$-linked mannose (24).

biosynthesis: What determines the relative frequency of either triantennary glycan type at the two glycosylation sites?

IN VIVO OBSERVATIONS WITH HUMAN TRANSFERRINS BEARING DIFFERENT GLYCANS

Similar to the situation with other plasma glycoproteins (25), there is reason to believe that the microheterogeneity of transferrin glycans, as surveyed in the foregoing section, is referable to a complex biosynthetic "template" that operates with remarkable precision. Therefore, the composition of transferrin from healthy individuals is being maintained within narrow margins with regard to prosthetic allomerism. However, a relative increase in transferrin molecules possessing triantennary glycans has been observed in hypotransferrinemic sera from cirrhotic patients (19, 22).

It is unclear at present what biological significance transferrin glycan heterogeneity entails, if any. The presence of the third antenna affects neither the extra/intravascular distribution nor the catabolic rate of transferrin as evident from an animal experiment depicted in Figure 7-2. Similarly, the rate of uptake of ^{59}Fe by the rat liver from human diferric transferrin

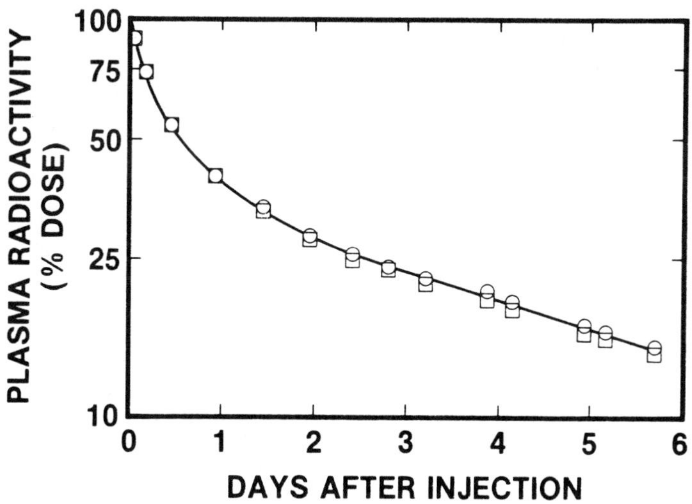

Figure 7-2. Clearance of human transferrins bearing different glycans from the plasma of a New Zealand white rabbit. Transferrin possessing two biantennary glycans (○) was labeled with ^{125}I and mixed with ^{131}I-transferrin possessing one bi- and one triantennary glycan (□) for injection. Dose, 20 μg of each protein. Points denote protein-bound radioactivities.

possessing two biantennary glycans is comparable to the rate obtained with transferrin that has one bi- and one triantennary glycan (Fig. 7-3). Clearly, more studies are required before the possible role of the transferrin glycans in physiology can be assessed in a comprehensive manner.

BEHAVIOR OF HUMAN ASIALOTRANSFERRINS IN THE RAT

In spite of their intrinsic differences with respect to affinity for the Gal/GalNAc-specific lectin (cf. Table 7-1), all human asialotransferrins studied so far (HAsTf-1, -2, and -3) exhibit properties that are fully comparable in a qualitative sense. The "asialotransferrin-like behavior" comprises of the following characteristic features.

1. The metabolic fate of asialotransferrins is markedly dependent on the dose (4). Thus, a small dose is catabolized very slowly, while a large dose is metabolized rapidly (Fig. 7-4). The three asialotransferrin types are interchangeable regarding their catabolism-promoting effect (17).

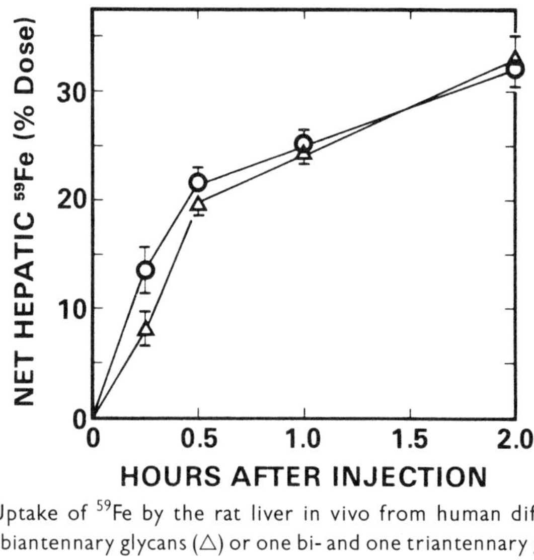

Figure 7-3. Uptake of ^{59}Fe by the rat liver in vivo from human diferric transferrin containing two biantennary glycans (△) or one bi- and one triantennary glycan (○). Dose, 20–25 μg of ^{59}Fe-labeled transferrin per 100 g of body weight. Each point is the mean from three animals; the bars are standard errors of the means. The term "net" signifies that the values were corrected for ^{59}Fe trapped in the post-mortem hepatic plasma volume by references to a differently labeled volume marker.

2. The hepatic half-life of asialotransferrin is long (approx. 2.8 hr; Ref. 26) as compared with other asialoglycoproteins (27, 28).
3. Subcellular asialotransferrin is present in a low-density organelle, having a median equilibrium density of 1.11 g/cm^3 in continuous sucrose density gradients (29).
4. Release of the subcellular asialotransferrin into the circulation is readily accomplished by a saturating amount of another asialoglycoprotein (asialo-orosomucoid, asialofetuin, asialomucin) (26). The release can be initiated at any time after the injection of the asialotransferrin dose.
5. A fraction of the injected asialotransferrin becomes partially resialylated (30). When this happens, the protein loses its affinity for the Gal/GalNAc-specific lectin and reappears spontaneously in the blood.
6. Asialotransferrins are potent donors of their iron to the liver (31).

It should be pointed out that of the features mentioned above, the study of numbers 2-5 is feasible only with a small dose.

Although the behavior of the three human asialotransferrins is qualitatively

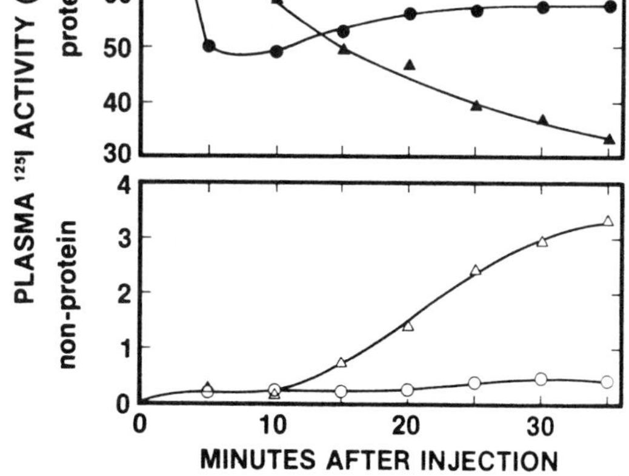

Figure 7-4. Plasma radioactivities following the injection of two different doses of ^{125}I-asialotransferrin, prepared by the enzymic desialylation of human whole transferrin, in rats. Doses: 5 μg/100 g (circles), and 390 μg/100 g (triangles). Closed symbols denote protein-bound, and open symbols nonprotein (acid-soluble) ^{125}I. Note the timing and extent of the rise in nonprotein ^{125}I that followed the injection of the large dose.

the same, their participation in certain biological processes (e.g., resialylation, iron deposition) is, nevertheless, quantitatively different. This is so because of the prevailing differences in lectin binding. The differences in affinity effect characteristic distribution patterns for each asialotransferrin type between the liver ("bound") and plasma ("unbound"). Since the fraction of radioligand that disappears from the plasma during the early phase (first 15 to 30 min) of an experiment is always fully recoverable with the liver in an undegraded form (4, 17), the radioactivity remaining in the plasma is a reliable indirect measure of liver binding. Thus, as can be seen in Figure 7-5, an association constant of 1.6×10^6 M^{-1} for HAsTf-1 (Table 7-1) translates into an in vivo situation in which approximately one-half of the dose is liver-associated and the other half is free; by using double-labeled (^{59}Fe, ^{125}I) HAsTf-1, it is possible to show that these two fractions are in a continuous exchange. By comparison, a K_A of 1.1×10^8 M^{-1} for HAsTf-3 affords a physiological distribution with 95-97% of the ligand bound by the liver and only 3-5% free. From the data in Figure 7-5, it is clear that HAsTf-3 ranks first among asialotransferrins (including those from other mammalian species) as a tool in cell biology.

ASIALOTRANSFERRIN DIACYTOSIS

All the features that constitute "asialotransferrin-like" behavior are explicable by the subcellular movement of asialotransferrin, termed *diacytosis*. The term has already been defined in the Introduction, but its usage requires a few explanatory comments. Etymologically, diacytosis means "movement through cell," and, hence, any ligand that passes through a cell undegraded qualifies for the term. (Of the synonyms currently in use, the Latin-Greek hybrid, *transcytosis* is equally descriptive. In contrast, *reverse endocytosis* seems a less accurate description because exocytosis—at least in the case of asialotransferrin—is mediated by a subcellular organelle that is not identical with the endosome.) Diacytosis is therefore not unique to asialotransferrin: Transferrin (32, 33), immunoglobulin A (34), lactoferrin (35), and even other asialoglycoproteins (36-40) are also diacytosed. Since ligands can be transported between different points of entry and exit in this manner, the vector is irrelevant to the concept of diacytosis. Last, it should be mentioned that the capability of diacytosing ligands is not an exclusive property of the hepatocyte; all eukaryotic cells investigated so far seem to be able to diacytose transferrin (7), and polymorphonuclear leukocytes diacytose a chemotactic peptide (41).

Regarding the hepatic diacytosis of HAsTf-3, the kinetics as understood today may be summarized as follows. Release studies with competing asialoglycoproteins in intact animals (26) suggest that HAsTf-3, after its initial capture by the Gal/GalNAc-specific lectin, is in a continuous movement in and out of the cell. During these repeated diacytic cycles, HAsTf-3 molecules are being slowly lost from the liver due to two simultaneous biochemical processes; namely, degradation and resialylation. Whether some HAsTf-3 may also be translocated to the bile is not yet known. As the combined result of these losses, the protein-bound radioactivity content of the liver decreases with a half-life of approximately 2.8 hr.

Diacytosis can be interrupted, and any HAsTf-3 still present in the pathway can be released, by the injection of a competing asialoglycoprotein. Release to the circulation is quantitative and is accompanied by no induction of HAsTf-3 degradation. Since the asialoglycoprotein-lectin complex dissociates after endocytosis (42, 43), it is a reasonable assumption that competing asialoglycoproteins exert their effect by occupying the free lectin units on the hepatocyte plasma membrane and thus prevent the recapture of HAsTf-3 following its exocytosis. Release of HAsTf-3 from isolated hepatocytes

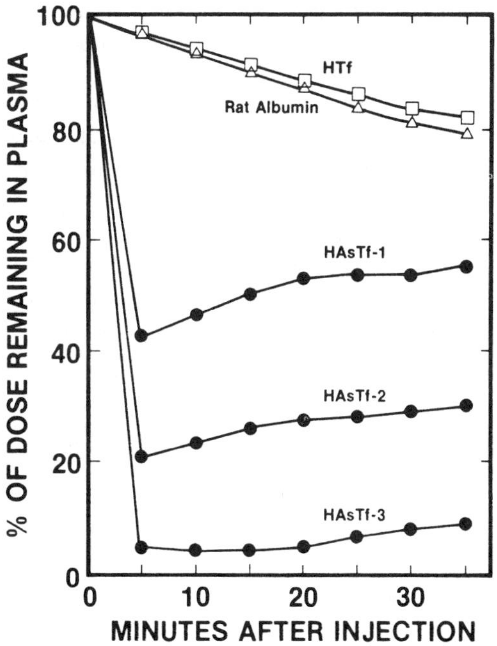

Figure 7-5. Initial plasma radioactivity curves of the human asialotransferrins in the rat. The ligands were labeled with ^{125}I and mixed with a ^{131}I-labeled plasma volume marker (rat albumin or human transferrin [HTf]). Dose, 1–2 μg of an asialotransferrin per 100 g of body weight. Points represent protein-bound radioactivities. Note that the magnitude of the rapid initial drop in the plasma level of each asialotransferrin type parallels the association constants given in Table 7-1.

can also be effected by resuspending the cells in a medium that contains a chelating agent (e.g., EGTA) or galactopyranoside (5).

Characteristically, the release curve observed in vivo can be resolved mathematically into a fast and a slow component, showing that 40% of the subcellular HAsTf-3 is discharged with a half-life of 0.94 min and the rest with a half-life of 4.9 min (26). The biological meaning of these kinetic compartments is unclear. They cannot be linked serially; otherwise, the pool represented by the faster component would be depleted to an insignificant residue in less than 6 min. Therefore, it must be assumed that they operate in parallel. It is interesting to note in this context that asialo-orosomucoid is also released from detergent-solubilized (44) or digitonin-treated (45) hepatocytes at two distinct rates, and so is [^{14}C]sucrose from macrophages and fibroblasts (46). On interrupting diacytosis of HAsTf-3 in suspended hepatocytes by EGTA, the discharge of the ligand is nearly (but not entirely) single-exponential, corresponding to a half-life of 20 min (5). It remains a task for the future to decide whether the segregation of these various macromolecules occurs in the same subcellular compartments or, perhaps, in unrelated ones.

The number of diacytic cycles that HAsTf-3 undergoes in an intact rat can be calculated from the half-life of the hepatic protein-bound radioactivity (2.8 hr) by reference to the respective release times of the compartments, assuming a steady state. These calculations indicate that HAsTf-3 molecules in the slow compartment are diacytosed 34 times, on the average, and those in the fast compartment are diacytosed 178 (!) times before their elimination from the pathway.

Deviations from the general kinetic picture drawn above for HAsTf-3 diacytosis are observed in both directions. Thus, in a group of 34 rats, we found the half-life of the hepatic protein-bound radioactivity to be significantly shorter than 2.8 hr in 12%, and significantly longer in 6% of the animals. It is clear from studies with the same dose solution in several recipients that we are dealing here with a true biological variation and not with fluctuations in the quality of the radioligand. And this poses a question: Why is HAsTf-3 being diacytosed in the first place?

THE MECHANISM OF DIACYTOSIS FOR HAsTf-3

Mention has already been made of a study by Young et al. (8), which was aimed at the clarification of the role that transferrin receptors may play in asialotransferrin evading degradation during its sojourn in the hepatocyte. These authors worked with suspended rat hepatocytes and used an asialoprotein prepared from the major component (47-50) of rat transferrin as the radioligand.

The preparation, when applied alone, was taken up by the cells, and 63% of the radioactivity released to the medium during a subsequent incubation period of 30 min was protein-bound. When the same experiment was carried out in the presence of an excess of asialo-orosomucoid, the portion of protein-bound ^{125}I in the medium increased to 81%, whereas in the presence of an excess of diferric transferrin it decreased to 21%. The conclusion drawn from this study was that either receptor (lectin or transferrin receptor) can afford the internalization of rat asialotransferrin; when uptake is channeled through the transferrin receptor, the ligand behaves as transferrin (i.e., it is salvaged), while in the alternative case it shares the fate of other asialoglycoproteins (i.e., it is degraded).

Dekker et al. (9) were unable to reproduce the above results on hepatocytes. Similarly, in vivo data available to this writer are at variance with the "either/or" theory of Young et al. (8). The core of the problem is that the single glycan attachment present in rat asialotransferrin is insufficient to accomplish endocytosis of this protein by the Gal/GalNAc-specific lectin alone; that is, without the assistance of a transferrin receptor. This may be concluded, among others, from experiments on animals in which more than 90% of the plasma transferrin was converted to the diferric form by the intraperitoneal injection of ferric-nitrilotriacetate complex before the ligand was injected. (The complex absorbing from the peritoneal cavity saturates the circulating transferrin with iron for several hours [51]. The saturation is needed because only 2Fe-transferrin binds strongly to transferrin receptors [52].) The failure of this treatment in inducing rapid catabolism of rat asialotransferrin is obvious from the nonprotein radioactivity curves in Figure 7-6 (see also Fig. 7-4 for comparison). The molar ratio of diferric transferrin to asialotransferrin in the study under consideration was close to 5,000:1. With this amount of competitor present for the transferrin receptor, any rat asialotransferrin uptake could have taken place only via the Gal/GalNAc-specific lectin. This would have led to prompt catabolism of the endocytosed protein, which was not the case. In agreement with these metabolic findings, rat asialotransferrin does not detectably interact with the immobilized Gal/GalNAc-specific lectin from rat liver, and its binding to purified rat liver plasma membranes (in the presence of diferric transferrin) is as insignificant as that of rabbit asialotransferrin (18).

It is hard to say why Young and his colleagues obtained the results they did. Freshly isolated hepatocytes are known to be in a catabolic state (53), which may affect their subcellular ligand sorting mechanism in favor of the lysosomal pathway. Furthermore, purified transferrin is prone to be contaminated with heme or heme-hemopexin (cf. "brown transferrin" [48]), particularly if the plasma used for fractionation was hemolytic. The removal of this contaminant cannot be accomplished by DEAE-chromatography, but only be an additional chromatographic step that utilizes carboxymethyl cellulose (54) or phenyl boronate (55). Neither of these separation media was deployed in the study under consideration. Hepatocytes do possess a heme binding site (56).

Clearly, then, the mechanism for the diacytosis of HAsTf-3 is not inferable from observations with rat asialotransferrin. A working hypothesis that accommodates all the experimental data gathered so far with the human protein has been introduced recently (57) and is presented in Figure 7-7. The underlying concept is that in order to be transported back to the surface, the ligand must remain membrane-bound while inside the cell. The vesicles containing endocytosed asialoglycoproteins, including HAsTf-3 (58), are rapidly acidified (59, 60), and this results in the dissociation of the ligand-receptor complex (44). Unlike asialo-orosomucoid and asialofetuin, HAsTf-3 has an additional binding site; that is, the one for the transferrin receptor, which is active at a low pH (52, 61, 62). If the cellular compartment in which the dissociation occurs contains a free transferrin receptor, HAsTf-3 binds to

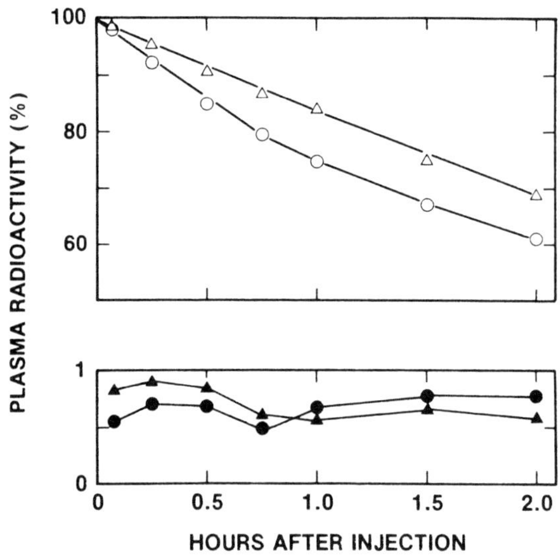

Figure 7-6. Plasma clearance of rat ^{131}I-labeled transferrin that contains a triantennary glycan (triangles; 1 μg/100 g) and its ^{125}I-labeled asialo derivative (circles; 3 μg/100 g) in a rat pretreated with ferric nitrilotriacetate complex. The complex, corresponding to 0.75 mg of Fe/100 g, was administered intraperitoneally 1 hr before the labeled proteins. Open symbols in the top section denote protein-bound radioactivities and closed symbols in the lower diagram nonprotein radioactivities, both expressed as percentages of the respective doses. For further explanations see the text.

it and is brought back to the plasma membrane. Diacytosis of HAsTf-3 is therefore a *dual*-receptor event, in which the lectin and the transferrin receptor come into action sequentially, the former being responsible for the endocytic and the latter for the exocytic aspect of the cycle. The receptor

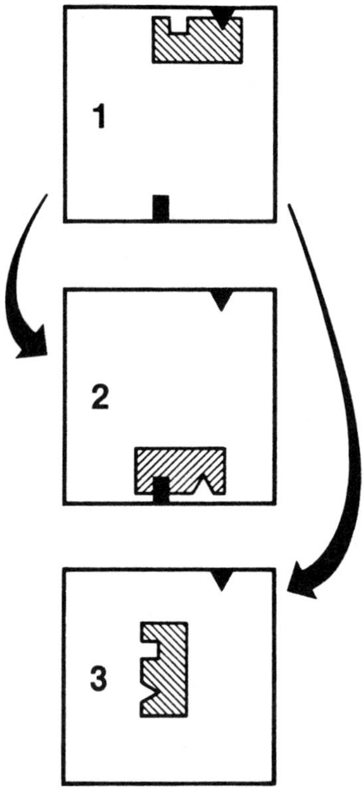

Figure 7-7. Schematic illustration of the writer's hypothesis of how HAsTf-3 escapes lysosomal degradation following its uptake by the rat liver. Big, numbered squares denote a plasmalemma-derived subcellular organelle (e.g., an endosome), closed triangles the Gal/GalNAc-specific lectin, and closed squares the transferrin receptor. Two kinds of indentation on the HAsTf-3 molecule (shaded body) symbolize the dual nature (i.e., the two different binding sites) of this ligand. In square 1, a freshly endocytosed HAsTf-3 is seen bound by the lectin. When the HAsTf-3/lectin complex dissociates as the result of a decrease in the pH, two possibilities arise. One is that a transferrin receptor is available in the same organelle to which HAsTf-3 can bind (square 2). If this is the case, HAsTf-3 follows the route of the membrane flow and is returned to the cell surface. The other possibility is that a transferrin receptor is not available (it is absent or is occupied by a transferrin molecule), in which case HAsTf-3 remains free in the endosomal fluid (square 3) and is degraded. For further explanations see the text.

switch takes place at a subcellular site (57). The above model, established in studies in vivo, is also supported by the results of experiments with isolated hepatocytes (5, 63).

There is ample evidence from studies with competitors to postulate that endocytosis of HAsTf-3 is mediated by the lectin and not by the transferrin receptor. For example, the hepatic binding of the ligand is not affected by establishing a 5×10^3-fold molar excess of endogenous diferric transferrin, or a 2.4×10^4-fold molar excess of human diferric transferrin in the host's circulation. Also, the apo form of HAsTf-3, which binds poorly to transferrin receptors at physiological pH (52), is taken up by the liver as efficiently as the diferric form. In sharp contrast, hepatic HAsTf-3 binding is abolished by pretreating the animal with an excess of asialo-orosomucoid (17).

The role of transferrin receptors in HAsTf-3 exocytosis is perceived from the effect of diferric transferrin on the hepatic half-life of HAsTf-3. The point has already been made in a previous section of the mass dependence of human asialotransferrin catabolism as one of the fundamental features of "asialotransferrin-like" behavior (4). The mass dependence manifests itself in the circumstance that diacytosis is observed only if the dose is kept very small (approximately 1 μg/100 g). The molecular mechanism underlying this phenomenon has remained elusive for years (26). Recently, however, we found that not only a larger dose of asialotransferrin, but also the injection of diferric transferrin (homologous and heterologous alike), can shorten the lifetime of a small dose of HAsTf-3 (57). Nevertheless, the effect of diferric transferrin is less spectacular: It reduces the hepatic half-life of HAsTf-3 from 2.8 hr to approximately 43 min as compared with asialotransferrin, which does so to 8-9 min (26). Furthermore, diferric transferrin, in sharp contradistinction to asialotransferrin, is effective only in very large quantities (>5 mg/100 g)—and, even then, not invariably.

Before considering the implications of the findings presented for HAsTf-3 exocytosis, the point should be made that the phenomenon of mass-dependent catabolism effectively eliminates the possibility of two receptors (lectin *and* transferrin receptor) participating simultaneously in the endocytosis of HAsTf-3. Iron uptake studies (see next section) suggest, however, that such a receptor cooperation may be operational with rat asialotransferrins. Of further relevance to this issue is that rat asialotransferrin are not subject to mass-dependent catabolism.

In all probability, mass-dependent catabolism of HAsTf-3 is due to a numerical imbalance between Gal/GalNAc-specific lectin units and transferrin receptors on the hepatocyte. According to a recent study (64), there are 1.8×10^5 surface lectin units per cell, which is contrasted with 3.6×10^4 (65)—or possibly even much fewer (66, 67)—transferrin receptors. For HAsTf-3 this means that the endocytic machinery of the cell has an excess capacity relative to the number of intraendosomal transferrin receptors to which the ligand can switch for exocytosis. Since the blood always contains some diferric transferrin, it seems reasonable to assume that only a fraction of the transferrin receptors en route through the cell at a given time are unoccupied. This factor imposes a further limit on the diacytic potential of the cell with respect to HAsTf-3. Other ligands, the survival of which does not depend on a receptor switch (e.g., lactoferrin [35]), can be diacytosed en mass.

A prerequisite to the postulated receptor switch is that all components of the reaction sequence are present in the same subcellular organelle. Results of recent cell subfraction studies suggest that this is indeed the case (68).

One observation that seemingly conflicts with the role envisaged for transferrin receptors in the exocytic phase of the diacytic cycle is the inferior efficacy of diferric transferrin itself in producing the expected metabolic change. Transferrin is a less potent inducer of HAsTf-3 catabolism than asialotransferrin (see above), and sometimes it fails altogether. (In our hands, the failure rate is 6/43 for a dose of >2 mg/100 g). The likely reason for this result is that the bulk of transferrin uptake by the hepatocyte takes place through fluid-phase pinocytosis rather than receptor-mediated endocytosis. Sibille and co-workers (69) and Octave and co-workers (70) were the first to draw attention to this fact, which has so far received inadequate attention—perhaps because of the current preoccupation with the idea that everything is mediated by receptors. In vivo data are available in the writer's laboratory in support of the observations by these researchers. An inevitable consequence of fluid-phase pinocytosis is the spread of the injected diferric transferrin over a large number of pinosomes. Therefore, the probability of blocking a transferrin receptor in the *same* endosome that contains an HAsTf-3 molecule is by this technique correspondingly small. When, however, asialotransferrin is used to induce catabolism, spreading is prevented by the lectin and the ligand is channeled to a comparatively small number of transport vesicles in which its number outweighs the transferrin receptors available for a "switch" and exocytosis.

Recently, the question of which hepatic cell types possess transferrin receptors was raised by Soda and Tavassoli (66) and Kishimoto and Tavassoli (67). These investigators found that, in contradistinction to the sinusoidal epithelium, hepatocytes and Kupffer cells did not adhere to latex particles coated with transferrin, nor did they bind ^{125}I-transferrin in vitro. Since sinusoidal cells are devoid of the Gal/GalNAc-specific lectin (71), it is interesting to ask which cells attract asialotransferrin. J. Rudolph in this laboratory injected rats with rat ^{125}I-transferrin or ^{125}I-HAsTf-2, and separated the liver cells subsequently on a metrizamide gradient; in preliminary experiments, he found that the transferrin was associated with both parenchymal and nonparenchymal cells, whereas >90% of HAstf-2 was confined to the parenchymal cell population.

ASIALOTRANSFERRIN AND BODY IRON DISTRIBUTION

It is a common feature to all asialotransferrins investigated so far that they deposit their iron in the liver more efficiently than their nondesialylated counterparts (9, 31, 72). This also holds true for those asialotransferrins that possess a single glycan attachment and therefore interact very weakly with the Gal/GalNAc-specific lectin (e.g., rat and rabbit asialotransferrins; cf. previous section). In the case of the human protein, it suffices to remove two out of four sialic acid residues to induce increased hepatic deposition of the iron.

The enhanced hepatic accumulation of iron from whole (unfractionated) rat asialotransferrin is illustrated in Figure 7-8, together with the results from a control group in which iron uptake from native transferrin was in the same order of magnitude as also found by others (73). Furthermore, Figure 7-8, in conjunction with Figure 7-3, demonstrates that native transferrins from different species are unequal iron donors when injected into a given species (72). The clearance of iron to extrahepatic tissues is, at least in the rat, the same regardless of whether the radioiron is administered as transferrin or asialotransferrin.

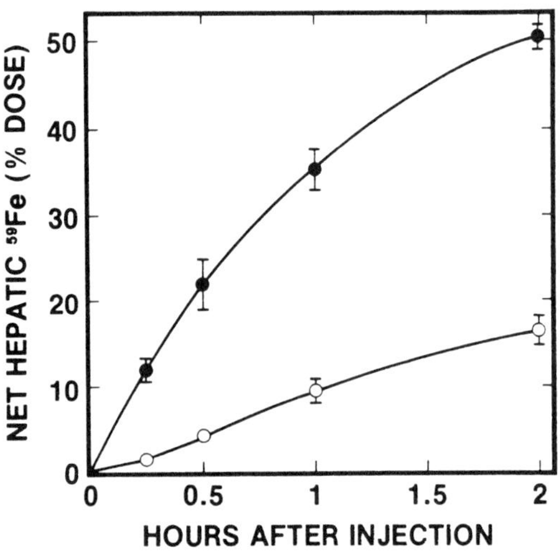

Figure 7-8. Uptake of ^{59}Fe by the liver of the intact rat from homologous diferric whole transferrin (○) and homologous diferric whole asialotransferrin (●). Dose, 20–25 μg ^{59}Fe-labeled protein per 100 g of body weight. Each point is the mean from three rats, the bars denoting standard errors. The meaning of the term "net" is explained in the legend to Figure 7-3.

In equilibrium dialysis experiments, Jarritt and Charlwood (74) found that desialylation did not affect the Fe-binding constants of human transferrin. It is a reasonable assumption, therefore, that iron transfer from asialotransferrin involves the same mechanism as from transferrin, and that more favorable disposition of asialotransferrin to surrender iron stems from its additional affinity for the liver as provided by the Gal/GalNAc-specific lectin. Whether the iron separates from transferrin first inside the cell, as many workers think, or already at the plasma membrane, as others believe (75, 76, 77), is immaterial in the present context.

Lectin binding alone seems sufficient to effect hepatic iron delivery from the human asialotransferrin types in the rat. HAsTf-3 and HAsTf-2, by virtue of their high affinities for the lectin, are capable of almost exclusively homing the attached ^{59}Fe to the liver (i.e., with a minimal spill to extrahepatic tissues) in a matter of minutes; if an excess of asialomucin is injected 5 min later, the liver retains the iron and releases the apoprotein.

More intriguing is the situation with rat asialotransferrin. When this protein, labeled with ^{59}Fe, is injected in rats in which the lectin is blocked (e.g., by pretreatment with asialomucin), the iron uptake curve seen in Figure 7-8 for asialotransferrin assumes the position of that for transferrin. The same effect is also achieved, however, by pretreating the rat with a large excess (23 mg/100 g) of diferric transferrin (J. Rudolph and E. Regoeczi, unpublished work). In other words, enhanced iron uptake from a weak asialotransferrin, such as the rat protein, is observed only if free lectin and free transferrin receptor are available on the liver simultaneously, thus implying that the enhanced iron uptake is mediated by a synergism (cooperation) between the two types of receptor. In any other case, rat asialotransferrin functions as ordinary transferrin. (One might ask by reference to the massive injection of diferric transferrin mentioned above, how a trace of asialotransferrin can deliver ^{59}Fe at the same rate as does control transferrin in the absence of a transferrin overload. The likely answer is fluid-phase pinocytosis; this form of transport obeys first-order kinetics, whereby the fractional transfer rate is independent of substrate concentration.)

The foregoing paragraphs show that transferrin, when wholly or partially deficient in sialic acid, distributes its iron in the body distinguishably from the normal protein. Sialic acid deficiency modulates the physiological interaction of transferrin with the liver and thus promotes preferential accumulation of iron in that organ. From model experiments in animals, the loss of sialic acid from circulating transferrin (or from α_1-antitryspin, for that matter [78]) seems unlikely (79). Nevertheless, incompletely sialylated transferrin may appear in the blood because of deficient glycosylation during protein biosynthesis, and in several laboratories (80, 81), such transferrin molecules have been detected in the sera of alcohol abusers. Because the terminating sugar of the incomplete antennae in transferrin from alcoholics is galactose (82), these molecules may be expected to have iron-donating properties similar to those of transferrins modified in vitro. The long survival of homologous asialotransferrin in humans (2) on the one hand, and the serum iron elevating effect of ethanol (83) on the other, prompt one to hypothesize that a common complication of alcoholism, hepatic siderosis, might causally be linked to abnormal transferrin glycosylation (72). An added impetus to this line of thinking is given by the recent study of Potter and his co-workers showing that the fractional catabolic rate of transferrin is normal in alcoholic liver disease (84).

IN VIVO RESIALYLATION OF ASIALOTRANSFERRIN

For the first 20 to 30 min after the injection of a small dose of ^{125}I-HAsTf-3 in a rat, only a few percent of the dose remains detectable in the plasma. Subsequently, however, the plasma protein-bound radioactivity level rises again, usually peaking between 15% and 20% of the dose approximately 3 hr after the injection (30). The gel-filtration profile of the labeled protein recovered at this stage from the plasma is the same as that of the injected material, suggesting the absence of any gross proteolytic modifications. Nevertheless, the recovered protein no longer binds to columns of the immobilized Gal/GalNAc-specific lectin. Lectin binding is restored by treatment with neuraminidase, demonstrating that HAsTf-3 has undergone resialylation in vivo.

The degree of resialylation can be gauged electrophoretically from the increase in electronegativity relative to a reference preparation of human transferrin that is comprised of all forms of the protein with respect to sialic acid content. The extent of resialylation thus found is two to three residues per molecule of protein, which corresponds to the number of residues required to restitute one of the glycan attachments. There is no evidence obtained by this technique of resialylation going to completion. The likely reason for the partial, instead of a complete, resialylation is that access of the sialyltransferase to one of the glycans is conceivably hindered sterically because of HAsTf-3 binding to a transferrin receptor (see the section on the mechanism of diacytosis for HAsTf-3). This view gains support from the recent demonstration by Snider and Rogers (85) of the resialylation of desialylated transferrin receptors in K562 cells. Resialylation of the transferrin receptor, which seems kinetically very similar to HAsTf-3 resialylation in the rat liver, is impaired when the receptor is occupied by transferrin. It must be stressed that the view expressed previously regarding the restoration of one glycan—as opposed to the attachment of two to three sialic acids to either glycan at random—is at present an assumption that requires experimental verification. At any rate, the remaining exposed galactosyl residues are insufficient to secure the recapture of exocytosed HAsTf-3 by the lectin, whereby the protein is set free.

After a pulse injection, partially resialylated HAsTf-3 disappears from the plasma of rats with similar kinetics to control transferrin, suggesting that the tissue distribution and catabolic rate of both proteins are comparable. This observation provides the rationale for the use of deconvolution as the method of computing the overall portion of a dose of HAsTf-3 resialylated by the liver from the plasma reappearance curve of HAsTf-3 and the plasma disappearance curve of a differentially labeled transferrin, injected as a pulse. The original value of 25% was computed from a composite plasma curve derived from several rats, each contributing a single point (30). A modified version of the method, involving a set of measurements on individual animals over a period of 3 hr, was recently described (86). The new approach facilitates testing the effects of cytotropic agents on HAsTf-3 resialylation. With its help it was possible to show that microtubule inhibitors (colchicine, vinblastine, and taxol) increase the proportion of the dose undergoing resialylation by 35-39% (86).

HAsTf-1 and HAsTf-2 also undergo partial resialylation in the rat, though more slowly than HAsTf-3 (87). On the whole, the speeds of resialylation of the three asialotransferrin are in the reverse order of their affinities for the Gal/GalNAc-specific lectin. In the case of HAsTf-1, for instance, it takes about 2.5 days before the total body radioactivity curve assumes the slope of control transferrin (Fig. 7-9). This slowness contrasts sharply with the speed by which these asialotransferrins deposit their iron in the liver (80-90% in less than 1 hr). The interesting inference is that each asialotransferrin molecule probably passes through the liver more than once before the critical number of sialyl residues become attached for the protein to be no longer attracted by the hepatic lectin (87).

No other desialylated plasma glycoprotein is known at present to serve as an acceptor of sialic acid in vivo after reinjection. However, the mouse liver transforms a small portion (3-5%) of exogenously administered ganglioside G_{M1} to the disialosyl derivative, possibly (but not certainly) by sialyltransferase acting directly on the original probe rather than on its metabolites (88).

The physiological significance of asialotransferrin resialylation is unknown. The phenomenon could point to the existence of a general membrane repair process taking place in the Golgi as envisaged by Farquhar (89). Transferrin does reach the *trans*-Golgi in KB carcinoma cells (90), most probably because of binding to subcellular transferrin receptors (91). Asialotransferrin, routed to the Golgi by an analogous mechanism, could be "mistaken" for a membrane protein and processed accordingly; this view is strengthened by the observations of Snider and Rogers (85). Noteworthy in this connection is that membrane glycoproteins are apparently not as static as had been thought earlier: According to Tauber and co-workers (92), the half-lives of the terminal sugars in these proteins are significantly shorter than those of the core sugars and of the peptide moieties. This could be an indication of monosaccharide turnover in membrane glycoproteins (as opposed as a plain, stepwise degradation), in which case the resialylation of asialotransferrins may be perceived as the anabolic pendant to the finding of Tauber and others (92).

THE RECOGNITION SYSTEM FOR BIANTENNARY ASIALOGLYCANS OF THE RABBIT BONE MARROW

Since knowledge of the existence of the marrow recognition system came about with the help of asialotransferrins, this chapter would not be complete without briefly mentioning this subject.

The bone marrow, in rabbits, is a significant competitor against the liver for HAsTf-1 (3). As can be seen in Figure 7-10, accretion of the ligand by the

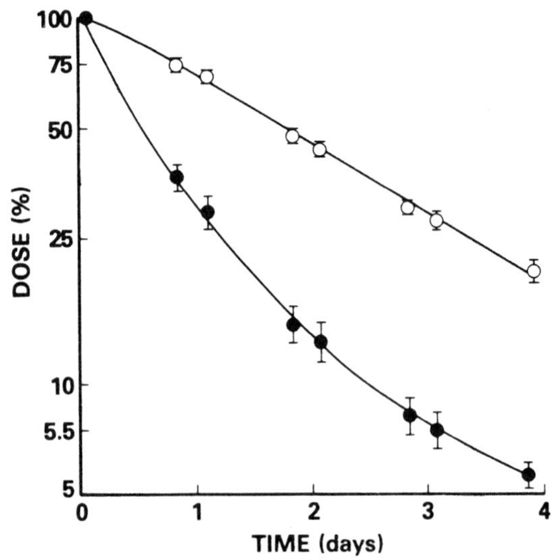

Figure 7-9. Catabolism of ^{131}I-HAsTf-1 (●) and its fully sialylated counterpart (○), labeled with ^{125}I, in the rat. A mixture of both proteins (approximately 2 μg/100 g of each) was injected intravenously at zero time, and the behavior of the radioactivities was followed by measuring total body radiation. Values are means and their standard error from five animals.

marrow is comparatively slow and leads to degradation. Pittman and his co-workers (93) injected human asialotransferrin labeled with ^{125}I-tyramine-cellobiose (a label that becomes trapped at the site of protein degradation) in rabbits, and found that 50% of the label was accumulated by the liver and 31% by the bone marrow.

As regards to the glycan specificity of this recognition system, optimal interaction seems to require at least two biantennary asialoglycans as concluded from the following observations. Human transferrin, rabbit asialotransferrin, galactose-oxidized HAsTf-1, and asialofetuin are not accreted by the marrow; HAsTf-1 with one residual sialic acid is a marginal case; HAsTf-1 is taken up well, whereas a mixture of HAsTf-2 and HAsTf-3 is taken up less well but still to a significant extent. The affinity of this asialotransferrin mixture for the marrow is markedly enhanced by retaining one sialic acid in the glycans ("pseudobiantennary" asialotransferrin). Chicken α_1-acid glycoprotein, containing galactose-terminating chains (94), and human asialo-orosomucoid block the marrow-HAsTf-1 interaction; the blocking power of these proteins resides with their subfractions, which are retained by Sepharose-concavalin-A; that is, which contain biantennary glycans (3). The latter findings suggest that the recognition system under consideration is not primarily concerned with transferrin.

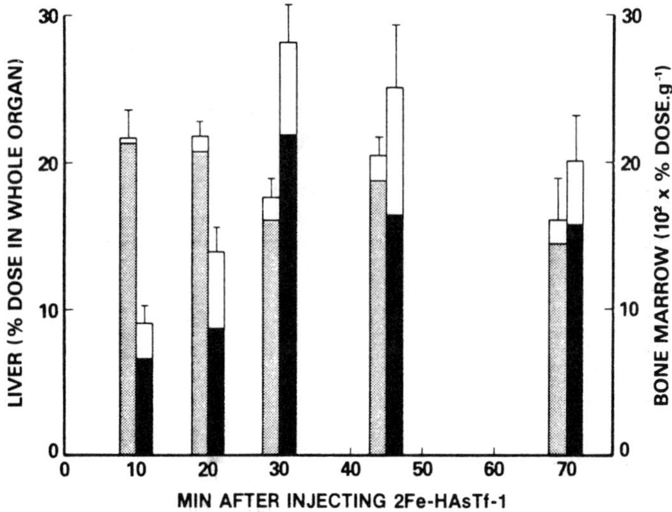

Figure 7-10. Clearance of ^{125}I-HAsTf-1 by the liver and bone marrow in the rabbit. Protein-bound ^{125}I in the liver is shown as shaded bars and in the marrow as closed bars, the open bars representing acid-soluble ^{125}I. Each value is the mean and its standard error from five to seven animals.

Judged by the comparatively long survival of human asialotransferrin in other species—notably in man (2), and the rat (Fig. 7-9), dog, goat, pig, sheep, and rhesus (E. Regoeczi, unpublished observations), as well as in the chicken and duck (95)—the bone marrow of the rabbit seems, so far, unique. However, it is impossible to say without direct marrow assays whether the activity in the above species is absent altogether or is just very weak. Kataoka and Tavassoli (96) reported recently the binding of galactosylated bovine albumin to the sinusoidal endothelium of rat bone marrow. The probe contained 38-50 galactose residues per molecule as compared with four in HAsTf-1. Thus the rat marrow may well have a recognition system that is galactose specific, although its glycan specificity remains unclear.

REFERENCES

1. A. G. Morell, G. Gregoriadis, I. H. Scheinberg, J. Hickman, and G. Ashwell, 1971, The role of sialic acid in determining the survival of glycoproteins in the circulation, *J. Biol. Chem.* **246:**1461-1467.
2. E. Regoeczi, M. W. C. Hatton, and K. -L. Wong, 1974, Studies of the metabolism of asialotransferrins: Potentiation of the catabolism of human asialotransferrin in the rabbit, *Can. J. Biochem.* **52:**155-161.
3. E. Regoeczi, P. A. Chindemi, M. W. C. Hatton, and L. R. Berry, 1980, Galactose-specific elimination of human asialotransferrin by the bone marrow in the rabbit, *Arch. Biochem. Biophys.* **205:**76-84.
4. E. Regoeczi, P. Taylor, M. W. C. Hatton, K. -L. Wong, and A. Koj, 1978, Distinction between binding and endocytosis of human asialo-transferrin by the rat liver, *Biochem. J.* **174:**171-178.
5. H. Tolleshaug, P. A. Chindemi, and E. Regoeczi, 1981, Diacytosis of human asialotransferrin type 3 by isolated rat hepatocytes, *J. Biol. Chem.* **256:**6526-6528.
6. D. T. Connolly, R. R. Townsend, K. Kawaguchi, W. R. Bell, and Y. C. Lee, 1982, Binding and endocytosis of cluster glycosides by rabbit hepatocytes. Evidence for a short-circuit pathways that does not lead to degradation, *J. Biol. Chem.* **257:**939-945.
7. E. Regoeczi, 1984, Role of cell-surface receptors in the behavior in vivo of transferrin, in *Pathophysiology of Plasma Protein Metabolism*, G. Mariani, ed., Macmillan, London, pp. 373-395.
8. S. P. Young, A. Bomford, and R. Williams, 1983, Dual pathways for the uptake of rat asialotransferrin by rat hepatocytes, *J. Biol. Chem.* **258:**4972-4976.
9. C. J. Dekker, M. J. Kroos, C. Van der Heul, and H. G. Van Eijk, 1985, Uptake of sialo and asialo transferrins by isolated rat hepatocytes. Comparison of a heterologous and a homologous system, *Int. J. Biochem.* **17:**701-706.
10. M. W. C. Hatton, E. Regoeczi, and K. -L. Wong, 1974, Studies of the metabolism of asialotransferrins: Relationship between the carbohydrate composition of bovine, canine, and porcine asialotransferrins and their metabolic behavior in the rabbit, *Can. J. Biochem.* **52:**845-853.
11. G. A. Jamieson, M. Jett, and S. L. DeBernardo, 1971, The carbohydrate sequence of the glycopeptide chains of human transferrin, *J. Biol. Chem.* **246:**3686-3693.

12. G. Spik, B. Bayard, B. Fournet, G. Strecker, S. Bouquelet, and J. Montreuil, 1975, Studies on glycoconjugates. LXIV. Complete structure of two carbohydrate units of human serotransferrin, *FEBS Lett.* **50:**296-299.
13. E. Regoeczi, K.-L. Wong, M. Ali, and M. W. C. Hatton, 1977, The molecular components of human transferrin type C, *Int. J. Pept. Protein Res.* **10:**17-26.
14. K.-L. Wong, and E. Regoeczi, 1977, Some observations on the carbohydrate composition of purified transferrin, *Int. J. Pept. Protein Res.* **9:**241-248.
15. K.-L. Wong, M. T. Debanne, M. W. C. Hatton, and E. Regoeczi, 1978, Human transferrin, asialotransferrin and the intermediate forms, *Int. J. Pept. Protein Res.* **12:**27-37.
16. M. W. C. Hatton, L. März, L. R. Berry, M. T. Debanne, and E. Regoeczi, 1979, Bi- and tri-antennary human transferrin glycopeptides and their affinities for the hepatic lectin specific for asialo-glycoproteins, *Biochem. J.* **181:**633-638.
17. E. Regoeczi, P. Taylor, M. T. Debanne, L. März, and M. W. C. Hatton, 1979, Three types of human asialo-transferrin and their interactions with the rat liver, *Biochem. J.* **184:**399-407.
18. M. T. Debanne, P. A. Chindemi, and E. Regoeczi, 1981, Binding of asialotransferrins by purified rat liver plasma membranes, *J. Biol. Chem.* **256:**4929-4933.
19. G. Spik, V. Debruyne, and J. Montreuil, 1983, *Alterations of the Carbohydrate Structure of Human Serotransferrin in Liver Disease,* H. Popper, W. Reutter, E. Köttgen, and F. Gudat, eds., MTP Press, Lancaster (U.K.), pp. 477-483.
20. L. März, M. W. C. Hatton, L. R. Berry, and E. Regoeczi, 1982, The structural heterogeneity of the carbohydrate moiety of desialylated human transferrin, *Can. J. Biochem.* **60:**624-630.
21. T. Krusius, and J. Finne, 1981, Use of the Smith degradation in the study of the branching pattern in the complex-type carbohydrate units of glycoproteins, *Carbohydr. Res.* **90:**203-214.
22. V. Debruyne, J. Montreuil, and G. Spik, 1984, Crossed immunoaffinity electrophoresis of human transferrin in normal and cirrhotic sera, *Protides Biol. Fluids* **31:**63-68.
23. R. L. Hudgin, W. E. Pricer, Jr., G. Ashwell, R. J. Stockert, and A. G. Morell, 1974, The isolation and properties of a rabbit liver binding protein specific for asialoglycoproteins, *J. Biol. Chem.* **249:**5536-5543.
24. M. Lowe, and B. Nilsson, 1983, The structure of the complex type oligosaccharide from rabbit hepatic binding protein. A re-examination, *J. Biol. Chem.* **258:**1885-1887.
25. M. W. C. Hatton, L. März, and E. Regoeczi, 1983, On the significance of heterogeneity of plasma glycoproteins possessing *N*-glycans of the complex type: A perspective, *Trends Biochem. Sci.* **8:**287-291.
26. E. Regoeczi, P. A. Chindemi, M. T. Debanne, and M. W. C. Hatton, 1982, Dual nature of the hepatic lectin pathway for human asialotransferrin type 3 in the rat, *J. Biol. Chem.* **257:**5431-5436.
27. W. A. Dunn, J. H. LaBadie, and N. N. Aronson, Jr., 1979, Inhibition of ^{125}I-asialofetuin catabolism by leupeptin in the perfused rat liver and in vivo, *J. Biol. Chem.* **254:**4191-4196.
28. P. A. Charlwood, E. Regoeczi, and M. W. C. Hatton, 1979, Hepatic uptake and degradation of trace doses of asialofetuin and asialo-orosomucoid in the intact rat, *Biochim. Biophys. Acta* **585:**61-70.
29. M. T. Debanne, and E. Regoeczi, 1981, Subcellular distribution of human asialotransferrin type 3 in the rat liver, *J. Biol. Chem.* **256:**11266-11272.
30. E. Regoeczi, P. A. Chindemi, M. T. Debanne, and P. A. Charlwood, 1982, Partial resialylation of human asialotransferrin type 3 in the rat, *Proc. Natl. Acad. Sci. (USA)* **79:**2226-2230.
31. E. Regoeczi, P. A. Chindemi, and M. T. Debanne, 1983, Altered hepatic uptake of iron from asialotransferrins in *Structure and Function of Iron Storage and Transport Proteins,* I. Urushizaki, P. Aisen, I. Listowsky, and J. W. Drysdale, eds., Elsevier Science Publishers, Amsterdam, pp. 271-272.

32. F. M. Van Bockxmeer, and E. M. Morgan, 1979, Transferrin receptors during rabbit reticulocyte maturation, *Biochim. Biophys. Acta* **584:**76-83.
33. A. Ciechanover, A. L. Schwartz, A. L. Dautry-Varsat, and H. F. Lodish, 1983, Kinetics of internalization and recycling of transferrin and the transferrin receptor in a human hepatoma cell line, *J. Biol. Chem.* **258:**9681-9689.
34. J. M. Schiff, M. M. Fisher, and B. J. Underdown, 1984, Receptor-mediated transport of immunoglobulin A and asialoglycoprotein: Sorting and missorting of ligands revealed by two radiolabeling methods, *J. Cell Biol.* **98:**79-89.
35. E. Regoeczi, P. A. Chindemi, M. T. Debanne, and J. -P. Prieels, 1985, Lactoferrin catabolism in the rat liver, *Am. J. Physiol.* **248:**G8-G14.
36. U. Gärtner, R. J. Stockert, A. G. Morell, and A. W. Wolkoff, 1981, Modulation of the transport of bilirubin and asialoorosomucoid during liver regeneration, *Hepatology* **1:**99-106.
37. F. G. M. Russel, J. G. Weitering, R. Oosting, G. M. M. Groothuis, M. J. Hardonk, and D. K. F. Meier, 1983, Influence of taurocholate on hepatic clearance and biliary excretion of asialo intestinal alkaline phosphatase in the rat in vivo and in the isolated perfused rat liver, *Gastroenterology* **85:**225-234.
38. R. R. Townsend, D. A. Wall, A. L. Hubbard, and Y. C. Lee, 1984, Rapid release of galactose-terminated ligands after endocytosis by hepatic parenchymal cells: Evidence for a role of carbohydrate structure in the release of internalized ligand from receptor, *Proc. Natl. Acad. Sci. (USA)* **81:**466-470.
39. P. H. Weigel, and J. A. Oka, 1984, Recycling of the hepatic asialoglycoprotein receptor in isolated rat hepatocytes. Receptor-ligand complexes in an intracellular slowly dissociating pool return to the cell surface prior to dissociation, *J. Biol. Chem.* **259:**1150-1154.
40. T. -M. Chang, and D. W. Kullberg, 1984, Diacytosis of ^{125}I-asialoorosomucoid by rat hepatocytes. A non-lysosomal pathway insensitive to inhibition by inhibitors of ligand degradation, *Biochim. Biophys. Acta* **805:**268-276.
41. G. Daukas, D. A. Lauffenburger, and S. Zigmond, 1983, Reversible pinocytosis in polymorphonuclear leukocytes, *J. Cell Biol.* **96:**1642-1650.
42. K. Bridges, J. Harford, G. Ashwell, and R. D. Klausner, 1982, Fate of receptor and ligand during endocytosis of asialoglycoproteins by isolated hepatocytes, *Proc. Natl. Acad. Sci. (USA)* **79:**350-354.
43. J. Harford, A. W. Wolkoff, G. Ashwell, and R. D. Klausner, 1983, Monensin inhibits intracellular dissociation of asialoglycoproteins from their receptor, *J. Cell Biol.* **96:**1824-1828.
44. J. Harford, K. Bridges, G. Ashwell, and R. D. Klausner, 1983, Intracellular dissociation of receptor-bound asialoglycoproteins in cultured hepatocytes. A pH-mediated nonlysosomal event, *J. Biol. Chem.* **258:**3191-3197.
45. J. A. Oka, and P. H. Weigel, 1983, Recycling of the asialoglycoprotein receptor in isolated rat hepatocytes. Dissociation of internalized ligand from receptors occurs in two kinetically and thermally distinguishable compartments, *J. Biol. Chem.* **258:**10253-10262.
46. J. M. Besterman, J. A. Airhard, R. C. Woodworth, and R. B. Low, 1981, Exocytosis of pinocytosed fluid in cultured cells: Kinetic evidence for rapid turnover and compartmentation, *J. Cell Biol.* **91:**716-727.
47. G. H. Beaton, A. E. Selby, and A. M. Wright, 1961, Starch gel electrophoresis of rat serum proteins. I. Procedure and designation of components, *J. Biol. Chem.* **236:**2001-2004.
48. A. H. Gordon, and L. N. Louis, 1963, Preparation and properties of rat transferrin, *Biochem. J.* **88:**409-414.
49. H. Huebers, E. Huebers, W. Rummel, and R. R. Crichton, 1976, Isolation and characterization of iron-binding proteins from rat intestinal mucosa, *Eur. J. Biochem.* **66:**447-455.
50. S. Okada, B. Jarvis, and E. B. Brown, 1979, In vivo evidence for the functional heterogeneity

of transferrin-bound iron. V. Isotransferrins: An explanation of the Fletcher-Huehns phenomenon in the rat, *J. Lab. Clin. Med.* **93:**189-198.

51. Y. Yamanoi, R. Matsuura, and M. Awai, 1982, Mechanism of iron toxicity in the liver and pancreas after a single injection of ferric nitrilotriacetate, *Acta Haematol. Jpn.* **45:**1229-1235.
52. E. H. Morgan, 1983, Effect of pH and iron content of transferrin on its binding to reticulocyte receptors, *Biochim. Biophys. Acta* **762:**498-502.
53. P. O. Seglen, 1977, Protein-catabolic state of isolated rat hepatocytes, *Biochim. Biophys. Acta* **496:**182-191.
54. M. W. C. Hatton, E. Regoeczi, K. -L. Wong, and G. J. Kraay, 1977, Bovine serum transferrin phenotypes AA, D_1D_1, D_2D_2, EE: Their carbohydrate compositions and electrophoretic multiplicity, *Biochem. Genet.* **15:**621-640.
55. N. D. Cook, R. J. Simpson, K. Osterloh, and T. J. Peters, 1985, Rapid preparation of highly purified human transferrin, *Anal. Biochem.* **149:**349-353.
56. A. Smith, and W. T. Morgan, 1981, Hemopexin-mediated transport of heme into isolated rat hepatocytes, *J. Biol. Chem.* **256:**10902-10909.
57. E. Regoeczi, and A. Koj, 1985, Diacytosis of human asialotransferrin type 3 in the rat liver is due to the sequential engagement of two receptors, *Exp. Cell Res.* **160:**1-8.
58. M. J. Geisow, and W. H. Evans, 1984, pH in the endosome. Measurements during pinocytosis and receptor-mediated endocytosis, *Exp. Cell Res.* **150:**36-46.
59. B. Tycko, C. H. Keith, and F. R. Maxfield, 1983, Rapid acidification of endocytic vesicles containing asialoglycoprotein in cells of a human heptoma line, *J. Cell Biol.* **97:**1762-1776.
60. F. R. Maxfield, 1985, Acidification of endocytic vesicles and lysosomes, in *Endocytosis*, I. Pastan, and M. C. Willingham, eds. New York, Plenum Press, pp. 235-257.
61. A. Dautry-Varsat, A. Ciechanover, and H. F. Lodish, 1983, pH and recycling of transferrin during receptor-mediated endocytosis, *Proc. Natl. Acad. Sci. (USA)* **80:**2258-2262.
62. R. D. Klausner, G. Ashwell, J. Van Renswounde, J. B. Harford, and K. R. Bridges, 1983, Binding of apotransferrin to K652 cells: Explanation of the transferrin cycle, *Proc. Natl. Acad. Sci. (USA)* **80:**2263-2266.
63. H. Tolleshaug, 1984, Intracellular segregation of asialo-transferrin and asialofetuin following uptake by the same receptor system in suspended hepatocytes, *Biochim. Biophys. Acta* **803:**182-190.
64. P. H. Weigel, and J. A. Oka, 1983, The large intracellular pool of asialoglycoprotein receptors functions during the endocytosis of asialoglycoproteins by isolated rat hepatocytes, *J. Biol. Chem.* **258:**5095-5102.
65. S. P. Young, and P. Aisen, 1980, The interaction of transferrin with isolated hepatocytes, *Biochim. Biophys. Acta* **633:**145-153.
66. R. Soda, and M. Tavassoli, 1984, Liver endothelium and not hepatocytes or Kupffer cells have transferrin receptors, *Blood* **63:**270-276.
67. T. Kishimoto, and M. Tavassoli, 1985, Endothelial binding of transferrin in fractionated liver cell suspensions, *Biochim. Biophys. Acta* **846:**14-20.
68. M. T. Debanne, M. Bolyos, J. Gauldie, and E. Regoeczi, 1984, Two populations of prelysosomal structures transporting asialoglycoproteins in rat liver, *Proc. Natl. Acad. Sci. (USA)* **81:**2995-2999.
69. J. -C. Sibille, J. -N. Octave, Y. -J. Schneider, A. Trouet, and R. R. Crichton, 1982, Transferrin protein and iron uptake by cultured hepatocytes, *FEBS Lett.* **150:**365-369.
70. J. -N. Octave, Y. -J. Schneider, A. Trouet, and R. R. Crichton, 1983, Iron uptake and utilization by mammalian cells. I. Cellular uptake of transferrin and iron, *Trends Biochem. Sci.* **8:**217-220.
71. C. J. Steer, and R. Clarenburg, 1979, Unique distribution of glycoprotein receptors on parenchymal and sinusoidal cells of rat liver, *J. Biol. Chem.* **254:**4457-4461.

72. E. Regoeczi, P. A. Chindemi, and M. T. Debanne, 1984, Transferrin glycans: A possible link between alcoholism and hepatic siderosis, *Alcoholism: Clin. Exp. Res.* **8:**287-292.
73. J. W. Wilms, and R. G. Batey, 1983, Effect of iron stores on hepatic metabolism of transferrin-bound iron, *Am. J. Physiol.* **244:**G138-G144.
74. P. H. Jarritt, and P. A. Charlwood, Competitive binding of iron by transferrin from different vertebrates, *Biochim. Biophys. Acta* **453:**319-331.
75. E. S. Cole, and J. Glass, 1983, Transferrin binding and iron uptake in mouse heptocytes, *Biochim. Biophys. Acta* **762:**102-110.
76. C. G. D. Morley, K. A. Rewers, and A. Bezkorovainy, Iron metabolism pathways in the rat hepatocytes, *Clin. Physiol. Biochem.* **1:**3-11.
77. C. G. D. Morley, and A. Bezkorovainy, 1983, The behavior of transferrin receptors in the rat hepatocyte plasma membranes, *Clin. Physiol. Biochem.* **1:**318-323.
78. K. Omichi, G. Savic, and A. Yoshida, 1983, Preparation of neuraminidase-resistant human α_1-protease inhibitor and its clearance in blood circulation, *Int. J. Biochem.* **15:**1345-1351.
79. K. -L. Wong, P. A. Charlwood, M. W. C. Hatton, and E. Regoeczi, 1974, Studies of the metabolism of asialotransferrins: Evidence that transferrin does not undergo desialylation in vivo, *Clin. Sci. Mol. Med.* **45:**763-774.
80. H. Stibler, and S. Borg, 1981, Evidence of a reduced sialic acid content in serum transferrin in male alcoholics, *Alcoholism: Clin. Exp. Res.* **5:**545-549.
81. H. G. Van Eijk, W. L. Van Noort, M. -L. Dubelaar, and C. Van der Heul, 1983, The microheterogeneity of human transferrins in biological fluids, *Clin. Chim. Acta* **132:**167-171.
82. E. Cervén, H. Stibler, and S. Borg, 1981, Determination of terminal sugars in transferrin by radio-lectin immunoassay (RLIA)—A new microanalytical procedure, *Ups. J. Med. Sci.* **86:**39-53.
83. L. W. Sullivan, and V. Herbert, 1964, Suppression of hematopoiesis by ethanol, *J. Clin. Invest.* **43:**2048-2062.
84. B. J. Potter, R. W. G. Chapman, R. M. Nunes, D. Sorrentino, and S. Sherlock, 1985, Transferrin metabolism in alcoholic liver disease, *Hepatology* **5:**714-721.
85. M. D. Snider, and O. C. Rogers, 1985, Intracellular movement of cell surface receptors after endocytosis: Resialylation of asialo-transferrin receptor in human erythroleukemia cells, *J. Cell Biol.* **100:**826-834.
86. E. Regoeczi, P. A. Charlwood, and P. A. Chindemi, 1985, The effects of cytotropic compounds on the resialylation of human asialotransferrin type 3 in the rat, *Exp. Cell Res.* **157:**495-503.
87. E. Regoeczi, P. A. Chindemi, and M. T. Debanne, 1984, Partial resialylation of human asialotransferrin types 1 and 2 in the rat, *Can. J. Biochem. Cell Biol.* **62:**853-858.
88. R. Ghidoni, S. Sonnino, V. Chigorno, B. Venerando, and G. Tettamanti, 1983, Occurrence of glycosylation and deglycosylation of exogenously administered ganglioside G_{m1} in mouse liver, *Biochem. J.* **213:**321-329.
89. M. G. Farquhar, 1983, Multiple pathways of exocytosis, endocytosis, and membrane recycling: Validation of a Golgi route, *Fed. Proc.* **42:**2407-2413.
90. M. C. Willingham, J. A. Hanover, R. B. Dickson, and I. Pastan, 1984, Morphologic characterization of the pathway of transferrin endocytosis and recycling in human KB cells, *Proc. Natl. Acad. Sci. (USA)* **81:**175-179.
91. J. E. Lamb, F. Ray, J. H. Ward, J. P. Kushner, and J. Kaplan, 1983, Internalization and subcellular localization of transferrin and transferrin receptors in HeLa cells, *J. Biol. Chem.* **258:**8751-8758.
92. R. Tauber, C. -S. Park, and W. Reutter, 1983, Intramolecular heterogeneity of degradation in plasma membrane glycoproteins: Evidence for a general characteristic, *Proc. Natl. Acad. Sci. (USA)* **80:**4026-4029.

93. R. C. Pittman, T. E. Carew, C. K. Glass, S. R. Green, C. A. Taylor, and A. D. Attie, 1983, A radioiodinated, intracellularly trapped ligand for determining the sites of plasma protein degradation in vivo, *Biochem. J.* **212:**791-800.
94. E. Regoeczi, M. W. C. Hatton, and P. A. Charlwood, 1975, Carbohydrate-mediated elimination of avian plasma glycoprotein in mammals, *Nature* **254:**699-701.
95. M. W. C. Hatton, and E. Regoeczi, 1976, Studies of the metabolism of asialotransferrins: Nonspecific changes in the metabolic behaviour of human asialotransferrin in avians, *Can. J. Biochem.* **54:**336-340.
96. M. Kataoka, and M. Tavassoli, 1985, Identification of lectin-like substances recognizing galactosyl residues of glycoconjugates on the plasma membrane of marrow sinus endothelium, *Blood* **65:**1163-1171.

8

Potential Role of Cell Surface Lectin in Lymphocyte Recirculation

Steven D. Rosen
University of California, San Francisco

Lloyd M. Stoolman
University of Michigan

THE PHENOMENOLOGY OF LYMPHOCYTE RECIRCULATION*

Lymphocytes are an extraordinarily mobile population of cells in the body. As was first demonstrated by Gowans (1) and Gowans and Knight (2), most of the small lymphocytes in lymphoid organs (lymph nodes, spleen, Peyer's patches, tonsils) are migratory cells that have recently arrived from the blood and will shortly return to the blood via collecting lymphatics. The process by which lymphocytes travel from the blood into lymphoid organs and back to the blood is referred to as lymphocyte recirculation. The extent of this recirculation is considerable. For example, in the rat the flux of lymphocytes from the thoracic duct (a major lymphatic duct) to the blood is sufficient to replace the blood lymphocytes 11 times daily (1, 3). Approximately an equal number of lymphocytes must leave the blood in order to maintain a fixed level in the blood. The rate of lymphocyte recirculation varies with the class of lymphocytes as well as with their state of differentiation. For example, in the rodent, mature B cells take somewhat longer than T cells to travel from the blood to the thoracic lymph (4, 5). Immature B and T cells, located in the bone marrow and thymus, respectively, show a very limited ability to recirculate (6, 7, 8). Figure 8-1 illustrates the pathway of lymphocyte recirculation through a lymph node.

*Literature review completed May, 1984.

Lymphocyte recirculation serves to distribute immunocompetent lymphocytes throughout the body. This phenomenon has been demonstrated experimentally in rodents by documenting the accumulation of antigen-specific lymphocytes in lymphoid organs challenged with the specific antigen. Specifically, Thrush and Emeson (9, 10) radiolabeled lymphocytes in animals primed with either sheep red blood cells (SRBC) or chicken red blood cells (CRBC). The former were labeled with ^{3}H-thymidine, while the latter were labeled with ^{14}C-thymidine. These labeled cells were then injected intravenously into animals that had received injections of CRBC and SRBC into the left and right feet, respectively. The ratios of ^{3}H and ^{14}C were determined in the lymph nodes draining the sites of inoculation at various

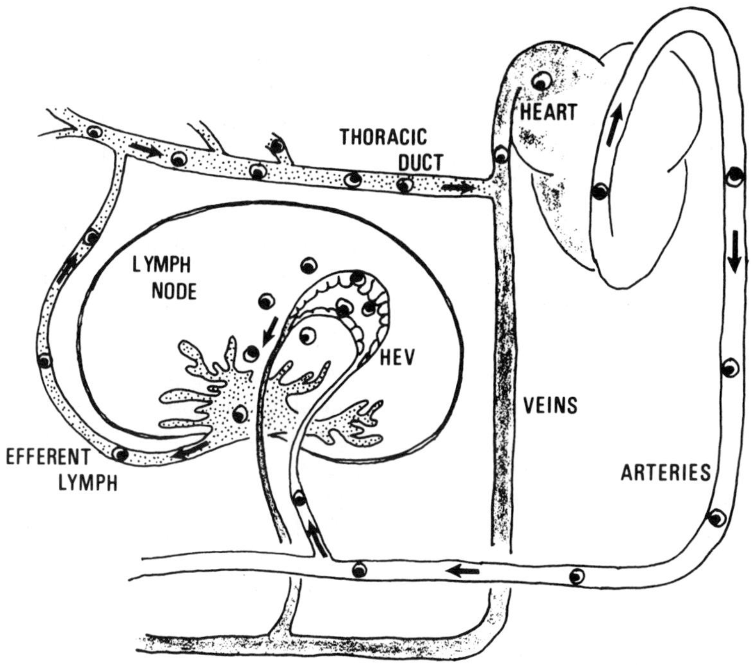

Figure 8-1. The pathway of lymphocyte recirculation. Blood-borne lymphocytes are carried to lymphoid organs such as lymph nodes, bind to the walls of HEV, and pass into the parenchyma of the organ. The extravasated lymphocytes percolate through the node, where they may encounter processed antigens. Recirculating lymphocytes that do not meet appropriate antigens eventually leave the node through the efferent lymphatics. These merge with larger lymphatic vessels, which collect into the major lymphatic ducts (e.g., thoracic duct) that empty into large veins at the base of the neck, thereupon returning the lymphocytes to the blood-vascular system. On the average, the recirculation time for lymphocytes in rodents is about 24 hr. *(Diagram by Ted Yednock.)*

intervals after the challenge. Cells obtained from animals primed with SRBC accumulated preferentially in the nodes draining the site of SRBC injection, while those obtained from animals primed with CRBC accumulated in nodes draining the site of CRBC inoculation. Thus, specific lymphocytes appear to be recruited to the lymphoid organs where their cognate antigens are sequestered. These results and others (8) are compatible with the hypothesis that continuous recirculation of lymphocytes between the bloodstream and lymphoid organs provides a means of assuring the maximal exposure of memory T and B cells to antigens that are trapped in lymphoid organs.

The anatomy of lymphocyte recirculation has been elucidated only relatively recently. In 1964, Gowans and Knight reported (2) that in the rat, lymphocytes migrate from blood into lymph nodes across a specialized microvascular network, the postcapillary high endothelial venules (HEV). HEV are found in several mammalian species, including man (2, 11-15) and are distinguished by their plump, cuboidal endothelial lining cells in contrast to the flattened endothelium typical of most microvasculature. Circulating T and B cells first bind to the luminal surface of these HEV and then migrate between adjacent endothelial cells, crossing a thin basement membrane and the surrounding perivascular sheath before entering the cortex of the lymph node (16-20). HEV serve as the portal of lymphocyte entry into lymph nodes and all lymphoid organs except the thymus, spleen, and bone marrow. These include Peyer's patches, tonsils, adenoids, and the specialized submucosal lymphoid nodules in the respiratory and gastrointestinal (GI) tracts known as bronchus-associated lymphoid tissue (BALT) and gut-associated lymphoid tissue (GALT), respectively. In addition, the transient appearance of HEV may be associated with the accumulation of small lymphocytes at sites of local inflammation (21). Figure 8-2 depicts the architecture of a rat lymph node, showing the location and organization of the HEV.

The fact that relatively few blood leukocytes (monocytes, granulocytes) other than lymphocytes adhere to and traverse HEV suggested to the earliest investigators that lymphocyte-HEV adherence must involve highly selective cell surface recognition mechanisms (2, 11, 12). In addition, the adhesive interactions between HEV and lymphocytes must develop rapidly and be relatively strong, since the blood-borne lymphocytes have only a few seconds to bind to the HEV and then must resist strong hydrodynamic forces as the blood streams past. A further indication of the strength of this interaction is the fact that HEV-bound lymphocytes are not dislodged by vigorous vascular perfusion of isolated lymph nodes, which is sufficient to flush out the plasma and the other blood cells (19). Detailed morphological studies in the rodent (19, 20) have revealed that the HEV endothelial cells extend broad, domelike protrusions into the lumen of the vessel (Fig. 8-3); the lymphocytes appear to bind to the endothelium via microvilli that project into shallow pits on the luminal HEV surface. Preservation of surface coast material with alcian blue (which binds to acidic glycoconjugates) reveals electron-dense material in the 12- to 26-nm intercellular spaces between adherent lymphocytes and endothelial cells (19). Another distinctive feature of HEV is the presence of a very prominent Golgi apparatus, which is usually found close to the luminal surface (20). Andrews et al. (23, 24) have reported that the Golgi and associated vesicles are sites of extensive ^{35}S-sulfate incorporation.

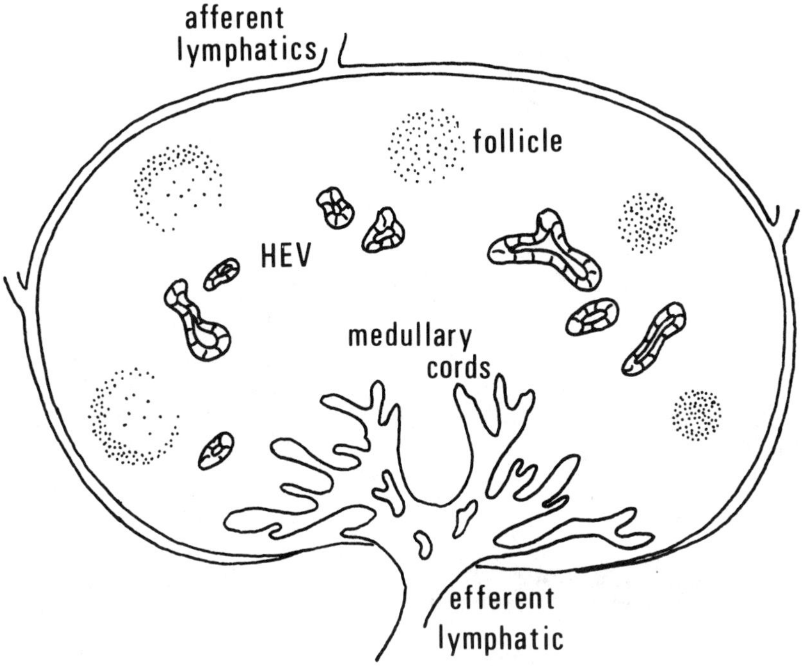

Figure 8-2. Architecture of a rat lymph node. Afferent lymph containing some lymphocytes and antigen arrives at the node through the afferent lymphatics. The lymph percolates through the node, passing through the cortex and the medullary cords and carbohydrate before exiting through the efferent lymphatic vessels. The HEV, which are the portals of entry of blood-borne lymphocytes into the node, are located in the paracortical region of the node, between and deep to the follicles. T and B lymphocytes bind to the luminal surface of these cuboidal vessels and then pass between the endothelial cells to enter the nodal parenchyma. B cells segregate into follicle areas and T cells move into the paracortical regions. If these lymphocytes do not encounter antigens to which they can respond, they eventually depart with the efferent lymph. A germinal center arises in a follicle when antigen-stimulated B cells differentiate and proliferate. Germinal center cells and probably T cell blasts are unable to recirculate (87, 88). However, cells that have been activated in a particular lymphoid tissue may be reprogrammed with respect to their migratory specificities so that their progeny (effector cells or memory cells) tend to return to the same general sites where they first encountered antigen (reviewed in ref. 86). *(Diagram by Ted Yednock.)*

The predominant product appears to be a glycolipid that is secreted. Since this component is unique to HEV among the vascular endothelia, it has been suggested that it may play a role in the diapedesis of lymphocytes.

Work in recent years has been directed at the elucidation of the molecular basis for the adhesive interaction between lymphocytes and HEV during recirculation. These studies are reviewed below with special emphasis on the possible role of carbohydrates as recognition determinants in the process.

IN VIVO STUDIES OF LYMPHOCYTE-HEV INTERACTION

Gesner and Ginsberg (25) were the first investigators to suggest a possible role for glycoconjugates as cell surface receptors in lymphocyte recirculation.

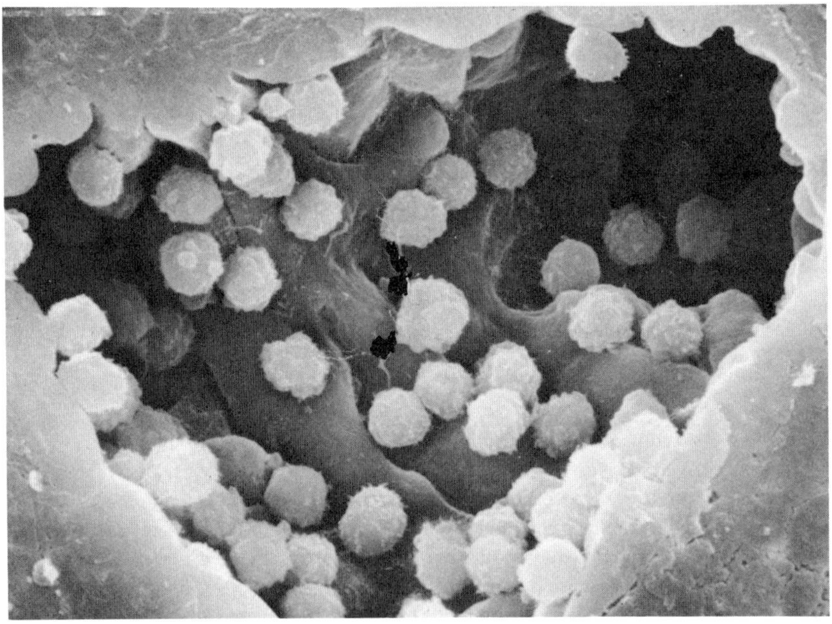

Figure 8-3. Lymphocytes bound to luminal surface of HEV. Shown is an SEM view of an HEV exposed by freeze-fracturing of a rat cervical lymph node. The nonadherent blood cells were flushed out of the blood vessels by perfusing the rat with phosphate buffered saline (PBS) and glutaraldehyde. Plump, bulging endothelial cells extend into the lumen of the HEV. The bound lymphocytes have microvilli that are used for attachment to the endothelial cells. After binding, the lymphocytes migrate between adjacent endothelial cells, traverse a thin basement membrane, and move into the node. Bar indicates 20 μm. *(SEM by Julie Emerson and Ted Yednock.)*

They observed that treatment of rat lymphocytes with mixed glycosidases (from *Clostridium perfringens*) before infusion into animals markedly reduces the ability of the lymphocytes to accumulate in the spleen and lymph nodes ("lymphocyte homing") but increases their accumulation in the liver. This redistribution can be prevented if specific sugars (L-fucose or N-acetyl-D-galactosamine) are included in the incubation medium, indicating that removal of sugars from the lymphocyte surface is responsible for the altered lymphocyte homing. It is now recognized that the liver contains a variety of carbohydrate-binding proteins, which are capable of trapping cells or glycoproteins displaying appropriate carbohydrate groups (26). Thus, it is likely that the glycosidase treatment of the lymphocytes does not impair their ability to interact with HEV but rather causes the lymphocytes with altered cell surface carbohydrates to be recognized and sequestered by the liver; consequently, fewer lymphocytes are available for entry into the lymphoid organs. Neuraminidase treatment of lymphocytes also augments accumulation of the cells in the liver, apparently at the expense of entry into lymphoid organs (27). These experiments with glycosidases illustrate an important caveat for the interpretation of in vivo studies of lymphocyte recirculation: A decrease in the ability of treated lymphocytes to localize in lymphoid organs may be the consequence rather than the cause of an increase in the localization in other organs (e.g., liver, lung).

Trypsin treatment of lymphocytes also dramatically impairs homing into lymph nodes, but its mechanism of action is distinctly different from that of the glycosidases described previously (28). The decreased lymph node homing is not accounted for by an excessive sequestration of the lymphocytes in the lung or liver, although there is a modest increase of treated lymphocytes in these organs as well as in the blood. A specific deficiency in homing is indicated by the observation that trypsin-treated lymphocytes do not accumulate in lymph nodes when perfused through the blood supply of an isolated chain of nodes (29). Furthermore, vascular perfusion of trypsin through a lymph node results in the release of lymphocytes bound to HEV (19). This evidence strongly suggests that trypsin has a direct effect on the adhesive interaction between lymphocytes and HEV. In the last several years, with the advent of an in vitro assay of lymphocyte binding to HEV, it has been possible to verify this conclusion directly (see the following).

IN VITRO STUDIES OF LYMPHOCYTE-HEV INTERACTION

The complexities of interpreting in vivo experiments illustrated in the discussion above led Stamper and Woodruff (30) to develop an in vitro assay of lymphocyte-HEV binding. This assay has made it possible to study cell adhesion events and their underlying molecular mechanisms in isolation from other complicated physiologic processes (e.g., blood flow to lymphoid organs, sequestration of lymphocytes by the liver, migration of lymphocytes across HEV wall, etc.).Thus, one can determine if a particular treatment of lymphocytes or HEV (e.g., enzymes, antibodies, lectins, etc.) designed to alter or mask specific cell surface sites, affects the lymphocyte-HEV adhesive interaction. The in vitro assay has dramatically enhanced our understanding of the mechanisms involved in lymphocyte-HEV adherence (see the following). Detailed molecular questions have been asked, and there are now candidates for lymphocyte adhesion factors as well as information on the role of carbohydrates in the adhesive interaction. In studies of lymphocyte-HEV binding, in vivo assays have now largely assumed the role of validating the conclusions based on the in vitro assay.

The Stamper-Woodruff assay employs frozen sections of lymph node, usually obtained from the rat. After fixation with glutaraldyde, the sections are overlaid with a suspension of viable lymphocytes and then gyrated at 7-10°C. The sections are then washed free of nonadherent lymphocytes, fixed, and stained. Upon examination with the light microscope, exogenous lymphocytes are observed predominantly in association with the HEV (Fig. 8-4*A*). The exogenous lymphocytes are readily identified, because they lie above the plane of the section, and because they stain more intensely than the cells within the plane of the section. The HEV are distinguished as relatively large blood vessels with plump endothelial cells that are well delineated from the surrounding nodal parenchyma. Figure 8-4*B* shows a scanning electron microscopy (SEM) view of lymphocytes bound to an HEV in a frozen section. The lymphocytes appear to attach to the endothelial cells via microvilli in correspondence with in vivo observations (see above). Morphological studies have not yet established that the lymphocytes are actually binding to remnants of HEV plasma membrane exposed in the sections, although this possibility seems likely.

Stamper and Woodruff (30) noted a remarkable degree of selectivity in the binding of thoracic duct lymphocytes to lymph node HEV in their frozen section assay. The density of binding is approximately 100 times greater to HEV sites as compared to non-HEV sites. Furthermore, there is no significant association of the overlaid lymphocytes with other vascular structures in the node, such as capillaries, lymphatic vessels, or thin-walled venules.

Butcher, Scollay, and Weissman (31), working primarily with the mouse, have adopted a modified version of the frozen section assay in which air-dried rather than aldehyde-fixed sections are employed. These workers have been able to control for variability in binding from section to section by including a reference population of lymphocytes (usually labeled with fluorescent dyes) with the various test populations of lymphocytes. When bind-

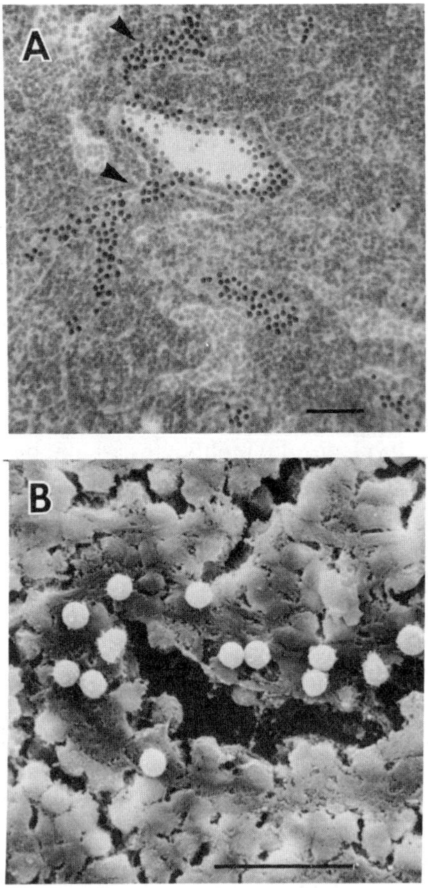

Figure 8-4. Frozen section assay of lymphocyte–HEV binding. *A:* Light microscope view of lymphocytes bound to HEV exposed in frozen section of rat cervical lymph node. A frozen section was air-dried and then overlaid with a suspension of peripheral node lymphocytes. After 50 min of gyration at 7°C, the section was washed free of unbound lymphocytes, fixed with glutaraldehyde, and then stained with toluidine blue. The large lumen belongs to an HEV. Many exogenous lymphocytes (which stain darkly) are associated with the luminal surface of this vessel. Tangentially sectioned HEV (indicated by arrowheads) also bind lymphocytes. Bar indicates 50 μm. *B:* SEM view of lymphocytes bound to HEV in a frozen section of rat cervical lymph node. The light, spherical cells are the exogenous lymphocytes, which have bound to remnants of HEV cells surrounding a large lumen. Bar indicates 20 μm *(SEM taken by Julie Emerson and Ted Yednock.)*

ing on each section is normalized to the internal standard, variability among replicates becomes minimal, and comparison of binding among various test populations of lymphocytes is greatly facilitated.

Validation of the in vitro system requires the demonstration that the ability of lymphocytes to bind to HEV in the intact animal correlates with lymphocyte-HEV adherence measured on frozen sections. A considerable body of evidence from Woodruff's laboratory and from Butcher, Weissman, and associates, has convincingly established this correlation in the rat and mouse. For example, it is well known that lymphocytes obtained from the thoracic duct, lymph nodes, and spleen include a high proportion of cells that are capable of recirculation, whereas thymocytes and bone marrow lymphocytes show a limited ability to recirculate. In the in vitro assay, the lymphocytes from the former sources exhibit substantial binding to lymphocyte HEV, while those from the latter sources bind poorly (30). As was reviewed previously, trypsinized lymphocytes are unable to home to lymph nodes, apparently as a result of their diminished capacity to bind to HEV. Woodruff and co-workers demonstrated that trypsinized lymphocytes also are incapable of binding to lymph node HEV in the frozen section assay (32). The most satisfying correlations are based on quantitative comparisons of in vitro vs. in vivo results. Butcher and co-workers (31) infused labeled lymphocytes from various species into mice and directly measured their binding to mesenteric lymph node HEV. When the same lymphocytes were tested for adherence to HEV in frozen sections of mouse mesenteric nodes, the correlation coefficient between the two assays was 1.00. Thus, it is clear from these results, as well as from others reviewed in the material following, that the frozen section assay accurately measures a physiologically relevant interaction between circulating lymphocytes and HEV.

The frozen section assay has been instrumental in revealing a extraordinary degree of specificity in the adherence of various lymphocyte populations to HEV. Stevens and co-workers (33) showed in the mouse that T lymphocytes bind about 1.5 times better than B lymphocytes to peripheral lymph node HEV, whereas B lymphocytes bind about two to three times better than T lymphocytes to Peyer's patch HEV. These adhesive specificities are proposed to underlie the selective homing in vivo of T vs. B cells to peripheral lymph nodes and B vs. T cells to Peyer's patches. While the organ specificity of these normal lymphocyte subsets is partial, many lymphoma lines are capable of nearly absolute organ specificity. Thus, Butcher and co-workers (34) identified several lymphomas that bind with extreme preference to peripheral node HEV and others that bind almost exclusively to Peyer's patch HEV. The existence of these node-specific and Peyer's patch-specific lymphomas strongly suggests that there are at least two sets of complementary recognition molecules available for expression on lymphocytes (and their transformed derivatives) and HEV (see model [86]).

Studies with lymphocytes from different animal species have revealed a significant degree of species specificity. The binding of lymphocytes to HEV in frozen sections of mouse mesenteric lymph node decreases exponentially with increasing evolutionary distance separating the source of the lymphocytes from the mouse (35). This result implies that recognition sites on the lymphocyte have changed continuously through evolution, resulting in an ever-decreasing degree of cross-species complementarity; presumably the HEV recognition sites have evolved in parallel to maintain maximum compatibility within each species (35).

The frozen section assay has also been useful in defining several critical requirements for lymphocyte adherence to HEV. In the first place, the lymphocytes must be viable: Fixed or metabolically poisoned lymphocytes do not bind (36). In contrast, the HEV seem to be passive partners in the interaction, since in the frozen sections they are dead (and can even be fixed), and yet they still will support binding. Attachment of lymphocytes is inhibited by cytochalasin B, indicating that functionally intact microfilaments are essential (32). Perhaps these microfilaments are required for the maintenance of the microvilli, which make the actual contacts with the endothelial cells. Calcium must be present for the initial binding of lymphocytes to HEV (32). Once attachment has occurred to HEV in frozen sections, removal of calcium by EDTA chelation does not release the adherent lymphocytes. In apparent conflict is the finding that vascular perfusion of EDTA through a lymph node removes HEV-bound lymphocytes (19). Calcium may be important in the initial stabilization of cell contact, possibly exerting its role through microfilaments or some other cytoskeleton system (32). Alternatively, calcium may directly modulate the binding activity of an adhesive molecule.

A still unexplained feature of the frozen section assay is the fact that optimal lymphocyte attachment takes place between 7°C and 10°C (36). Raising or lowering the temperature markedly impairs binding. Preliminary experiments (36) indicate that attached lymphocytes may elute from HEV in frozen section at the physiological temperature of 37°C. Stamper and Woodruff speculate that dissociation of adherent lymphocytes from HEV sites may reflect a physiological process involved in the passage of the lymphocytes through the endothelium.

IDENTIFICATION OF POTENTIAL ADHESIVE MOLECULES

Chin, Woodruff, and colleagues (37-41) have described two factors in rat thoracic duct lymph that affect lymphocyte attachment to HEV in frozen sections of peripheral lymph node. One of these, the adherence enhancing factor (AEF), dramatically augments binding of mature lymphocytes but not thymocytes to HEV. The functional significance of the AEF is unclear, as its biological and biochemical characterizations are very limited to date.

The other lymph factor, known as high endothelial binding factor (HEBF), inhibits the attachment of lymphocytes to HEV. HEBF achieves this inhibition by reacting with sites on the HEV, presumably through a binding interaction. Purification of HEBF reveals that it is a trypsin-sensitive, 160,000-dalton glycoprotein capable of binding to the lentil lectin (α-methyl-D-mannoside specificity). HEBF is a surprisingly prominent component of the lymph, with a concentration of about 14 micrograms per ml. Anti-HEBF antibody binds to the cell surface of the majority of lymphocytes from thoracic lymph, spleen, and lymph nodes, suggesting that HEBF present in lymph may arise from cell surface shedding from lymphocytes. The antigen is present in very much reduced amounts on the cell surface of thymocytes and bone marrow cells, which corresponds with the poor HEV binding exhibited by these cells. The key result implicating HEBF as an adhesive factor is the finding that anti-HEBF antibody or Fab fragments substantially inhibits the binding of lymphocytes to peripheral lymph node HEV in the frozen section assay. Furthermore, the antibodies block the entry of intravenously injected lymphocytes into lymph nodes, whereas homing of lymphocytes into Peyer's patches is not affected. SDS-PAGE analysis of immunoprecipitated antigen from surface-labeled lymphocytes revealed a major band at 70,000 daltons and two minor bands at 92,000 and 45,000 daltons. These antigens are not class specific, since they are present on both T and B cell populations. It is proposed that the HEBF on the surface of lymphocytes is directly involved in the binding of the lymphocytes to lymph node HEV.

Gallatin, Weissman, and Butcher (42) have used a monoclonal antibody approach to define a potential adhesive molecule on mouse lymphocytes. A monoclonal antibody called MEL-14 was developed that binds preferentially to the surface of lymphocytes with the ability to attach to lymph node HEV in frozen sections; specifically, MEL-14 binds strongly to normal lymphocytes and lymph node-specific lymphocytes (cells that attach to HEV of lymph nodes), but it binds poorly or not at all to Peyer's patch-specific lymphomas, bone marrow cells, thymocytes, germinal center blasts, or T-cell clones (cells that do not attach to lymph node HEV). In addition, MEL-14 blocks the attachment of lymphocytes to HEV in frozen sections of lymph node and prevents the homing of lymphocytes to lymph nodes in vivo. However, the antibody does not inhibit lymphocyte attachment to Peyer's patch HEV, despite the fact that most of the lymphocytes that attach express the MEL-14 antigen. Immunoprecipitation of surface-labeled cells with MEL-14 revealed an 80,000-dalton subunit (by SDS-PAGE) from normal lymphocytes and a 90,000-dalton component from a lymph node-specific lymphoma.

There are a number of similarities and differences between the HEBF from rat and the MEL-14 antigen from mouse. Specific antibodies, directed against each, block lymphocyte attachment to lymph node HEV but not to Peyer's patch HEV. The antigens are not class specific (B vs. T), although it is proposed that the MEL-14 antigen may be more highly represented on mouse T cells than B cells (42). HEBF isolated either from lymph or from lymphocyte membranes has been demonstrated to have biological activity: It inhibits lymphocyte-HEV attachment, apparently by binding to HEV sites. There have been no reports as yet of biological activity for the MEL-14 antigen. The biochemical characteristics of the two components appear to be very different: Subunit molecular weights do not correspond; subunits of the MEL-14 antigen increase in apparent molecular weight with reduction on SDS-PAGE, whereas the HEBF components decrease in apparent molecular weight; HEBF is a relatively basic component, with a pI of 8.3-8.6 (41), while MEL-14 antigen has a pI of 4.2 (Eugene Butcher, personal communication). It seems very unlikely that these differences can be attributed to species variation between the mouse and rat. It is more likely that HEBF and the MEL-14 antigen are components of separate (i.e., parallel) adhesive systems or possibly distinct elements of a single multicomponent adhesive system.

POTENTIAL ROLE OF CELL SURFACE LECTIN AS ADHESIVE MOLECULE

Our approach to the problem of lymphocyte-HEV adherence has been to focus on a potential mechanism of the cell recognition event. Specifically, we have asked if this adhesion event is based on the interaction of cell surface carbohydrate-binding proteins (hereafter referred to as cell surface lectins) with complementary glycoconjugates on the adjacent cells. This hypothesis is worthy of consideration for several reasons. First, on theoretical grounds, oligosaccharide chains are excellent candidates for recognition molecules, since the combination of different monosaccharides and linkages provides the potential for an enormous diversity of structures; proteins are plausible candidates for the other member of the complementary pair, since there are many known examples of highly specific recognition of carbohydrates by proteins (e.g., plant lectins, blood group-specific antibodies, glycosidases, and glycosyltransferases). A second and more significant basis for considering this hypothesis is that cell surface lectins have been implicated as cell recognition molecules in a large number of systems (although this is clearly not their exclusive function (70, 71)). These systems include microbe-host cell interactions (43, 44); stage-specific aggregation of cellular slime molds (45) and myxobacteria (46); attachment of nitrogen-fixing bacteria to the roots of legumes (47); the cohesion of teratocarcinoma stem cells (48); the bind-

Table 8-1. Lymphocyte Attachment in the Presence of Monosaccharides

Sugar[a]	*Binding Relative to Control*[b]	*Significance*[c]
D-galactose	244 ± 45	N.S.
N-acetyl-D-glucosamine	216 ± 50	N.S.
D-fucose	177 ± 15	—
No added sugar (control)	100 ± 14	$p < 0.05$
D-mannose	107 ± 25	$p < 0.01$
L-fucose	94 ± 14	$p < 0.05$

Note: For further details consult reference 53, from which data were taken.
[a]Lymphocyte attachment to peripheral lymph node HEV (rat) was measured in the presence of various monosaccharides (150 mM).
[b]Binding is shown relative to the no-added-sugar control (binding in presence of inhibitor/binding in control X 100). Means and SEMs shown are based on at least eight independent replicates.
[c]The significance (Student's t-test) of the difference between binding in the presence of D-fucose and each of the other conditions is shown. N.S. denotes no significant difference.

ing of desialylated erythrocytes by Kupffer cells in the liver (49); and sperm-egg binding in invertebrates (50, 51) and vertebrates (52).

To investigate the relevance of the lectin hypothesis of intercellular adhesion to lymphocyte-HEV adherence, we employed a modification of the Stamper-Woodruff frozen section assay. Generally, we used rat cervical lymph node sections, which were either unfixed or paraformaldehyde-fixed with freshly prepared paraformaldehyde (0.5-1%). The lymphocyte suspensions were usually teased from cervical lymph nodes. Further details of the assay are given elsewhere (53, 54).

Our first experiments (53) examined the effects of a series of neutral monosaccharides, commonly found in cell surface glycoconjugates, on the binding of lymphocytes to HEV. At 150 mM, those sugars with an axial 2-OH and an equatorial 4-OH (D-mannose and L-fucose) inhibit binding relative to sugars that lacked these features (Table 8-1). However, the binding in the presence of D-mannose or L-fucose is not significantly different from that in the absence of added sugar. One interpretation of these data is that hypertonic solutions nonspecifically enhance the binding of lymphocytes to HEV unless the solute has an independent inhibitory activity. Therefore, the lack of a significant difference in binding in the presence of D-mannose or L-fucose vs. the no-added-sugar control may indicate that the inhibitory activities of these sugars are countered by nonspecific effects of hypertonicity induced by the high sugar concentration.

We found that we could enhance the suspected inhibitory activities of the simple sugars by altering the ionic strength of the buffer used in the binding assay. Increasing the ionic strength by adding NaCl both inhibits lymphocyte attachment directly and augments the inhibitory potencies of specific neutral monosaccharides (Fig. 8-5, Table 8-2). In the presence of 60 mM added

NaCl, L-fucose, D-mannose, and α-methyl-D-mannoside inhibit attachment 10- to 40-fold relative to D-galactose, L-rhamnose, and N-acetyl-D-glucosamine and 5- to 20-fold relative to D-fucose, D-glucose, D-ribose, and the no-added-sugar control.

The dose-response curves for D- and L-fucose under "permissive conditions" (elevated ionic strength) lend further support to the hypothesis that hypertonic conditions augment the adherence of lymphocytes to the HEV (Fig. 8-6). Relative to the no-added-sugar control, D-fucose produces a

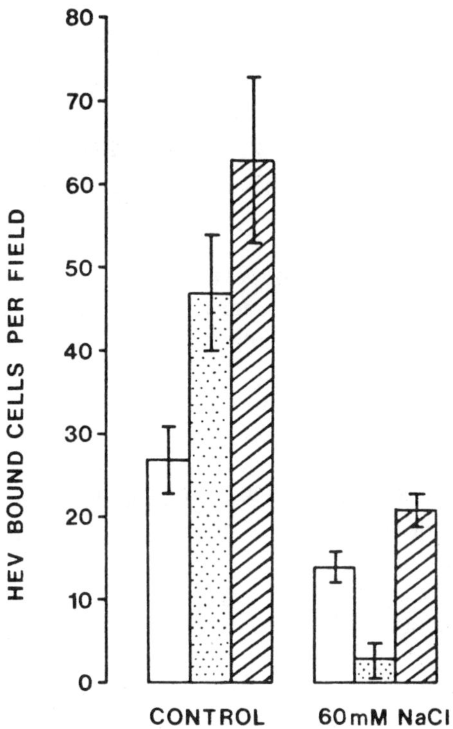

Figure 8-5. Lymphocyte attachment in the presence of L- and D-fucose with or without added NaCl. The ordinate indicates the number of lymphocytes bound to rat peripheral lymph node HEV per microscopic field (200X). Clear bar, no added sugar; stippled bar, L-fucose (150 mM); crosshatched bar, D-fucose (150 mM). Two sets of assays were done: one in isoionic medium (control), and the other in the presence of supplementary NaCl (60 mM). Note that the addition of salt reduces the level of binding in the no-added-sugar control and allows significant inhibition by L-fucose relative to the no-added-sugar control as well as D-fucose. Means and standard error of the means (SEMs) are based on four independent replicates. Further details are given in reference 53. *(From Stoolman and Rosen [53], with permission.)*

Table 8-2. Lymphocyte Attachment Under Permissive Conditions to HEV in the Presence of Neutral Monosaccharides

Sugar[a]	*Binding Relative to Control*[b]
D-galactose	217 ± 60
L-rhamnose	202 ± 60
N-acetyl-D-glucosamine	159 ± 46
D-ribose	102 ± 36
No added sugar (control)	100 ± 34
D-glucose	109 ± 41
D-fucose	101 ± 42
α-methyl-D-mannoside	26 ± 8
L-fucose	12 ± 8
D-mannose	4.5 ± 3.5

Note: For further details consult reference 53, from which data were taken.
[a]Lymphocyte binding to peripheral lymph node HEV (rat) was measured in the presence of various neutral monosaccharides (150 mM) with 60 mM NaCl added (permissive condition).
[b]Binding is shown relative to the no-added-sugar control (with NaCl): binding in the presence of sugar/binding in control X 100. Means and SEMs are based on four replicates.

dose-dependent increase in lymphocyte attachment that plateaus at a level of 150%. In contrast, L-fucose produces no significant increase in binding at low concentrations followed by significant inhibition at 75 mM and 150 mM. In light of these data, we conclude that the most appropriate "controls" for judging the inhibitory potency of an active sugar are to compare it to equimolar concentrations of other sugars. If one uses D-fucose as the control for L-fucose, then one-half maximal inhibition of binding (I-50) requires approximately 50 mM L-fucose. The inhibitory activities of L-fucose and D-mannose do not appear to be due to metabolic suppression of lymphocyte function, since neither trypan blue exclusion nor the rate of protein synthesis is selectively reduced by the active sugars.

The modulation of lymphocyte-HEV binding by ionic strength indicated a possible role for charge in the interaction. Therefore, we examined the effects of a variety of phosphorylated monosaccharides on lymphocyte binding. We found that specific phosphorylated monosaccharides are very potent inhibitors in isoionic media; that is, without the addition of supplementary NaCl (54). At 10 mM, the structurally related sugars D-mannose 6-phosphate (M6P) and D-fructose 1-phosphate (F1P) produce 80-90% inhibition of lymphocyte binding (Fig. 8-7). In contrast, D-mannose 1-phosphate (M1P), D-fructose 6-phosphate (F6P), D-galactose-phosphates (1 and 6), and D-glucose-phosphates (1 and 6) have little or no effect at this concentration. The I-50 for M6P is 2-3 mM. Therefore, phosphorylation of D-mannose at a specific position (carbon 6) significantly increases its inhibitory activity (25- to 50-fold).

This initial series of experiments established three independent modulators of the lymphocyte-HEV binding interaction: (1) inhibition by neutral and phosphorylated carbohydrates with an axial 2-OH and equatorial 4-OH; (2) accentuation by hypertonic solutions; and (3) inhibition by hyperionic solutions. The hypertonic effect may result from the reported increase in the

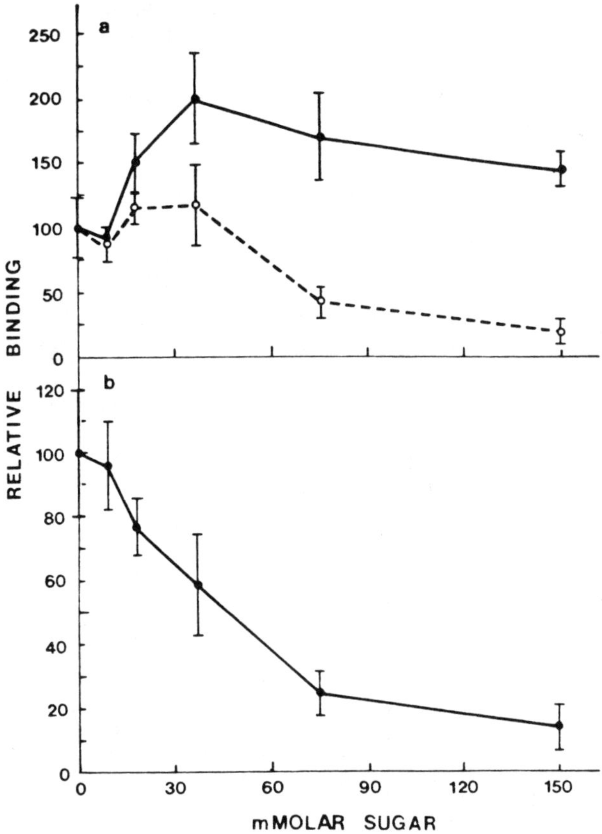

Figure 8-6. Dose-response curves for D- and L-fucose with added NaCl. Lymphocyte binding to rat peripheral lymph node HEV was measured in the presence of L-fucose or D-fucose at varying concentrations (abscissa) with 60 mM supplementary NaCl added. *a:* Binding relative to a no-added-sugar control. Relative binding (ordinate) = (binding in the presence of fucose)/(binding in the control) X 100. D-fucose, solid line; L-fucose, dashed line. Means and SEMS are based on four independent replicates. Significance differences are seen between L-fucose and D-fucose at 37 mM and higher concentrations. *b:* Inhibition of L-fucose relative to D-fucose. Relative binding (ordinate) = (binding in the presence of L-fucose)/(binding in the presence of D-fucose) X 100. Further details are given in reference 53. *(From Stoolman and Rosen [53].)*

number and prominence of microvilli on the surface of lymphocytes exposed to hypertonic media (55). Since circulating lymphocytes anchor to the HEV at the tips of their microvilli, conditions that enhance these structures might strengthen the intercellular adhesive bond. The impact of hyperionic solutions suggests that specific or nonspecific charge interactions contribute to the adhesion of lymphocytes to HEV. Finally, inhibition by several structurally related neutral and phosphorylated monosaccharides under conditions that minimize intracellular transport (7-10°C incubations) is consistent with the

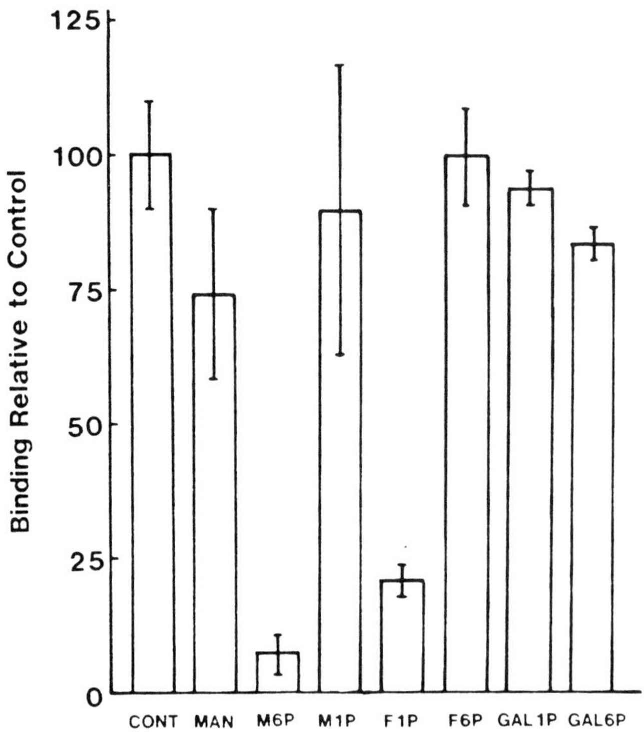

Figure 8-7. Lymphocyte attachment in the presence of various monosaccharides. The ordinate indicates binding of lymphocytes to rat peripheral lymph node HEV in the presence of monosaccharides (10 mM) relative to a no-added-sugar control. Relative binding (ordinate) = (binding in the presence of sugar)/(binding in the control) X 100. The assay was done in isoionic medium (no supplementary NaCl). CONT (no-added-sugar control); MAN (D-mannose); M6P (D-mannose-6-phosphate); M1P (D-mannose-1-phosphate); F1P (D-fructose-1-phosphate); F6P (D-fructose-6-phosphate); GAL1P (D-galactose-1-phosphate); GAL6P (D-galactose-6-phosphate). Glucose phosphates (1 and 6) show no significant inhibitory activity at 10 mM. Further details are given in reference 54. *(From Stoolman et al. [54].)*

involvement of a specific cell surface lectin in the adhesive interaction between lymphocytes and HEV. In addition, the particular carbohydrate specificity of inhibition suggests that the putative adhesive lectin may be related to one of the recently described phosphomannosyl receptors (see discussion below).

To explore these hypotheses further, we examined the effects of a variety of structurally defined polysaccharides and glycoproteins on the attachment of lymphocytes to HEV. Frequently, the affinity of lectins for polysaccharides greatly exceeds their affinity for monosaccharides. Such enhanced activity may be due either due to a valency effect in which multiple sugar determinants in the polysaccharide can be bound simultaneously to multiple sites on the lectin (56) or to the presence of a specific oligosaccharide structure in the polysaccharide with high affinity for the lectin (57). Consistent with the carbohydrate specificity defined with the monosaccharides, we found that a L-fucose-rich polysaccharide (fucoidin) and a M6P-rich core polysaccharide (PPME) from a yeast mannan (isolated from *Hansenula holstii*) have potent inhibitory activity, whereas an extensive list of diverse polysaccharides and glycoproteins have little or no activity (Table 8-3). Fucoidin is a 100,000-dalton, sulfated polysaccharide consisting predominantly of α1,2- and α1,3-linked L-fucose (58). PPME has a molecular weight of approximately 10^6 daltons and consists exclusively of mannose and phosphate (ratio of 6:1), with the latter found principally in the 6 phosphomonoester form (59, 60). The I-50's for fucoidin and PPME are both approximately 1 X 10^{-8} M. Recently, Spangrude et al. (61) have shown that fucoidin is also a potent inhibitor of in vitro lymphocyte-HEV binding in the mouse. Furthermore, fucoidin has an equally impressive in vivo activity in the mouse: when it is infused intravenously, there is rapidly developing and pronounced inhibition of lymphocyte entry into both lymph nodes and Peyer's patches. It should be pointed out that PPME and M6P are also excellent inhibitors of lymphocyte binding to peripheral lymph node HEV in the mouse (Ted Yednock, unpublished data).

Fucoidin and PPME appear to act at the lymphocyte surface to inhibit attachment of the cells to the HEV. Preincubation of the lymphocytes, but not the frozen sections, with either fucoidin or PPME results in persistent inhibition of lymphocyte attachment (53, 54). Since both the preincubation and subsequent binding assays are carried out at temperatures that minimize pinocytosis (4°C and 10°C, respectively), it is likely that inhibition results from an interaction with the lymphocyte surface. Ted Yednock (unpublished observations) has obtained more direct evidence for the presence of a lymphocyte cell surface lectin with the appropriate carbohydrate-binding specificity by utilizing PPME-conjugated latex beads. These beads adhere to the surface of lymphocytes derived from lymph nodes but not to the surface

Table 8-3. Lymphocyte Attachment in the Presence of Glycoconjugates

Substance[a]	*I-50 (μg/ml)*[b]
Fucoidin	1-5
PPME	10-20
Dextran sulfate	>25
Dermatan sulfate	>25
Chondroitin sulfate	>50
Hyaluronic acid	>50
Milk oligosaccharides	>100
Heparin	>100
Egg jelly coat fucan	>100
Desulfated fucoidin	>100
Dephosphorylated PPME	>200
Native mannan	>1000
Horseradish peroxidase	>1000
Invertase	>1000
Thyroglobulin	>1000
Glycogen	>1000

Note: For further details consult references 53 and 54, from which data were taken.
[a]Various glycoconjugates were tested for their ability to inhibit the attachment of lymphocytes to peripheral lymph node HEV (rat). Assays were done under nonpermissive conditions (no supplementary NaCl).
[b]I-50 denotes the concentration required to achieve 50% inhibition of binding relative to a control with no glycoconjugate added.

of thymocytes. Thus, cells that exhibit a very low degree of binding to HEV (thymocytes) apparently lack binding sites for PPME. In addition, the binding of conjugated beads to lymphocytes could be selectively inhibited by M6P as compared with M1P, suggesting that PPME and M6P interact with the same receptor on the cell surface.

The inhibitory activities of fucoidin and PPME depend upon their anionic groups, since desulfated fucoidin and dephosphorylated PPME are both inactive in the frozen section assay. It is therefore important to determine if the charge imparted to the lymphocyte surface (which is the apparent site of action of both fucoidin and PPME) is sufficient to account for their activities. Using cellular electrophoresis techniques, we demonstrated (Fig. 8-8) marked increases in lymphocyte surface charge density in the presence of charged polysaccharides that lack significant inhibitory activity (e.g., heparin or dextran sulfate at 25 μg per ml). In contrast, fucoidin, PPME, and M6P produce maximal inhibition at concentrations that increase the surface charge density significantly less than either dextran sulfate or heparin.

It should be noted that heparin and dextran sulfate are not totally inactive in the frozen section assay. At high concentrations (100 μg/ml and 25-50 μg/ml, respectively), inhibition is observed (53 and unpublished observations). Furthermore, the infusion of large quantities of either heparin or dextran sulfate into rats results in partial inhibition of lymphocyte entry into lymph nodes and a concomitant lymphocytosis (62, 63, 64). The cell electrophoresis experiments suggest that the very large increases in lymphocyte surface charge produced by these polysaccharides may be responsible for both their in vivo and in vitro activities.

The specific inhibitory activities of a variety of structurally related monosaccharides and polysaccharides, and the preliminary experiments with PPME beads, suggest that a lectin on the surface of lymphocytes is involved in lymphocyte attachment to HEV in peripheral lymph nodes. Our inhibition studies indicate that the carbohydrate-binding specificity of this lectin is

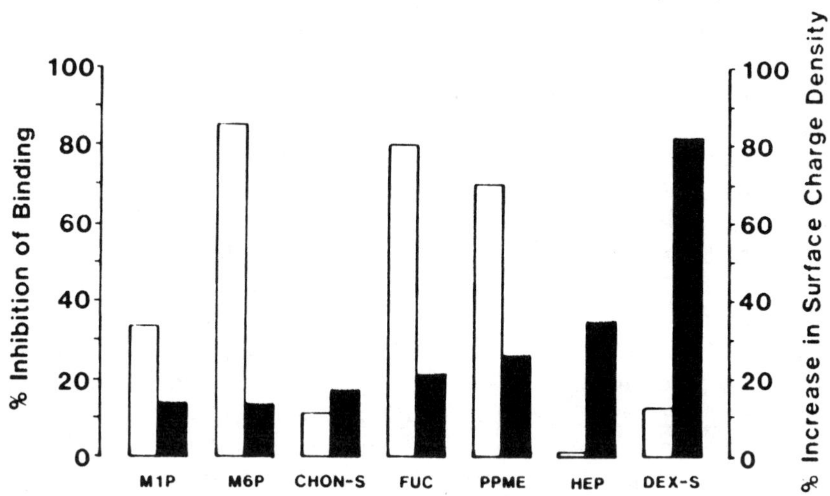

Figure 8-8. Effects of monosaccharides and polysaccharides on negative surface charge density of lymphocytes. The increase in surface charge density on the lymphocyte surface, due to the binding of charged carbohydrates, was determined by measuring the electrophoretic mobility of rat lymphocytes in the presence of sugars or polysaccharides. Clear bars depict percent inhibition of lymphocyte binding to HEV in the presence of carbohydrates relative to a control with no added carbohydrate. Solid bars depict the percent increase in negative surface density in the presence of carbohydrates relative to a control with no added carbohydrates. M1P (mannose-1-phosphate, 10 mM); M6P (mannose-6-phosphate, 10 mM); CHON-S (chondroitin sulfate, 25 μg/ml); FUC (fucoidin, 10 μg/ml); PPME (core phosphomannan from *H. holstii*, 200 μg/ml); HEP (heparin, 25 μg/ml); DEX-S (dextran sulfate, 25 μg/ml). SEMs on electrophoresis measurements ranged from 3% to 4%. SEMs for lymphocyte-HEV binding ranged from 10% to 20%. The differences between fucoidin and PPME, on the one hand, and heparin and dextran sulfate, on the other hand, are statistically significant. Further details are given in reference 54. *(From Stoolman et al. [54].)*

for phosphomannosyl determinants or structurally related groups. There have been several other reports (65-69) of cell surface lectins on lymphocytes, but none with a phosphomannosyl specificity. However, one of these lectins, described in a preliminary report by Kieda and Monsigny (69), is of particular interest, because it has been implicated in lymphocyte-HEV adherence. In this study (carried out in the mouse), a series of neoglycoproteins were tested for their ability to inhibit the attachment of splenocytes to mesenteric node HEV in vitro. Neoglycoproteins containing β-D-galactoside residues inhibit binding, but those containing alpha-linked D-mannose, D-glucose, D-galactose, N-acetyl-D-galactosamine, or N-acetyl-D-glucosamine are inactive. The apparent discrepancy between our results and those of Kieda and Monsigny may be attributable to the fact that these investigators employed splenocytes and mesenteric node HEV in contrast to the peripheral node lymphocytes and HEV used in our studies. It is conceivable that carbohydrate specificity varies with the source of the lymphocytes and/or HEV used in the adhesion assay. This is a provocative possibility, especially in light of the organ specificity manifested in lymphocyte-HEV attachment (see previous material).

The phosphomannosyl specificity of the putative "adhesive lectin" that we have detected on lymphocytes is similar to that described for membrane receptors implicated in the intracellular targeting of M6P-containing acid hydrolases to the lysosome. This class of receptor is widely distributed in mammalian tissues on both plasma and internal membranes (72, 73). The cell surface receptor on human diploid fibroblasts appears to mediate the absorptive pinocytosis of M6P-containing glycoproteins and polysaccharides (74). The receptor has been identified as a 215,000-dalton membrane protein in several sources, including fibroblasts (75), bovine hepatocytes (76), Swarm rat chondrosarcoma cells (77), and a murine lymphoma (78). Recently, phosphomannosyl-binding activity has been detected in a 57,000-dalton component purified from liver (79). This component is immunologically cross-reactive with the 215K molecule, suggesting that the larger molecule may be a macromolecular complex.

The adsorptive pinocytosis of acid hydrolases by fibroblasts, like the binding of lymphocytes to HEV, is selectively inhibited by M6P, F1P, and PPME; in contrast, M1P, F6P, the glucose phosphates, the galactose phosphates, and alkaline phosphatase-treated PPME show little activity in either system (53, 54, 74, 80). A further parallel is that only those neutral monosaccharides bearing an axial 2-hydroxyl and equatorial 4-hydroxyl (i.e., L-fucose, D-mannose, and others) inhibit in these systems, although their activities are significantly less that those of the phosphorylated monosaccharides (M6P and F1P). The only apparent discrepancy is that fucoidin, the most potent inhibitor of lymphocyte-HEV binding so far identified (53), does not inhibit the binding of an acid hydrolase (α-mannosidase from *Dictyostelium discoideum*) to the surface of alveolar macrophages (81). Clearly, both the macrophage receptor and the putative adhesive lectin from lymphocytes must be purified before their structural and functional relationship to the 215K (or 57K) phosphomannosyl receptor can be defined.

Ligatin is a plasma membrane lectin with a binding specificity for phosphorylated monosaccharides (82-85). It appears to serve as a "baseplate" for the cell surface attachment of peripheral glycoproteins in a wide variety of tissues. In the embryonic chick neural retina, ligatin and its associated glycoproteins may play a role in intercellular adhesion. Interestingly, the carbohydrate-binding specificity of ligatin appears to vary with the tissue source. Thus, M6P is the best inhibitor of ligatin binding to β-N-acetylhexosaminidase when the source of ligatin is ileum. However, D-glucose 1-phosphate is superior to M6P when macrophage ligatin is used. The comparison of ligatin with the lymphocyte lectin is complicated by this tissue variation.

The adhesive site on the HEV surface is less amenable to study than that on the lymphocyte. Until endothelium with the structural and functional characteristics of HEV can be brought into culture, the adhesive site must be characterized in situ or in frozen sections, thus limiting the experimental methodologies that one can apply. Despite these limitations, several lines of evidence suggest that carbohydrate-containing moieties on the HEV may be involved in the attachment of lymphocytes. (This result would, of course, be compatible with the lectin model of intercellular adhesion, given the presence of an adhesive lectin on the lymphocyte surface.) Anderson and Anderson (18) found that the alcian blue-positive glycocalyx of rat HEV is three- to fourfold greater in thickness than the glycocalyces of microvascular beds that do not bind circulating lymphocytes. The fact that the frozen sections can be fixed with aldehydes without loss of specific binding activity is also consistent with the possibility that the recognition sites on the HEV are carbohydrate in nature. While most proteins would be denatured by aldehyde fixation, carbohydrates should be chemically unaltered. Direct evidence for the involvement of carbohydrates comes from our preliminary experiments with periodate. When paraformaldehyde-fixed frozen sections are treated with periodate under conditions favoring carbohydrate oxidation (low temperature, low pH, in the dark), lymphocytes are no longer able to bind to the sections, although the morphology of the sections is well preserved (Table 8-4, Fig. 8-9). In the presence of excess substrate (ethylene glycol or glucose), periodate does not destroy adhesive sites on the HEV. While strongly suggesting that a carbohydrate-containing moiety on the HEV forms part of the attachment site for lymphocytes, this evidence does not constitute rigorous proof, since periodate can oxidize amino acids under

Table 8-4. Inhibition of Lymphocyte Binding to HEV by Periodate Oxidation of Sections

Condition[a]	*Binding Relative to Control*[b]
Control	100 ± 14
Periodate	2 ± 1
Periodate + glucose	93 ± 33
Periodate + ethylene glycol	106 ± 22

Note: See Figure 8-9 for micrographs treated and control sections.
[a]Lymphocyte binding was measured to HEV in frozen sections (rat peripheral lymph node) that had been treated with periodate plus or minus glucose or ethylene glycol. Conditions of the periodate oxidation are given in the legend to Figure 8-9.
[b]Binding is shown relative to a control that was not exposed to periodate but was otherwise treated identically. Means and SEMs are based on seven independent replicates. Exposing the sections to glucose or ethylene glycol alone did not alter binding relative to the control.

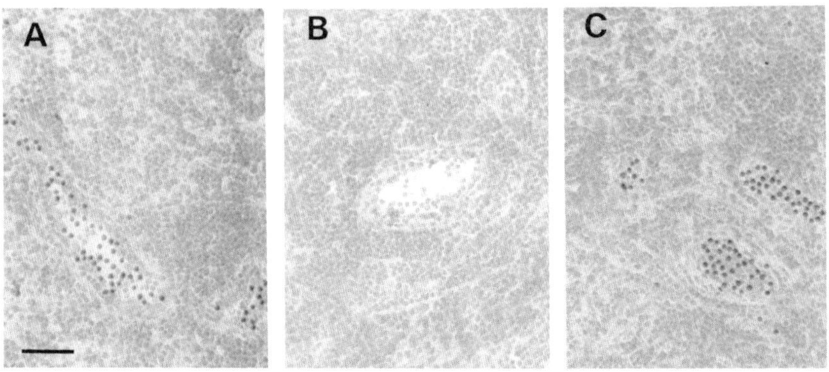

Figure 8-9. Effect of periodate oxidation of lymph node sections on lymphocyte binding to HEV. Frozen sections of rat peripheral lymph node were cut and fixed with paraformaldehyde (1%) as described in reference 54. Sections were exposed to 50 mM sodium metaperiodate in sodium acetate buffer (50 mM, 100 mM NaCl, pH 5.0) plus or minus D-glucose (100 mM) or ethylene glycol (250 mM) for 1 hr at 4°C in the dark. Control sections were treated identically except that periodate was omitted. Binding of lymphocytes to the sections was carried out by standard procedures (54), except that 10 mM sodium borohydride was included in the lymphocyte suspension. The sodium borohydride treatment prevented nonspecific binding of lymphocytes to aldehyde groups generated by periodate oxidation of the sections. *A:* Binding of lymphocytes to control section, bar indicates 50 M. *B:* Binding to periodate-oxidized section; *C:* Binding to section treated with periodate plus excess substrate (ethylene glycol). Numerical data for this experiment are shown in Table 8-4.

some conditions. To define further the nature of the recognition sites on the HEV, future experiments might be directed at determining whether specific enzyme treatments (mannosidases, fucosidases, phosphatases, sulfatases, etc.) of the frozen sections impair lymphocyte attachment. Ultimately, the HEV adhesive sites will have to be isolated before detailed characterization is possible.

The precise contribution of the phosphomannosyl-specific lectin identified in our studies to the adhesive interaction between lymphocytes and HEV remains to be determined. Three possible mechanisms for lectin-mediated adhesion are depicted in Figure 8-10. In A, the membrane lectin forms a direct bridge to a carbohydrate-containing site on the HEV. Presumably, this site would contain M6P, sulfated fucose or a structurally related carbohydrate. Alternatively, as illustrated in B, the phosphomannosyl lectin may serve as a baseplate for another adhesive molecule that forms the actual bridge to the HEV. The extrinsic bridging molecule may also be a lectin in

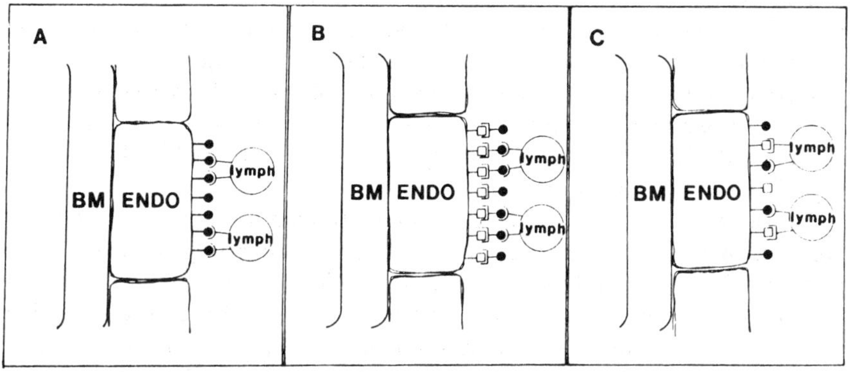

Figure 8-10. Hypothetical models of lymphocyte-HEV binding. In all three models, the semicircles represent the binding site of the phosphomannosyl-specific lectin on the surface of lymphocytes. In model A, this lectin directly mediates attachment of the lymphocytes ("lymph") to the HEV ("ENDO") by recognizing carbohydrate determinants (closed circles) on the luminal surface. This carbohydrate determinant is presumed to be structurally related to mannose-6-phosphate or fucose sulfate. In model B, the phosphomannosyl lectin serves as a baseplate molecule (a ligatin-like molecule) for another adhesive molecule that forms the actual bridge to the HEV. This bridging molecule has a carbohydrate determinant that is recognized by the phosphomannosyl lectin, and it also has a receptor function that allows it to recognize another class of determinant (indicate by squares) on the HEV. In model C, two parallel systems of cell adhesion are shown, one of which is based on the phosphomannosyl lectin.

light of the evidence that the HEV site is carbohydrate in nature. Finally, as depicted in C, the binding interaction may be based on multiple independent systems, one of which involves the phosphomannosyl lectin. Additional models can be postulated, and the schematics shown are presented only as a framework for evaluating the possible relationships amongst the growing number of putative adhesive molecules.

Finally, additional study is required to define the relationship between the phosphomannosyl-specific lectin and the lymphocyte membrane antigens (MEL-14 antigen and HEBF) implicated as adhesive factors in binding to HEV. Either of these suspected adhesive molecules may be identical to the membrane lectin we have characterized. Alternatively, either antigen could be a secondary bridging molecule such as shown in Figure 8-10*B* or part of independent adhesive mechanisms such as those suggested in Figure 8-10*C*.

CONCLUSIONS

The work reviewed clearly documents that the attachment of lymphocytes to a specialized microvasculature (the HEV) constitutes the first step in the migration of lymphocytes from the bloodstream into lymphoid organs. The development of in vitro techniques for measuring the adhesive interaction has led to the identification of two adhesive molecules: the MEL-14 antigen in the mouse and HEBF in the rat. Our studies suggest that a lectin on the surface of the lymphocyte and possibly a glycoconjugate on the HEV surface contribute to this adhesive interaction.

Elucidation of the binding interaction between normal lymphocytes and HEV in lymphoid organs may add to our understanding of lymphocyte migration into nonlymphoid tissues and provide insights into the process by which malignant lymphocytes spread (metastasize) from the organ in which they arise. Specifically, the structural similarity between the morphologically distinctive HEV and the microvasculature that transiently develops at sights of chronic inflammation suggests that the molecular basis of cellular migration into both lymphoid and nonlymphoid tissues may be similar. Futhermore, the demonstration that lymphoid malignancies recapitulate the adhesive and migratory characteristics of their normal counterparts raises the provocative possibility that a selective adhesive interaction between circulating malignant lymphoid cells and the microvasculature may facilitate the spread of lymphoid malignancies.

It is clear that the study of lymphocyte recirculation and migration offers a variety of exciting avenues for future investigation.

REFERENCES

1. J. L. Gowans, 1959, The recirculation of lymphocytes from blood to lymph in the rat, *J. Physiol. (Lond.)* **146:**54–69.

2. J. L. Gowans, and E. J. Knight, 1964, The route of re-circulation of lymphocytes in the rat, *Proc. R. Soc. B.* **159:**257-282.
3. J. L. Gowans, 1957, The effect of the continuous re-infusion of lymph and lymphocytes on the output of lymphocytes from the thoracic duct of unanaesthetized rats, *Br. J. Exp. Pathol.* **38:**67-78.
4. W. L. Ford, and S. J. Simmonds, 1972, The tempo of lymphocyte recirculation from blood to lymph in the rat, *Cell Tissue Kinet.* **5:**175-189.
5. J. C. Howard, 1972, The life-span and recirculation of marrow-derived small lymphocytes from rat thoracic ducts, *J. Exp. Pathol.* **135:**185-199.
6. F. Brahim, and D. G. Osmond, 1970, Migration of bone-marrow lymphocytes demonstrated by selective bone marrow labelling with thymidine-H3, *Anat. Rec.* **168:**139-160.
7. H. Cantor, and I. Weissman, 1976, Development and function of subpopulations of thymocytes and T lymphocytes, *Prog. Allergy* **20:**1-64.
8. W. L. Ford, 1975, Lymphocyte migration and immune responses, *Prog. Allergy* **19:**1-59.
9. D. R. Thrush, and E. E. Emeson, 1972, The immunologically specific retention of recirculating long lived lymphocytes in lymph nodes stimulated with xenogeneic erythrocytes, *J. Exp. Med.* **135:**754-763.
10. D. R. Thrush, and E. E. Emeson, 1973, Selective DNA synthesis by cells localizing in response to xenogenic erythrocytes, *J. Exp. Med.* **138:**659-671.
11. V. T. Marchesi, and J. L. Gowans, 1964, The migration of lymphocytes through the endothelium of venules in lymph-nodes: An electron microscopic study, *Proc. R. Soc. B.* **159:**282-290.
12. I. Goldschneider, and D. D. McGregor, 1968, Migration of lymphocytes and thymocytes in the rat. I. The route of migration from blood to spleen and lymph nodes, *J. Exp. Med.* **127:**155-168.
13. J. Sprent, 1973, Migration of T and B lymphocytes in the mouse. I. Migratory properties, *Cell Immunol.* **7:**10-39.
14. G. A. Gutman, and I. L. Weissman, 1973, Homing properties of thymus-independent follicular lymphocytes, *Transplantation* **16:**621-629.
15. J. J. Miller, III, 1969, Studies of the phylogeny and ontogeny of the specialized lymphatic tissue venules, *Lab. Invest.* **21:**484-490.
16. G. I. Schoefl, 1972, The migration of lymphocytes across the vascular endothelium in lymphoid tissue; A reexamination, *J. Exp. Med.* **136:**568-588.
17. E. J. Wenk, D. Orlic, E. J. Reith, and J. A. G. Rhodin, 1974, The ultrastructure of mouse lymph node venules and the passage of lymphocytes across their walls, *J. Ultrastruc. Res.* **47:**214-241.
18. A. O. Anderson, and N. D. Anderson, 1975, Studies on the structure and permeability of the microvasculature in normal rat lymph nodes, *Am. J. Pathol.* **80:**387-412.
19. A. O. Anderson, and N. D. Anderson, 1976, Lymphocyte emigration from high endothelial venules in rat lymph nodes, *Immunology* **31:**731-748.
20. D. Anderson, A. O. Anderson, and R. G. Wyllie, 1976, Specialized structure and metabolic activities of high endothelial venules in rat lymphatic tissues, *Immunology* **31:** 455-473.
21. J. B. Smith, G. H. McIntosh, and B. Morris, 1970, The migration of cells through chronically inflamed tissues, *J. Pathol.* **100:**21-29.
22. M. H. Claesson, O. Jorgensen, and C. Roepke, 1971, Light and electron microscopic studies of the paracortical post-capillary high-endothelial venules, *Z. Zellforsch.* **119:**195-207.
23. P. Andrews, D. W. Milsom, and W. L. Ford, 1982, Migration of lymphocytes across specialized vascular endothelium. V. Production of a sulphated macromolecule by high endothelial cells in lymph nodes, *J. Cell Sci.* **57:**277-292.
24. P. Andrews, D. W. Milsom, and R. W. Stoddart, 1983, Glycoconjugates from high endothelial cell. I. Partial characterization of a sulphated glycoconjugate from the high endothelial cells of rat lymph nodes, *J. Cell Sci.* **59:**231-244.

25. B. M. Gesner, and V. Ginsberg, 1964, Effect of glycosidases on the fate of transfused lymphocytes, *Proc. Natl. Acad. Sci. (USA)* **52:**750-755.
26. G. Ashwell, and J. Harford, 1982, Carbohydrate-specific receptors of the liver, *Annu. Rev. Biochem.* **51:**531-554.
27. J. J. Woodruff, and B. M. Gesner, 1969, Effect of neuraminidase on the fate of transfused lymphocytes, *J. Exp. Med.* **129:**551-567.
28. J. J. Woodruff, and B. M. Gesner, 1968, Lymphocytes: Circulation altered by trypsin, *Science* **161:**176-178.
29. W. L. Ford, M. Sedgley, S. M. Sparshott, and M. E. Smith, 1976, The migration of lymphocytes across specialized vascular endothelium. II. The contrasting consequences of treating lymphocytes with trypsin or neuraminidase, *Cell Tissue Kinet.* **9:**351-361.
30. H. B. Stamper, Jr., and J. J. Woodruff, 1976, Lymphocyte homing into lymph nodes: *In vitro* demonstration of the selective affinity of recirculating lymphocytes for high-endothelial venules, *J. Exp. Med.* **144:**828-833.
31. E. C. Butcher, R. G. Scollay, and I. L. Weissman, 1979, Lymphocyte adherence to high endothelial venules: Characterization of a modified *in vivo* assay, and examination of the binding of syngeneic and allogeneic lymphocyte populations, *J. Immunol.* **123:**1996-2003.
32. J. J. Woodruff, I. M. Katz, L. E. Lucas, and H. B. Stamper, Jr., 1977, An *in vitro* model of lymphocyte homing. II. Membrane and cytoplasmic events involved in lymphocyte adherence to specialized high-endothelial venules of lymph nodes, *J. Immunol.* **119:**1603-1610.
33. S. K. Stevens, and E. C. Butcher, 1982, Differences in the migration of B and T lymphocytes: Organ-selective localization *in vivo* and the role of lymphocyte-endothelial cell recognition, *J. Immunol.* **128:**844-851.
34. E. C. Butcher, R. G. Scollay, and I. L. Weissman, 1980, Organ specificity of lymphocyte migration: Mediation by highly selective lymphocyte interaction with organ-specific determinants on high endothelial venules, *Eur. J. Immunol.* **10:**556-561.
35. E. C. Butcher, R. G. Scollay, and I. L. Weissman, 1979, Evidence of continuous evolutionary change in structures mediating adherence of lymphocytes to specialized venules, *Nature* **280:**496-498.
36. H. B. Stamper, and J. J. Woodruff, 1977, An *in vitro* model of lymphocyte homing. I. Characterization of the interaction between thoracic duct lymphocytes and specialized high endothelial venules of lymph nodes, *J. Immunol.* **119:**772-780.
37. Y. -H. Chin, G. D. Carey, and J. J. Woodruff, 1980, Lymphocyte recognition of lymph node high endothelium. I. Inhibition of *in vitro* binding by a component of thoracic duct lymph, *J. Immunol.* **125:**1764-1769.
38. Y. -H. Chin, G. D. Carey, and J. J. Woodruff, 1980, Lymphocyte recognition of lymph node high endothelium. II. Characterization of an *in vitro* inhibitory factor isolated by antibody affinity chromatography, *J. Immunol.* **125:**1770-1774.
39. G. D. Carey, Y. -H. Chin, and J. J. Woodruff, 1981, Lymphocyte recognition of lymph node high endothelium. III. Enhancement by a component of thoracic duct lymph, *J. Immunol.* **127:**976-979.
40. Y. -H. Chin, G. D. Carey, and J. J. Woodruff, 1982, Lymphocyte recognition of lymph node high endothelium. IV. Cell surface structures mediating entry into lymph nodes, *J. Immunol.* **129:**1911-1915.
41. Y. -H. Chin, G. D. Carey, and J. J. Woodruff, 1983, Lymphocyte recognition of lymph node high endothelium. V. Isolation of adhesion molecules from lysates of rat lymphocytes, *J. Immunol.* **131:**1368-1374.
42. W. M. Gallatin, I. L. Weissman, and E. C. Butcher, 1983, A cell-surface molecule involved in organ-specific homing of lymphocytes, *Nature* **304:**30-34.
43. I. Ofek, and E. H. Beachey, 1980, Bacterial adherence, *Adv. Intern. Med.* **25:**503-531.

44. J. C. Paulson, J. E. Sadler, and R. L. Hill, 1979, Restoration of specific myxovirus receptors to asialoerythrocytes by incorporation of sialic acid with pure sialyltransferases, *J. Biol. Chem.* **254:**2120-2124.
45. S. D. Rosen, J. E. Rose, D. L. Simpson, and S. H. Barondes, 1974, Carbohydrate-binding protein form *Polysphondylium pallidum* implicated in intercellular adhesion, *Nature* **252:**128, 149-151.
46. M. Cumsky, and D. R. Zusman, 1979, Nonfusing hemagglutinin: A development specific lectin of *Myxococcus xanthus, Proc. Natl. Acad. Sci. (USA)* **76:**5505-5509.
47. F. B. Dazzo, 1981, Bacterial attachment as related to cellular recognition in the rhizobium-legume symbiosis, *J. Supramol. Struct.* **16:**29-41.
48. L. B. Grabel, S. D. Rosen, and G. R. Martin, 1979, Teratocarcinoma stem cells have a cell surface carbohydrate-binding component implicated in cell-cell adhesion, *Cell* **17:**477-484.
49. H. Kolb, and V. Kolb-Bachofen, 1978, A lectin-like receptor on mammalian macrophage, *Biochem. Biophys. Res. Commun.* **85:**678-683.
50. G. P. Bolwell, J. A. Callow, and L. V. Evans, 1980, Fertilization in brown algae. Preliminary characterization of putative gamet receptors from eggs and sperm of *Fucus serratus, J. Cell Sci.* **43:**209-224.
51. C. G. Glabe, L. B. Grabel, V. D. Vacquier, and S. D. Rosen, 1982, Carbohydrate specificity of sea urchin sperm binding: A cell surface lectin mediating sperm-egg adhesion, *J. Cell Biol.* **94:**123-128.
52. T. T. F. Huang, E. Ohzu, and R. Yanagimachi, 1982, Evidence suggesting that L-fucose is part of a recognition signal for sperm zona pellucida attachment in mammals, *Gamete Res.* **5:**355-361.
53. L. M. Stoolman, and S. D. Rosen, 1983, Possible role for cell-surface carbohydrate-binding molecules in lymphocyte recirculation, *J. Cell Biol.* **96:**722-729.
54. L. M. Stoolman, T. S. Tenforde, and S. D. Rosen, 1984, Phosphomannosyl receptors may participate in the adhesive interaction between lymphocytes and high endothelial venules, *J. Cell Biol.* **99:**1535-1540.
55. F. Loor, 1977, Structure and dynamics of the lymphocyte surface, in relation to differentiation, recognition, and activation, *Prog. Allergy* **23:**1-153.
56. M. J. Krantz, N. A. Holzman, C. P. Stowell, and Y. C. Lee, 1976, Attachment of thioglycosides to proteins—Enhancement of liver membrane binding, *Biochemistry* **15:**3963-3968.
57. I. J. Goldstein, and C. E. Hayes, 1978, The lectins: Carbohydrate-binding proteins of plants and animals, *Adv. Carbohydr. Chem. Biochem* **35:**127-340.
58. A. J. Mian, and E. Percival, 1973, Carbohydrates of brown sea weeds *Himanthalia lorea* and *Bifurcaria bifurcata.* Part II. Structural studies of the "fucans," *Carbohydr. Res.* **26:**147-161.
59. M. E. Slodki, R. M. Ward, and J. A. Boundy, 1973, Concanavalin A as a probe of phosphomannan molecular structure, *Biochem. Biophys. Acta* **304:**449-456.
60. H. D. Fischer, M. Natowicz, W. S. Sly, and R. K. Bretthauer, 1980, Fibroblast receptor for lysosomal enzymes mediates pinocytosis of multivalent phosphomannan fragment, *J. Cell Biol.* **84:**77-86.
61. G. J. Spangrude, B. A. Braatan, and R. A. Daynes, 1984, Molecular mechanisms of lymphocyte extravasation. I. Studies of two selective inhibitors of lymphocyte recirculation, *J. Immunol.* **132:**354-362.
62. S. Sasaki, 1967, Production of lymphocytosis by polysaccharide polysulphates (heparinoids), *Nature* **214:**1041-1042.
63. J. W. B. Bradfield, and G. V. R. Born, 1974, Lymphocytosis produced by heparin and other sulphated polysaccharides in mice and rats, *Cell. Immunol.* **14:**22-32.
64. W. L. Ford, M. E. Smith, and P. Andrews, 1978, Possible clues to the mechanism underlying the selective migration of lymphocytes from the blood, *Proc. Soc. Exp. Biol.* **32:**359-392.

65. C. Kieda, D. J. Bowles, A. Ravid, and N. Sharon, 1978, Lectins in lymphocyte membranes, *FEBS Lett.* **94:**391-396.
66. H. Kolb, 1982, A ganglioside receptor on lymphocytes mediates recognition of self, *Biochem. Biophys. Res. Commun.* **105:**1488-1495.
67. J. R. Agpar and P. Cresswell, 1982, Expression of cell surface lectin on activated human lymphoid cells, *Eur. J. Immunol.* **12:**570-576.
68. M. Barzilay, M. Monsigny, and N. Sharon, 1982, Interaction of soybean agglutinin with human peripheral blood lymphocyte subpopulations: Evidence for the existence of a lectin-like substance on the lymphocyte surface, in *Lectins, Biology, Biochemistry, and Clinical Biochemistry,* T. C. Madsen, ed., Walter de Gruyter, Berlin, pp. 67-68.
69. C. Kieda, and M. Monsigny, 1983, Lymphocyte membrane lectin: Evidence for the involvement of a beta-galactoside receptor in cell-cell interaction, in *Intercellular Communication in Leukocyte Function,* J. W. Parker, and R. L. O'Brien, eds., John Wiley and Sons, Ltd., London, pp. 649-652.
70. S. H. Barondes, 1981, Lectins: Their multiple endogenous cellular functions, *Annu. Rev. Biochem.* **50:**207-231.
71. S. H. Barondes, 1984, Soluble lectins: A new class of extracellular proteins, *Science* **223:**1259-1264.
72. H. D. Fischer, A. Gonzales-Noriega, W. S. Sly, and D. J. Morre, 1980, Phosphomannosyl-enzyme receptors in rat liver, *J. Biol. Chem.* **255:**9608-9615.
73. W. J. Brown, and M. G. Farquar, 1984, The mannose-6-phosphate receptor for lysosomal enzymes is concentrated in cis golgi cisternae, *Cell* **36:**297-307.
74. A. Kaplan, D. T. Achord, and W. S. Sly, 1977, Phosphohexosyl components of a lysosomal enzyme are recognized by pinocytosis receptors on human fibroblasts, *Proc. Natl. Acad. Sci. (USA)* **74:**2026-2030.
75. K. E. Creek, and W. S. Sly, 1983, Biosynthesis and turnover of the phosphomannosyl receptor in human fibroblasts, *Biochem. J.* **214:**353-360.
76. G. G. Sahagian, J. Distler, and G. W. Jourdian, 1981, Characterization of a membrane-associated receptor from bovine liver that binds phosphomannosyl residues of bovine testicular beta-galactosidase, *Proc. Natl. Acad. Sci. (USA)* **78:**4289-4293.
77. A. W. Steiner, and L. H. Rome, 1982, Assay and purification of a solubilized membrane receptor that binds the lysosomal enzyme alpha-L-iduronidase, *Arch. Biochem. Biophys. Acta* **214:**681-687.
78. D. E. Goldberg, C. A. Gabel, and S. Kornfeld, 1983, Studies of the biosynthesis of the mannose-6-phosphate receptor in receptor-positive and -deficient cell lines, *J. Cell Biol.* **97:**1700-1706.
79. D. C. Mitchell, T. Mahler, and G. W. Jourdian, 1984, Detergent dissociation of bovine liver phosphomannosyl binding protein, *J. Cell. Biochem.* **24:**319-330.
80. A. Kaplan, D. Fischer, and W. S. Sly, 1978, Correlation of structural features of phosphomannan with their ability to inhibit pinocytosis of human beta-glucuronidase by human fibroblasts, *J. Biol. Chem.* **253:**647-650.
81. V. L. Shepard, A. L. Freeze, A. L. Miller, and P. D. Stahl, 1984, Identification of mannose-6-phosphate receptors in rabbit alveolar macrophages, *J. Biol. Chem.* **259:**2257-2261.
82. E. R. Jakoi, G. Zampighi, and J. D. Robertson, 1976, Regular structures in unit membranes. II. Morphological and biochemical characterization of two water-soluble proteins isolated from the suckling rat ileum, *J. Cell Biol.* **70:**97-111.
83. R. B. Marchase, P. Harges, and E. R. Jakoi, 1981, Ligatin from embryonic chick neural retina inhibits retinal cell adhesion, *Dev. Biol.* **86:**250-255.
84. E. R. Jakoi, K. Kempe, and S. M. Gaston, 1981, Ligatin binds phosphohexose residues on acidic hydrolases, *J. Supramol. Struct. Cell. Biochem.* **16:**139-153.

85. R. B. Marchase, L. A. Koro, C. M. Kelly, and D. R. McClay, 1982, Retinal ligatin recognizes glycoproteins bearing oligosaccharides terminating in phosphodiester-linked glucose, *Cell* **28:**813-820.
86. E. C. Butcher, and I. L. Weissman, 1984, Lymphoid tissues and organs, in *Fundamental Immunology,* W. E. Paul, ed., Raven Press, New York, pp. 109-127.
87. R. A. Reichert, W. M. Gallatin, I. L. Weissman, and E. C. Butcher, 1983, Germinal center B cells lack homing receptors necessary for normal lymphocyte recirculation, *J. Exp. Med.* **157:**813-827.
88. M. O. Dailey, C. G. Fathman, E. C. Butcher, E. Pillemer, and I. L. Weissman, 1982, Abnormal migration of T lymphocyte clones, *J. Immunol.* **128:**2134-2136.

9

Role of Lectins in Immune Recognition

Sandra L. White
Howard University Medical School

The problem of cellular recognition is of fundamental importance in a wide range of biological systems, not the least of which is the immune system. There exists the need to understand the complex relationships between immune cells, both phagocytic and nonphagocytic, and their ability to recognize and react to foreign materials. It is clear that the immunoglobulins, the classical adaptive recognition molecules produced by all vertebrates (1), cannot mediate all the recognition events and cellular interactions that are essential to vertebrate immunity. Furthermore, no evidence has yet been presented that animals below the phylogenetic level of the vertebrates can produce immunoglobulin molecules. However, a role in immunity and cell recognition, in both invertebrates and vertebrates, has been postulated for a wide variety of molecules, including immunoglobulins, products of the major histocompatibility gene complex (MHC) (2-5) and lectins (6). There is now substantial evidence that interactions between lectin-like receptors and the carbohydrate portion of glycoproteins and glycolipids on cell surfaces serve as the basis for the mechanisms of immune recognition (7-13) and nonimmune phagocytosis (14).

Lectins, a group of sugar-binding and cell-agglutinating proteins of non-immune origin, are ubiquitous in nature (15). An important characteristic of these lectins is that they have binding specificity for sugars that are commonly found in membrane and secretory glycoproteins (16). Several have binding specificity for terminal, nonreducing sugars (17, 18) and can agglutinate red blood cells, suggesting their multivalency (19, 20).

The concept that the carbohydrate moiety of a surface glycoprotein or glycolipid of one cell interacts with glycosyltransferase or a carbohydrate-binding protein on the surface of an adjacent cell as a basis for cellular interaction/adhesion was originally proposed by Tyler (21) and Weiss (2); the concept was more recently restored to prominence by Roseman, Roth and co-workers, and Den and co-workers (23, 24, 25). This hypothesis has been most exhaustively studied with cellular slime molds. (See refs. 6, 26, and 27 for recent reviews).

With respect to cells of the immune system, binding of lectins to them can result in dramatic effects on cellular function (28). Since Nowell's (29) original report in 1960 that phytohemagglutin (PHA) could stimulate mitotic division and proliferation of lymphocytes, the interactions of lectins and lectin-like molecules with cell-surface glycoproteins has become an area for intense investigation. The ability of simple or complex saccharides (for which the lectin is specific) to inhibit or abrogate lectin-induced functions has been used as evidence that lectins mediate these functions via binding to cell surface receptors whose structure is similar to, though not necessarily identical to, that of the inhibitory sugars. The binding sugar moieties are components of glycoproteins and glycolipids, important components of cell membranes. As informational molecules on cell surfaces, carbohydrates have distinctive advantages over proteins and nucleic acids. The potential for structural diversity is tremendous when one considers their ability to exist in a variety of spatial configurations, to be linked in different ways (α and β) and to have the ability to form branched molecules (e.g., by O- and N-glycosidic linkages). As many as 10^{24} structural combinations can exist for a 12-sugar carbohydrate composed of three mannose, three N-acetyl-glucosamine, three galactose, and three sialic acid residues (21).

This review is not intended to be encyclopedic; many important areas, such as the possible involvement of lectins in nonimmune phagocytosis and lymphocyte extravasation, are not considered in depth, as they are reviewed in other chapters of this book. The objective of this report is to critically evaluate the current state of knowledge on the role of lectin-cell surface glycoprotein interactions in immune recognition. Viewpoints or areas not considered in this review can be found in one of several recent surveys on this subject (11, 12, 14, 15, 30, 31, 32).

CYTOLYTIC TARGET CELL RECOGNITION

One of the most fascinating phenomena that still exists in immunobiology is the ability of the immune system to distinguish self from nonself. Lymphoid cells have a tremendous capacity to react with other lymphoid and nonlymphoid cells and respond to a vast array of foreign stimuli. Classically, these interactions have been mediated through highly specific immunoglobulins, cell-mediated responses, and cell surface glycoprotein determinants encoded for by genes of the MHC. However, over the last decade we have become increasingly aware of subpopulations of cells (natural killer and cytotoxic macrophages) with the ability to lyse a variety of target cells without prior sensitization (33, 34).

From the standpoint of phylogenetics, mononuclear phagocytes are probably the earliest cells demonstrating specialized host defense functions. There is evidence that an immune response more primitive than that concerned with humoral antibody formation must exist (35, 36). The question is, how do these cells distinguish or recognize foreign invaders so as to defend against them?

Blaese and co-workers (12) have investigated whether a lectin-sugar interaction process might be involved in the recognition of self from foreigners by primitive mononuclear phagocytes. They studied the ability of invertebrate mononuclear phagocytes to kill a variety of foreign targets in vitro and demonstrated not only that the mononuclear phagocytes were markedly cytotoxic toward the targets, but also that certain sugars could inhibit their cytotoxicity. These oligosaccharides exhibited target specificity in that certain sugars would inhibit cytotoxicity against some targets and not against others. Such findings were interpreted to not only rule out a nonspecific toxic effect of the sugars, but also to suggest that different sugar-lectin interactions were involved in the recognition of different target cells. In similar experiments, human monocytes spontaneously lysed a variety of xenogeneic erythrocytes without addition of exogenous lectins or antibody. A variety of simple sugars were capable of blocking the expression of cytotoxicity, again suggesting a lectin-oligosaccharide interaction as the mechanism of recognition (37). Other aspects of cell mediated functions such as lymphocyte transformation by PHA, monocyte phagocytosis of opsonized chicken erythrocytes, and PHA-induced cytotoxicity were not affected when carried out in the presence of identical concentrations of oligosaccharides.

Within the population of lymphoid cells in mammals, there exists another subpopulation of cells with the ability to lyse a variety of target cells without prior sensitization (33, 38, 39, 40). These cells have been designated natural killer (NK) cells on the basis of their in vitro killing capabilities and have been distinguished from conventional T and B lymphocytes. While it is true that NK cells have a broad degree of reactivity, it is equally apparent that there exists a fine degree of selectivity, since some target cells are highly sensitive to cytolysis whereas others are not (38). Most studies have suggested that NK activity is directed against multiple antigenic determinants, some of which may be shared by more than one target cell. The precise nature of the target antigens recognized by NK cells remains poorly understood, however. In order to elucidate the as yet unsolved question of the target recognition system of NK cells, several investigators have employed various mono- and oligosaccharides for inhibition of murine and human NK cells (41, 42).

Stutman and co-workers (41) demonstrated that murine natural cell-mediated cytotoxicity could be inhibited by several sugars, including D-mannose, D-galactose, and L-fucose, and suggested that the recognition structure may be lectin-like in nature. A receptor, specific for D-mannose-6-phosphate, on the surface of human fibroblasts mediates the pinocytosis of β-glucuronidase and α-1-iduronidase (43, 44). Mannose-6-phosphate (Man-6-P) has been implicated as the common recognition marker for intracellular transport of many glycosylated lysosomal hydrolases (45–48). Recently, Forbes and Oeltmann (48) have demonstrated that Man-6-P, fructose-1-phosphate, and fructose-6-phosphate can inhibit human natural cell-mediated cytotoxicity (NCMC) to K-562 when added directly to the incubation mixture of effector cells and target cells. When tested in the same manner, glucose-6-phosphate, galactose-1-phosphate, mannose-1-phosphate, galactose, glucose, mannose, and fucose had no effect. In further studies they established the presence of a Man-6-P receptor on the target cells and demonstrated that the ability of Man-6-P to inhibit lysis was not at the recognition stage (between target and effector), but suggested that the inhibition may occur with the recognition and uptake of lytic molecules, which contain Man-6-P residues. The fact that sugars more likely inhibit lysis events rather than the binding event is supported by recent reports that the target antigen for NK cells is the receptor for transferrin (49, 50).

In contrast to findings with inhibitory sugars and NK activity, conditions that exerted strong inhibitory effects on NK cells had no effect on cytotoxic T-lymphocytes (41). Furthermore, antibody-dependent cell-mediated cytotoxicity (ADCC) against K-562 and Chang tumor cells was not inhibited under these conditions (42). In separate experiments, Kaudewitz and associates (51) have reported contrasting findings. They show that under comparable conditions with various sugars, ADCC in the P-815 system is inhibited to a degree similar to that of NK cells. However, these investigators do not explore the possibility that rabbit antiserum that they used in the ADCC may be directed against similar carbohydrate determinants as the saccharides, thus inhibiting the cytotoxic reaction and possibly explaining their findings, which differ from those of others.

CELL-CELL INTERACTIONS

Blaese and co-workers have contributed significantly to our knowledge of the role of lectins in the recognition of "targets" by cytotoxic monocytes (12). As a result of some of their earlier studies using xenogeneic targets and analyzing self-xenogeneic recognition systems, they have also shed light on collaboration between autologous cells and the role of oligosaccharide specificities (8). The antigen-specific lymphocyte proliferation response to recall antigens using human peripheral blood monocyte cells as responders was examined in the presence of a large variety of sugars. Three major findings were the outcome of these studies: (a) sugars, such as cellobiose, lactose, maltose, and trehalose, which profoundly inhibited cytotoxic reactions had no effect on antigen-induced proliferation; (b) those sugars that did inhibit antigen-induced proliferation had no effect on PHA-induced proliferation, a finding used to rule out suggestions that the inhibition was due to nonspecific toxicity; and (c) certain sugars (e.g., N-acetylgalactosamine, L-rhamnose) inhibited the elicitation of a secondary response only if they were present at the initiation of the assay; if they were added 24 hr after the addition of antigen, there was no inhibition of antigen-induced proliferation. One possible explanation for this result is that the sugars could be interfering with antigen uptake by monocytes, either specifically of via some nonspecific toxic mechanism. However, data exist that argue against this idea (8). Another interesting interpretation has been offered by Parish and co-workers (52) and McKenzie and co-workers (53), who suggest that the sugars could be acting by blocking immune response (Ir) gene-restricted cellular interactions. These I region-associated (Ia) antigens express carbohydrate-defined antigenic determinants (54) and have been implicated in many immunological phenomena, such as macrophage-T cell interactions (55, 56), T cell help (57), T cell suppression (58, 59), and help for secondary IgG responses (60). The observations that certain monosaccharides selectively inhibit the secondary in vitro immunoglobulin (IgG) response (10); that the specificities of the inhibiting sugars differ depending on the strain of mouse; and that the differences in specificity can be mapped to the I region of the MHC (60), are consistent with the concept that carbohydrate-defined Ia antigens play a role in communication between lymphoid cells, and that their blockage can lead to loss of communication and thus immune function. In further support are observations that binding of antibodies directed at specific Ia (53) and DR (61) determinants on lymphoid cells may be blocked by specific sugars.

What is most intriguing is recent evidence that suggests there is homology in the primary structure of recognition molecules in the animal kingdom, including vertebrate immunoglobulins, T cell receptors, MHC products, C-reactive proteins (62), and some lectins (62, 63). It is therefore necessary to consider the possibility that immunoglobulins, lectins, and many membrane glycoproteins, with or without observed recognition functions, may be related structurally and functionally.

LYMPHOKINES AND IMMUNOREGULATORY FACTORS

Recently, increasing evidence has been provided for defined saccharide specificity of a variety of biologically important immunological mediators. The characteristics of the interaction between some lymphokines and their target cells appear to involve lectin-like reactions. Activity of migration inhibitory factor (MIF) is inhibited by L-fucose, L-rhamnose, and other 5-methylpentases (64, 65, 66), and MIF has been shown to bind to a fucosidase-sensitive receptor on the macrophage surface (64). Leukocyte inhibitory factor (LIF), however, is inhibited by N-acetyl-D-glucosamine (GluNAc; 65), whereas interferons (IFN) from humans show binding specificity for the sialyl-lactose portion of ganglioside GM (67, 68).

Of particular interest are the recently reported findings that interaction between HN glycoprotein of Sendai virus and receptors on the cell surface of mouse spleen cells can trigger production of IFN α/β (69, 70). In view of the functional similarity between the hemagglutinin-neuraminidase (HN) viral glycoproteins and lectins (71), Ito and co-workers (70) have recently reported findings suggesting that the HN glycoprotein of the Sendai virus may, through a mechanism similar to that of a lectin, stimulate spleen cells to produce IFN.

Other lymphocyte responses have also been shown to be modulated by lectin-like interactions. Muchmore and associates (72) have demonstrated that human T cell proliferation to soluble antigen is blocked by certain sugars, such as L-rhamnose, D-altrose, N-acetyl galactosamine, and D-mannose, whereas no effect was observed when cellobiose or arabinogalactan was used as the inhibiting sugar. Inhibition of T cell proliferation by some sugars occurred only if they were present from the initiation of culture. This would seem to suggest that some early event in the process of antigen-induced T cell proliferation is mediated via lectin binding to a cell surface receptor (thus the ability of the saccharides to inhibit).

Immunoregulatory factors may also have carbohydrate specificity. Greene and co-workers and Fleisher and co-workers have investigated suppression mediated by soluble factors in human cells. They have examined the supernatants of concanavalin-A (Con A)-activated human peripheral blood mononuclear cells for their effects on both T cell-dependent proliferation (73) and B cell-dependent immunoglobulin biosynthesis (74). The soluble suppressor of T cell proliferation was blocked by the monosaccharide (GluNAc), and at least some of the same surface glycoprotein receptors recognized by the inhibitory lectins wheat germ agglutinin (WGA) and *Agarius bisporus* lectin (which have GluNAc binding specificity), bound to the T suppressor factor also (73). The soluble suppressor factor of B immunoglobulin production could be blocked by L-rhamnose but was unaffected by GluNAc. It is interesting that the plant lectin WGA, which acts as a direct suppressor signal for T cell proliferation (73), binds strongly to GluNAc, the same sugar that inhibited the ability of the soluble suppressor of T cell proliferation.

In this regard, Barrett and his co-workers (75) used WGA as an in vitro model for saccharide-specific suppressor factor. The suppressive effect of WGA on lectin-stimulated blastogenesis and immunoglobulin production was studied. WGA inhibited PHA-, Con A-, and pokeweed mitogen (PWM)-induced mitogenic responses by 70-80% and suppressed PWM-driven immunoglobulin synthesis. The inhibitory effect of WGA was not due to cell death or to interference with lectin binding at the cell surface. This WGA-induced inhibition of blastogenesis was blocked by the addition of GluNAc, which prevents WGA binding to the cell surface. This WGA inhibition mimicked the suppressive effect of Con A-induced suppressor factor (64, 65, 73). They proposed that the biological relevance of their model was its ability to demonstrate how other endogenous inhibitory or suppressive lymphokines might bind to cells through GluNAc or other lectin receptors.

Other studies have described a T suppressor factor that is capable of inhibiting a variety of immune responses, nonspecifically and without H-2 restriction. This T suppressor factor is not inhibited by GluNAc, but rather by methyl-α-D-mannopyranoside (10).

Thus, one must consider the possibility that certain essential functions of immunocompetent cells, such as differentiation and proliferation, may be induced and regulated by sugar-binding protein cells.

There have been limited opportunities to study known endogenous lectins relative to their impact on immune functions. One group of lectins that have been studied are the low-molecular-weight lactose-binding proteins, electrolectins (ELs), found in lower animals (76, 77) and in humans (78, 79). The wide distribution of ELs in the animal kingdom and the conservation through evolution of their gross physicochemical properties strongly suggest an important physiological role. Many groups have attempted to demonstrate a role for ELs in the fusion of myoblasts into myotubes (80, 81, 82) but the results obtained remain controversial. Levi and associates (83) have carried out a systematic study of the possible interactions between ELs and cholinergic synapses, using experimental autoimmune myasthenia gravis as a model. They found that administration of ELs to the myasthenic rabbits resulted in complete recovery. Much to their surprise, they could find no evidence for an action of EL at the muscular level. However, ELs were found not only to bind lymphocytes but also to stimulate mitosis. As a result of further studies that demonstrated similarities between their saccharide binding specificity and ability to bind the same class of thymocytes, it has been proposed that one may assume that many findings reported in the literature concerning the interactions between peanut agglutinin (PNA) and lymphocytes could be extended to ELs. The most relevant observations are (a) PNA binding to mature lymphocytes with suppressor activity; and (b) PNA binding to immature thymocytes (84). These studies support, but do not prove, the proposition that an endogenous EL may play an important, but as yet undefined, role in the processes of T suppressor cell commitment.

SUMMARY

Only a small fraction of examples demonstrating the potential association between cellular lectin-like activity and recognition processes in the immune response have been discussed. As has been briefly mentioned, others include many recent studies that have provided evidence that the specific interaction between lymphocytes and high-endothelial venules (HEV), the basic mechanism for lymphocyte extravasation, is mediated through lectin-like cell surface binding molecules (85, 86). This dynamic process of lymphocyte recirculation is an essential component of the functional immune system. Lectins on bacteria have been useful in the differentiation of loosely related bacteria (87), as well as having a possible role in the stimulation of mast cells to release histamine (88). In addition, bacteria with distinctive cell surface carbohydrate structures bind specifically to different subsets of lymphoid cells (89), and these carbohydrate-lectin interactions have been shown to serve as the basis for recognition in nonimmune phagocytosis (14).

Indeed, the evidence presented in the studies reviewed here suggests, but by no means proves, that lectin-like interactions may be among the fundamental mechanisms in cellular immune recognition. However, one cannot deny that much more definitive research must be done beyond the scope of these sugar inhibition studies. In spite of the rapid advances in immunology, one of the most basic questions still remains unanswered: "How do immune cells talk to each other?" Perhaps future studies on lectin interactions between cells will resolve the question.

REFERENCES

1. J. J. Marchalonis, 1977, *Immunity in Evolution*, Harvard University Press, Cambridge, Mass.
2. B. Benacerraf, 1978, A hypothesis to relate the specificity of T lymphocytes and the activity of I-region-specific IR genes in macrophages and B lymphocytes, *J. Immunol.* **120:**1809-1812.
3. D. H. Katz, 1980, Adaptive differentiation of lymphocytes: Theoretical implications for

mechanisms of cell-cell recognition and regulation of immune responses, *Adv. Immunol.* **29:**137-207.

4. D. B. Amos, and F. E. Ward, 1975, Immunogenetics of the HL-A system, *Physiol. Rev.* **55:**206-246.
5. D. C. Schreffler, and C. S. David, 1975, The H-2 major histocompatibility complex and the I immune response region: Genetic variation, function, and organization, *Adv. Immunol.* **20:**125-195.
6. S. H. Barondes, 1981, Lectins: Their multiple endogenous cellular functions, *Annu. Rev. Biochem.* **50:**207-221.
7. T. Babu, T. Yoshida, S. Cohen, 1974, Activated T cells in the synovial fluid of arthritis patients: Characterization and comparison with in vitro activated human and murine T cells in cooperation with monocytes in cytotoxicity, *J. Immunol.* **122:**838-841.
8. A. V. Muchmore, and R. M. Blaese, 1980, Evidence that monocyte mediated cellular recognition phenomena are mediated by receptors with specificity for simple oligosaccharides, in *Macrophage Regulation of Immunity,* E. R. Unanue and A. S. Rosenthal, eds., Academic Press, New York, pp. 505-517.
9. T. K. Higgins, A. B. Sabatino, H. G. Remold, and J. R. David, 1978, Possible role of macrophage glycolipids as receptors for migration inhibitory factor (MIF), *J. Immunol.* **121:**880-886.
10. L. D. Tomaska, and C. R. Parish, 1981, Inhibition of secondary IgG responses by N-acetyl-D-galactosamine, *Eur. J. Immunol.* **11:**181-186.
11. D. M. Weir, 1980, Surface carbohydrates and lectins in cellular recognition, *Immunol. Today* **1:**45-51.
12. R. M. Blaese, G. Tosato, W. C. Greene, T. A. Fleisher, and A. V. Muchmore, 1983, The role of cell surface lectin-carbohydrate interactions in cellular recognition, cooperation, and regulation, *Am. J. Pediatr. Oncol. Hematol.* **5:**199-206.
13. U. H. Kaszenowski, and M. Karmer, 1981, Selective inhibition of T suppressor-cell function by a monosaccharide, *Nature* **289:**181-184.
14. N. Sharon, 1984, Surface carbohydrates and surface lectins are recognition determinants in phagocytosis, *Immunol. Today* **5:**143-147.
15. N. Sharon, 1982, Glycoproteins: Research booming on long-ignored ubiquitous compounds, *Mol. Cell. Biochem.* **42:**167-187.
16. R. Kornfeld, and S. Kornfeld, 1980, Structure of glycoproteins and their oligosaccharide units, in *The Biochemistry of Glycoproteins and Proteoglycans,* W. J. Lennarz, ed., Plenum Press, New York, pp. 1-27.
17. J. U. Baenziger, and Y. Maynard, 1980, Human hepatic lectin. Physiochemical properties and specificity, *J. Biol. Chem.* **255:**4607-4613.
18. E. B. Briles, W. Gregory, P. Fletcher, and S. Kornfeld, 1979, Vertebrate lectins from tissues of calf and chicken, *J. Cell Biol.* **81:**528-537.
19. V. I. Teichberg, I. Silman, D. D. Beitsch, and G. Resheff, 1975, A β-D-galactoside binding protein from electric organ tissue of *Electrophorus electricus, Proc. Natl. Acad. Sci. (USA)* **72:**1383-1387.
20. R. J. Stockert, A. G. Morele, and I. H. Schernberg, 1974, Mammalian hepatic lectin, *Science* **186:**365-366.
21. A. Tyler, 1946, An auto-antibody concept of cell structure, growth and differentiation, *Growth* **10:**7-19.
22. P. Weiss, 1947, Problem of specificity in growth and development, *Yale J. Biol. Med.* **19:**235-278.
23. S. Roseman, 1970, The synthesis of complex carbohydrates by multiglycosyltransferase systems and their potential function in intercellular adhesion, in *Chemistry and Physics of Lipids,* North Holland Publishing Co., Amsterdam, pp. 270-297.

24. S. Roth, E. J. McGuire, and S. Roseman, 1971, Evidence for cell-surface glycosyltransferases—Their potential role in cellular recognition, *J. Cell. Biol.* **51:**536-547.
25. H. Den, B. Kaufman, and S. Roseman, 1970, Properties of some glycosyltransferases in embryonic chicken brain, *J. Biol. Chem.* **245:**6607-6615.
26. S. H. Barondes, 1980, Developmentally regulated lectins in slime molds and chick tissue—Are they cell-adhesion molecules, in *Cell Adhesion and Motility,* A. S. G. Curtis and J. D. Pitts, eds., Cambridge University Press, Cambridge, Mass., pp. 309-328.
27. K. Olden, J. B. Parent, and S. L. White, 1982, Carbohydrate moieties of glycoproteins. A re-evaluation of their function, *Biochem. Biophys. Acta* **650:**209-232.
28. J. J. Oppenheim, and D. L. Rosenstreich, eds., 1976, *Mitogens in Immunobiology,* Academic Press, New York.
29. P. C. Nowell, 1960, Phytohemagglutinin: An initiator of mitosis in cultures of normal human leukocytes, *Cancer Res.* **20:**462-466.
30. N. Sharon, 1983, Lectin receptors and lymphocyte surface markers, *Adv. Immunol.* **34:**213-298.
31. M. E. A. Pereira, and E. A. Kabat, 1979, Immunochemical studies on lectins and their application to the fractionation of blood group substances and cells, *CRC Crit. Rev. Immunol.* **1:**33-78.
32. I. J. Goldstein, and C. E. Hayes, 1978, The lectins: Carbohydrate binding of plants and animals, *Adv. Carbohydr. Chem. Biochem.* **35:**127-340.
33. R. B. Herberman, M. E. Nunn, and D. H. Lavrin, 1975, Natural cytotoxic reactivity of mouse lymphoid cells against syngeneic and allogeneic tumors. I. Distribution of reactivity and specificity, *Int. J. Cancer* **16:**216-229.
34. R. Evans, and P. Alexander, 1976, Mechanisms of extracellular killing of nucleated mammalian cells by macrophages, in *Immunobiology of the Macrophage,* D. S. Nelson, ed., Academic Press, New York, pp. 535-576.
35. A. E. Nelstrop, G. Taylor, and P. Collard, 1968, Studies on phagocytosis: III. Antigen clearance studies in invertebrates and poikilothermic vertebrates, *Immunology* **14:**347-356.
36. N. Pearsall, and R. Weiser, eds., 1970, *The Macrophage,* Lea and Febiger, Philadelphia, pp. 17-20.
37. A. V. Muchmore, J. M. Decker, R. M. Blaese, 1981, Spontaneous cytotoxicity by human peripheral blood monocytes: Inhibition by monosaccharides and oligosaccharides, *Immunobiology* **158:**191-206.
38. J. C. Roder, K. Karre, and R. Kiessling, 1981, Natural killer cells, *Prog. Allergy* **28:**66-159.
39. R. Kiessling, and H. Wigzell, 1979, An analysis of the murine NK cell as to structure, function and biological relevance, *Immunol. Rev.* **44:**165-208.
40. R. B. Herberman, and H. T. Holden, 1978, Natural cell-mediated immunity, *Adv. Cancer Res.* **27:**305-377.
41. O. Stutman, P. Dien, R. E. Wisun, and C. Lattime, 1980, Natural cytotoxic cells against solid tumors in mice: Blocking of cytotoxicity by D-mannose, *Proc. Natl. Acad. Sci. (USA)* **77:**2895-2898.
42. R. P. McDermott, L. J. Kienker, M. J. Bertowich, and A. V. Muchmore, 1981, Inhibition of spontaneous but not antibody-dependent cell-mediated cytotoxicity by simple sugars: Evidence that endogenous lectins may mediate spontaneous cell-mediated cytotoxicity, *Immunology* **44:**143-152.
43. A. Kaplan, D. T. Achord, and W. S. Sly, 1977, Phosphohexosyl components of a lysosomal enzyme are recognized by pinocytosis receptors on human fibroblasts, *Proc. Natl. Acad. Sci. (USA)* **74:**2026-2030.
44. G. N. Sando, and E. F. Neufeld, 1977, Recognition and receptor-mediated uptake of a lysosomal enzyme, α-L-iduronidase, by cultured human fibroblasts, *Cell* **12:**619-627.
45. E. F. Neufeld, and G. Ashwell, 1980, Carbohydrate recognition systems for receptor-

mediated pinocytosis, in *The Biochemistry of Glycoproteins and Proteoglycans,* W. J. Lennarz, ed., Plenum, New York, pp. 241-266.

46. W. S. Sly, and P. Stahl, 1980, Receptor-mediated uptake of lysosomal enzymes, in *Transport of Macromolecules in Cellular Systems,* S. C. Silverstein, ed., Dahlein Konferenzem, Berlin, pp. 229-244.
47. K. Olden, B. A. Bernard, S. L. White, and J. B. Parent, 1982, Function of carbohydrate moieties of glycoproteins, *J. Cell. Biochem.* **18:**313-335.
48. J. T. Forbes, and T. N. Oeltmann, 1982, Carbohydrate receptors in natural cell-mediated cytotoxicity, in *NK Cells and Other Effector Cells,* R. B. Heberman, ed., Academic Press, New York, pp. 977-982.
49. L. Vondinelich, R. Sutherland, C. Schneider, R. Newman, and M. Greaves, 1983, Receptor for transferrin may be a target structure for natural killer cells, *Proc. Natl. Acad. Sci. (USA)* **80:**835-839.
50. R. Newman, J. Warner, and G. Dennert, 1984, NK recognition of target structures: Is the transferrin receptor the NK target structure?, *J. Immunol.* **133:**1841-1845.
51. P. Kaudewitz, H. W. Ziegler, G. R. Pape, and G. Riethmuller, 1982, Inhibition of spontaneous and antibody dependent cellular cytotoxicity using mono- and oligosaccharides, in *NK Cells and Other Natural Effector Cells,* R. Herberman, ed., Academic Press, New York, pp. 771-776.
52. C. R. Parish, T. J. Higgins, and I. F. C. McKenzie, 1978, Comparison of antigens recognized by xenogeneic and allogeneic anti-Ia antibodies: Evidence for two classes of Ia antigens, *Immunogenetics* **6:**343-354.
53. I. F. C. McKenzie, A. Clarke, and C. R. Parish, 1977, Ia antigenic specificities are oligosaccharides in nature: Hapten inhibition studies, *J. Exp. Med.* **145:**1039-1953.
54. C. R. Parish, D. C. Jackson, and I. F. C. McKenzie, 1977, Evidence that Ia antigenic specificities are defined by carbohydrates, in *Ir and Ia Antigens,* H. O. McDevitt, ed., Academic Press, New York, pp. 243-253.
55. R. H. Schwartz, A. Yano, and W. E. Paul, 1978, Interaction between antigen-presenting cells and primed T-lymphocytes, *Immunol. Rev.* **40:**153-180.
56. P. Erb, A. C. Stein, S. S. Alkan, S. Studer, E. Zoumbon, and R. H. Gisler, 1980, Characterization of accessory cells required for helper T-cell induction in vitro: Evidence for a phagocytic, Fc-receptor, and Ia-bearing cell type, *J. Immunol.* **125:**2504-2507.
57. R. J. Hodes, K. S. Hatchcock, and A. Singer, 1980, Major histocompatibility complex restricted self recognition, *J. Exp. Med.* **152:**1779-1794.
58. J. Theze, C. Waltenbaugh, M. E. Dorf, and B. Benacerraf, 1977, Immunosuppressive factor(s) specific for L-glutamic acid^{50}-L-tyrosine^{50} (GT) II. Presence of I-J determinants on the GT suppressive factor, *J. Exp. Med.* **146:**287-292.
59. T. Tada, and K. Okumura, 1979, The role of antigen specific T-cell factors in the immune response, *Adv. Immunol.* **28:**1-87.
60. L. D. Tomaska, and C. R. Parish, 1982, Inhibition of secondary IgG responses by monosaccharides: Evidence for I-region control, *J. Immunogenet.* **9:**63-68.
61. D. L. Mann, and A. V. Muchmore, 1980, Inhibition of DRw antisera reactivity by mono and oligosaccharides, *J. Immunol.* **124:**2218-2221.
62. G. W. Warr, G. R. Vasta, B. E. Ledford, and J. J. Marchalonis, 1983, Evidence for the existence of a diverse family of immunoglobulin-related recognition molecules, *J. Immunogenet.* **10:**17-23.
63. G. R. Vasta, J. J. Marchalonis, and H. Kohler, 1984, Invertebrate recognition protein cross-reacts with an immunoglobulin idiotype, *J. Exp. Med.* **159:**1270-1276.

64. H. G. Remold, 1973, Requirement for α-L-fucose on the macrophage membrane receptor for MIF, *J. Exp. Med.* **138:**1065-1076.
65. R. E. Rocklin, 1976, Role of monosaccharides in the interaction of two lymphocyte mediators with their target cells, *J. Immunol.* **116:**816-820.
66. G. Poste, R. Kirsh, and T. J. Fidler, 1979, Cell surface receptors for lymphokines. 1. Possible role of glycolipids as receptors for macrophage migration inhibitory factor (MIF) and macrophage activation factor (MAF), *Cell. Immunol.* **44:**71-88.
67. F. Besancon, H. Ankel, and S. Basu, 1976, Specificity and reversibility of interferon ganglioside interaction, *Nature* **259:**576-578.
68. V. E. Vengris, F. H. Reynolds, Jr., M. D. Hollenberg, and P. M. Pitha, 1976, Interferon action: Role of membrane gangliosides, *Virology* **72:**486-493.
69. Y. Ito, Y. Nishiyama, H. Shimokata, H. Takeyama, and A. Kunii, 1978, Active component of HVJ (Sendai virus) for interferon induction in mice, *Nature* **274:**801-802.
70. Y. Ito, M. Tsurudome, A. Yamada, M. Hishijama, 1984, Interferon induction in mouse spleen cells by mitogenic and nonmitogenic lectins, *J. Immunol.* **132:**2440-2444.
71. S. Kizaka, G. Snitokoff, and J. J. McSharry, 1983, Sendai virus glycoproteins are T-cell dependent B-cell mitogens, *Infect. Immunol.* **40:**592-600.
72. A. V. Muchmore, J. M. Decker, and R M. Blaese, 1980, Evidence that specific oligosaccharides block early events necessary for the expression of antigen specific proliferation by human lymphocytes, *J. Immunol.* **125:**1306-1311.
73. W. C. Greene, T. A. Fleisher, and T. A. Waldman, 1981, Soluble suppressor supernatants elaborated by concanavalin A-activated human mononuclear cells. I. Characterization of a soluble suppressor of T-cell proliferation, *J. Immunol.* **126:**1185-1191.
74. T. A. Fleisher, W. C. Greene, R. M. Blaese, and T. A. Waldman, 1981, Soluble suppressor supernatants elaborated by concanavalin A-activated human mononuclear cells. II. Characterization of a soluble suppressor of B-cell immunoglobulin production, *J. Immunol.* **126:**1192-1197.
75. D. J. Barrett, J. R. Edwards, B. A. Pietrantuono, and E. M. Ayoub, 1983, Inhibition of human lymphocyte activation by wheat germ agglutinin: A model for saccharide-specific suppressor factors, *Cell. Immunol.* **81:**287-297.
76. B. Barak, W. Gregory, P. Fletcher, and S. Kornfeld, 1979, Vertebrate lectins. Comparison of β-galactoside-binding lectins from tissues of calf and chicken, *J. Cell. Biol.* **81:**528-537.
77. D. Kobiler, E. C. Beyer, and S. H. Barondes, 1978, Developmentally regulated lectins from chick muscle, brain, and liver have similar chemical and immunological properties, *Dev. Biol.* **64:**265-272.
78. G. I. Franklin, F. S. Walsh, and E. J. Thompson, 1980, Endogenous lectins of human muscle, *FEBS Lett.* **118:**200-204.
79. J. T. Powell, 1980, Purification and properties of lung lectin. Rat lung and human lung β-galactoside-binding proteins, *Biochem. J.* **187:**123-129.
80. S. J. Kaufman, and M. L. Lawless, 1980, Thiodigalactosides binding lectin and skeletal myogenesis, *Differentiation* **16:**41-48.
81. R. G. MacBride, and R. J. Przybylski, 1980, Purified lectin from skeletal muscle inhibits myotube formation in vitro, *J. Cell. Biol.* **85:**617-625.
82. H. Den, D. A. Malingak, and A. Rosenberg, 1976, Lack of evidence for the involvement of a β-D-galactosyl specific lectin in the fusion of chick myoblasts, *Biochem. Biophys. Res. Commun.* **69:**621-627.
83. G. Levi, R. Tarrab-Hazdai, and V. I. Teichberg, 1983, Prevention and therapy with electrolectin of experimental autoimmune myasthenia gravis in rabbits, *Eur. J. Immunol.* **13:**500-507.

84. Y. Reisner, M. Biniaminov, E. Rosenthal, N. Sharon, and B. Ramot, 1979, Interaction of peanut agglutinin with normal human lymphocytes and with leukemic cells, *Proc. Natl. Acad. Sci. (USA)* **76:**447-451.
85. L. M. Stoolman, and S. D. Rosen, 1983, Possible role for cell-surface carbohydrate-binding molecules in lymphocyte recirculation, *J. Cell. Biol.* **96:**722-729.
86. G. J. Spangrude, B. A. Braaten, and R. A. Daynes, 1984, Molecular mechanisms of lymphocyte extravasation. I. Studies of two selective inhibitors of lymphocyte recirculation, *J. Immunol.* **132:**254-362.
87. H. B. Cole, J. W. Ezzell, K. F. Keller, and R. J. Doyle, 1984, Differentiation of *Bacillus anthracis* and other bacillus species by lectins, *J. Clin. Microbiol.* **19:**48-53.
88. C. Jensen, P. S. Stahl, C. Koch, H. Permin, T. C. Bog-Hansen, H. Lowenstein, N. Hoiby, and S. Norn, 1983, Intrinsic asthma and bacterial histamine release. Possible role of bacterial surface lectins, *Eur. J. Respir. Dis.* **64**(suppl. 128):391-393.
89. M. Feodorescu, E. Mayer, and S. Dray, 1977, Identification of five human lymphocyte subpopulations by their differential binding of various strains of bacteria, *Cell. Immunol.* **29:**352-357.

10

Developmentally Regulated Soluble Lectins

David Kobiler
Israel Institute for Biological Research

Lectins are divalent or multivalent carbohydrate-binding proteins that are characterized by their ability to cause carbohydrate-blockable agglutination of cells. In the last decade, several vertebrate lectins from different tissues of diverse species have been described, and they can be roughly classified into two groups. The first comprises integral membrane proteins of high molecular weight that can be solubilized by detergents and require divalent cations for their binding activity. These lectins have been implicated in the clearance of asialoglycoproteins from the serum (1).

The second class of vertebrate lectins comprises a group of soluble proteins, low in molecular weight and developmentally regulated. These lectins share a specific requirement for thiol-reducing groups and do not require divalent cations for agglutinating activity. The physiological roles of these lectins have not yet been deciphered; however, accumulating evidence suggests possible functions in intercellular interactions during tissue differentiation and in the secretion or organization of extracellular glycoconjugates (2, 3).

The following is an attempt to evaluate current knowledge concerning these soluble lectins and their possible functions.

OCCURRENCE AND CHARACTERIZATION OF SOLUBLE LECTINS

The recent work on vertebrate lectins was stimulated by Teichberg and co-workers (4), who reported the presences, isolation, and partial characterization of a β-D-galactoside-binding protein in the electric organ tissue of the eel *Electrophorus electricus.* This protein was named *electrolectin* because of its origin in electrogenic tissue and its lectin-like activity of agglutinating erythrocytes. The authors also described the occurrence of similar β-D-galactoside-binding activities in a variety of vertebrate tissues and tissue cultures (4). *Galaptins* is a generic term proposed for the β-galactoside-specific vertebrate lectins. The word is derived from their saccharide specificity and the Greek *hapto*—to touch or to join (5). However, operational designations, such as thiodigalactoside-binding (TDG-binding) lectin and chicken-lactose lectin, are more commonly used.

More recently, several lectins similar to the electrolectin have been identified, purified, and characterized from various vertebrate sources (Table 10-1). These agglutinins of trypsinized rabbit erythrocytes show similar saccharide specificity and are all inhibited by β-D-galactosyl-β-thiogalactopyranoside (thiodigalactoside-TDC) and lactose. They are also characterized by divalent cation-independent binding activity, a requirement for sulfhydryl reducing agents, a low molecular weight (Mr from 20,000 to 34,000), and an apparent dimeric structure (6-11).

The lectins are not tissue-specific, as within the same species, and lectins of a given class seem to be identical. Calf lectins, purified from several tissues, are essentially indistinguishable by SDS-polyacrylamide gel electrophoresis, isoelectric focusing, amino acid composition and radioimmunoassay (6, 12). Furthermore, these lectins show antigenic cross-reactivity with lectins isolated from human heart muscle and from human and rhesus money skeletal muscles (13).

Likewise, lectins isolated from embryonic and adult chicken tissues (chicken lactose lectin I [CLL-I]) (7, 8, 14-17) appear to be identical in molecular weight and dimeric structure. They also have the same isoelectric point and peptide maps and are antigenically indistinguishable. However, the chicken lectins are clearly different from the mammalian lectins, exhibiting a slightly larger molecular size and being antigenically distinct (12, 18).

A second type of lactose-binding lectin from embryonic and adult chicken has been identified and designated chicken lactose lectin II (CLL-II). They show lower specific hemagglutinating activity and differ in subunit size, in isoelectric points, and in tryptic peptide maps, and they are also immunologically distinct. Furthermore, while CLL-I has an apparent dimeric structure, CLL-II is isolated as a monomer of 14,000 Mr (16, 17). A similar monomeric lactose lectin from mammalian tissue was purified from erythroblast-enriched bone marrow (Mr 13,000) and termed *erythroid developmental agglutinin* (EDA) (19, 20).

The third type of soluble lectin that was identified in developing tissues was initially characterized as a N-acetylgalactosamine-sensitive agglutinin

Table 10-1. Some Vertebrate Soluble Lectins (Class I)

Source			
Species	*Tissues*	*Subunit Mr*	*References*
Electric eel	Electric organ	16,500	11
Chicken	Embryonic muscle Adult liver	15,000	7, 8, 15
Human	Lung	14,500	9
Rat	Lung	13,500	9
Calf	Heart, spleen, muscle	12,500	12 10

(21). However, since heparin is the most potent known inhibitor of this lectin, it has been referred to as chicken heparin lectin (CHL) (22, 23). This lectin would agglutinate trypsinized, glutaraldehyde-fixed rabbit erythrocytes only after they had been modified by aging (21), heat, or alcohol treatment (22). CHL has been purified from chicken liver and embryonic chick muscle, as well as from rat lung by affinity chromatography on heparin-Sepharose. Although the lectin forms large aggregates, it can be dissociated into subunits with apparent molecular weights of 13,000 and 16,000 (23, 24). A similar agglutinating activity was described for extracts of human muscle (25).

Soluble fucan-binding lectins, which also show a divalent cation-independent agglutinating activity and have been implicated in teratocarcinoma stem cell adhesion (26, 27), will be discussed elsewhere in this monograph.

A number of additional soluble lectins are beyond the limits of our oversimplified classification and will not be discussed here. Several of these lectins require divalent cations for activity (28, 29), while others show a hemagglutinating activity sensitive to hexosamines rather than to their N-acetyl derivatives (30, 31, 32); this characteristic raises the possibility that their hemagglutinating activity may not be due to their carbohydrate-binding capacity.

DEVELOPMENTAL REGULATION

Developmentally regulated species-specific lectins can be detected in cellular slime molds as they differentiate from a vegetative single-cell state to a social aggregation-competent state. The appearance of lectins on the cell surface parallels the development of cell cohesiveness and is also correlated with the presence of high-affinity species-specific glycoconjugate receptors on the cell surface. Extensive studies suggest that these lectins mediate specific cell-cell associations; however, the evidence is not yet conclusive (2).

It was suggested that the slime mold lectins are similar to embryonic lectins in that they play a role in cell-cell association and tissue organization. The demonstration that these lectins are developmentally regulated supports this contention. The appearance of a lectin at a specific stage in the development of a given tissue implies that it may participate in the specific process taking place at that stage. For example, CLL-I is found at low levels in extracts of muscle cells early in their development; the levels rise by two orders of magnitude between day 8 and day 12 and then decline to very low levels after day 16 of development. This striking increase in lectin hemagglutinating activity correlates with myoblast fusion into myotubes (33, 34, 35). A similar but less striking increase in lectin hemagglutinating activity was observed during differentiation of a myogenic cell line (34, 36). The chick β-D-galactoside lectin is already present in embryos already at the pregastrula and gastrula stages (37, 38). Developmentally regulated lectins were also found in embryonic chick skin and cartilage, correlating temporally with dermal condensation and chondriogenesis (39, 40).

Temporal changes in CLL-I were also found in the developing chick embryo nervous system, and manifested in different patterns for the lectins extracted from brain, optic tectum, retina, and spinal cord (35, 41, 42). High lectin levels are not correlated with cellular proliferation or cellular migration, but rather with neuronal differentiation. Similarly, circumstantial evidence would suggest that the developmentally regulated lectin of rat lung is involved in the development and regulation of this organ's cholinergic innervation.

It is inferred that these developmentally regulated lectins play a specific role in the embryonic differentiation of some tissues. However, they also appear to be especially abundant in other tissues at the adult stage, presumably mediating some functions of the fully mature tissue. Whereas the high levels of CLL-I in embryonic muscle tissue diminished to negligible amounts in the adult muscle, in adult liver the level of lectin activity far exceeds that found in embryonic liver. These findings suggest that the lectins could have been adapted for different functions in the different tissues.

The other two classes of soluble lectins, CLL-II and CHL, are also developmentally regulated, showing striking changes in concentration at different stages of development of some tissues. CLL-II is abundant in embryonic kidney, but falls to low levels in maturity; in contrast, the low levels of lectin activity found in embryonic intestine extracts rise strikingly in adult tissue (17). Changes during differentiation were also shown for EDA. The lectin is present on the erythroblast cell surface in bone marrow, is less pronounced on marrow reticulocytes, and is undetected on mature erythrocytes (19).

The changes in CHL activity during the development of pectoral muscle were followed (21), and they were found to correlate with muscle fusion and innervation. Also, the low levels of activity found in embryonic liver rise to abundance in the adult tissue (22, 23).

The three classes of chicken lectins were found to be present simultaneously in various tissues, but in differing concentrations. Furthermore, the relative levels of hemagglutinating activity showed marked changes during tissue development. The combined activity of several lectins in a given cell may participate in highly selective functions, which can be controlled by the relative concentrations of the lectins. On the other hand, each lectin can participate in a different process.

DIFFERENT ROLES IN DIFFERENT TISSUES AT DIFFERENT TIMES

The developmentally regulated nature of the lectins suggests that they may participate in a specific developmental function. However, the notion that lectin functions are confined solely to tissue differentiation was challenged by the finding of an identical lectin in adult tissue. Immunohistological studies to locate CLL-I in adult tissues indicate that the lectins play different roles in different tissues (15). The lectin was found to be concentrated in cells lining the liver sinusoids, in the secretory granules of the mucus-secreting intestinal goblet cells, and in the extracellular spaces surrounding pancreatic acini. The various locations of these extra lectins in the tissues examined argue against the possibility of a common function and suggest that the function of the lectin is different in each of these tissues (15).

The elucidation of the functional role of the lectins is further complicated by the fact that no specific endogenous receptor for CLL-I has been identified or isolated so far. The specificity of the binding sites of the lectins, inferred from hapten inhibition studies, suggests the existence of endogenous glycoconjugate receptors in the tissues from which they are derived. An attempt was made to determine the location of such potential receptors for exogenous CLL-I in several chicken tissues and their relation to the localization of endogenous CLL-I (44). Some tissues, such as intestine and kidney, showed abundant extracellular binding sites for the lectin, primarily in intercellular spaces in basement membranes and in material on the luminal surface. On the other hand, adult heart had neither endogenous lectin nor significant binding sites for CLL-I, and adult pancreas showed considerable endogenous lectin in extracellular sites surrounding exocrine lobules, but no binding sites for exogenously applied CLL-I. In tissues where binding sites have been demonstrated, they were not specific for CLL-I, since a nonrelated lactose-binding lectin, purpurin, binds to them as well (44). Theses results, although confirming the presence of appropriate complementary glycoconjugates that can serve as receptors, indicate that the binding capacity does not result in functional specificity.

Although a specific cellular function has not been proved unequivocally for any of these lectins, two main possibilities have been considered: (a) mediation of cell-cell interaction, and (b) organization of intra- or extracellular glycoconjugates.

The possible role of lectins in cell adhesion has been primarily supported by studies in differentiating systems. The working hypothesis was that specific cell adhesion could be achieved by a cell surface lectin with a particular recognition specificity for a carbohydrate-containing receptor on the surface of the adjacent cell. However, the carbohydrate-binding properties of the lectins can serve to crosslink intracellular or extracellular glycoconjugates. These two possible functions have been evaluated in several systems.

MYOBLAST FUSION

The observation that chick embryo pectoral muscle and cultured muscle cells contain β-D-galactoside-sensitive hemagglutinating activity at a high concentration (4) focused much interest on this system. Myoblast fusion into multinucleated myotubes involves at least two separate reactions: cell recognition and adhesion and subsequent membrane-membrane interactions that culminate in cell fusion (45). The striking increase in myoblast lectin levels, which correlates temporally with the fusion process of these cells, raised the possibility that the lectin plays a role in this event (34). However, as the fusion of myoblasts is strictly a calcium-requiring process, the role of the calcium-independent lectin, if any, could be limited to the early event of recognition.

The observation that TDG, potent inhibitor of the lectin activity, blocked the fusion of cultured myoblasts from an established cell line (36, 46) lent support to the hypothesis that the lectin plays a functional role in myoblast fusion. However, this conclusion was challenged, since TDG affects neither aggregation nor fusion of chick myoblasts in primary cultures (36, 45). It was proved that these conflicting observations were the consequence of varying experimental conditions (47). TDG markedly inhibited fusion of cell line myoblasts when the cells were grown in its presence for prolonged periods. However, this saccharide had no effect on development when it was present for restricted periods preceding fusion. These results, consistent with the previous positive and negative reports, were interrupted as not supporting the theory that this lectin plays a role in interactions of myoblasts immediately preceding or during fusion (47). In addition, it was shown that antilectin antibodies and their Fab preparations, which recognized the carbohydrate-binding site of CLL-I, had no effect on myotube formation when they were present in excessive amounts in muscle cultures throughout the fusion period (48).

In contrast, myotube formation was temporarily inhibited by the direct application of purified lectin in low amounts to skeletal muscle cultures. The endogenous CLL-I is assumed to react with glycoconjugates on the cell surface of adjacent cells, thus holding the two myoblasts in close apposition. When exogenous lactose-binding lectin is added, the adhesive sites (lectin receptors) are experimentally covered, blocking the bridging reaction and resulting in minimal lateral cell surface apposition, as was seen experimentally (49). However, these results have not been confirmed (48). Furthermore, the interpretation of these findings, if reproducible, is complicated by the finding that plant lectins (at higher concentrations) with different saccharide specificities also inhibit myotube formation (50), although the mechanism of action may be different (51).

More recently, the carbohydrate requirement for fusion of embryonic quail muscle cells was examined in culture by the use of tunicamycin to block glycosylation of N-linked glycoproteins (52). When the mononucleated, proliferating myoblasts were treated with tunicamycin at various times before fusion, fusion was strongly inhibited. However, fusion was partially restored when tunicamycin was administered in the presence of protease inhibitors. Olden and associates (52) interpreted these studies to indicate that fusion is mediated by glycoproteins with asparagine-linked oligosaccharides; however, the requirement for the carbohydrate portion of the glycoprotein appears to be indirect in that it acts to stabilize the protein moiety against proteolytic degradation (53, 54).

The cellular localization of the lectin is an important step in elucidation of the lectins' function. It is generally agreed that in the cultured myoblasts, lectins are primarily located in the cytoplasm (8, 12, 33, 47, 48, 55, 56), although some are detected on the surfaces of cultured cells (8, 55).

Immunohistochemical techniques have been used to study the localization of CLL-I in vivo in various stages of chick development (56). In myoblasts, the lectin was distributed throughout the cytoplasm, and some of it seemed to be associated with the cell surface. In maturing myotubes, the distribution of the lectin is similar to that of the sarcoplasmic reticulum and T tubules. Later in muscle development, CLL-I becomes localized extracellularly, and is associated with the surfaces of muscle fibers and surrounding connective tissue. The lectin may become externalized by migration through the T tubules, which are continuous with extracellular space (56). These results do not suggest that CLL-I participates in myoblast fusion, but rather that it may play a role in organizing complementary glycoconjugates in the intracellular tubules, on the muscle surface, or in extracellular material.

When all of the data accumulated so far on the participation of the lectin in this system are considered, it seems that the weight of evidence does not absolutely exclude such a possibility: however, other possible interpretations seem more adequate.

MATURATION OF THE NERVOUS SYSTEM

In a way that is similar to the behavior of muscle lectin, the striking changes in lectin activity during the development of the brain implied that this lectin may play a role in its differentiation (35). The peak of activity of β-galactoside-binding proteins has been correlated with synaptogenesis (56). It was suggested that the establishment of neuromuscular junctions has a regulatory effect on lectin concentration. The level of lectin activity increases upon denervation, and is lower in slow multiinnervated muscle than in fast moninnervated muscle (56). In the developing rat lung, where the temporal pattern of lectin activity was found closely parallel that of acetylcholinesterase, it was suggested that lectin activity is involved in the development and regulation of cholinergic innervation (43).

However, no direct evidence for a possible role of these lectins in synaptogenesis has been obtained. Moreover, lactose, which was shown to inhibit the hemagglutinating activity of the retinal lectin, had no effect on the in vitro formation of synapses between chick retinal neurons and rat myotubes (42).

In a different approach, the optic tectum, a well-characterized area of chick brain, was chosen for studying lectin distribution through immunohistochemical techniques (41). CLL-I is not detectable at very early stages, it becomes abundant in differentiating optic tectal neurons, and it is absent from most cells of 2-month-old chickens. The tectal lectin is abundant in developing neuronal cell bodies and their processes, and it was also detected on the cell surfaces of dissociated neurons. It was also observed in ependymal and glyal cells. The distribution of CLL-I in the tectum, like overall tectal development, was influenced by innervation by the optic nerve fibers. The peak of CLL-I activity does not seem to be correlated with extensive cellular proliferation or cellular migration, but rather with neuronal maturation, the stage at which neurons are sending out processes and receiving innervation. After innervation by incoming optic nerve fibers, there is a reduction in lectin concentration. Furthermore, the removal of optic vesicles causes the failure of innervation, which is correlated with a lack of depletion of CLL-I (41).

The circumstantial evidence presented supports a possible role for the lectin in the innervation process; however, no direct evidence has been presented so far.

ADHESION OF ENDODERM CELLS FROM GASTRULATING CHICK EMBRYO

Two lectin activities, specific for β-D-galactoside groups, were identified and purified from gastrulating chick blastoderms (37, 38). A soluble lectin, probably CLL-I, is present in tissue from the area pellucida and the area opaca, while particle-associated lectin activity is present mainly in the epiblast and endoderm of the area opaca (38).

TDG, the most potent inhibitor of these lectins, inhibits the aggregation of freshly prepared cells from the extra-embryonic endoderm of the gastrulating chick embryo (37, 57, 58). Aggregation is also inhibited when purified CLL-I is added to the assay system. Furthermore, cells maintained in culture secrete soluble lectin into the medium, which in turn decreases the tendency of the cells to aggregate (57, 58). The addition of exogenous purified blastoderm lectin or its inhibitor (TDG) to cultured blastoderm cells causes a delay and decrease of cell attachment to the substrate. The treated cells are fibroblast-like, their cell-cell contacts are limited to localized areas of cell surface, and many of them appear to migrate over the substrate. In contrast, in control cultures, cells appear epithelial in morphology and maximize their areas of apposition (57). Immunofluorescence studies with anti-CLL-I antibodies showed that in the majority of cells in culture, the lectin was present primarily as an irregular circular web in the central cytoplasm. In migrating cells, increased fluorescence was found in peripheral cytoplasm and retraction fibers, while in detaching cells, intense fluorescence was observed in the extracellular substrate-attached material (SAM "footprints") deposited by the cells (59).

These results are consistent with the hypothesis of a possible role of CLL-I in cell-cell aggregation and cell-substrate adhesion in this embryonal system. The adherence of cells, mediated by the endogenous lectin, can be inhibited by blocking the lectin (by TDG) or by blocking its apparent receptors (by exogenous CLL-I). Appealing as this interpretation may seem, more information is needed in order to prove a role for these lectins in the cell interactions of this system.

ERYTHROID DEVELOPMENT IN ADULT BONE MARROW

The inter-erythroblast aggregation in the bone marrow, forming "erythroblastic islands" around a central macrophage "nurse" cell, is a transient stage in erythroid development. EDA was implicated as a mediator of this process (19). This lectin, which has been purified from erythroblast-enriched rabbit bone marrow, was shown to have developmentally regulated and' tissue-specific in vivo and in vitro. The purified lectin, when added to a cell suspension, agglutinates the erythroblasts, while the mature erythrocytes are not affected. The interpretation that EDA functions in vitro as a bridging molecule is supported by the findings that the lectin agglutinates inert (glutaraldehyde-treated) erythroblasts as well as lactose-washed cells (from which all of the cell surface agglutinin has been removed). The agglutination of rabbit erythrocytes is calcium-independent and insensitive to disruption of cellular metabolism. Furthermore, anti-EDA Fab fragments block the in vitro agglutination of erythroblasts by EDA. It may be suggested that such a bridging mechanism may mediate inter-erythroblast association in vivo. This assumption is supported by the observation that EDA distribution on the cell surface of the erythroblasts and the susceptibility of lactose-washed cells to agglutination by EDA decline as a function of erythroid maturation, concomitant with reticulocyte release from the erythroblastic island (19, 20).

Further studies were done with MELC, a murine erythroleukemic cell line that can be chemically induced to differentiate along the erythroid developmental pathway in culture. The transformed cells were not readily agglutinated by EDA; however, upon induction, MELC acquired aggregation competence by EDA (20). The evidence obtained supports the hypothesis that EDA may be responsible for inter-erythroblast recognition and adhesion in vivo by forming simple bridges between cells, which can serve as short-term adhesive bonds. These contacts can be abolished when reticulocytes are released from the erythroblastic island prior to entering the circulation. However, further evaluation is needed in order to establish the role of this lectin in erythroid island formation.

INTRACELLULAR AND EXTRACELLULAR BINDING OF GLYCOCONJUGATES

The carbohydrate-binding properties of the lectin, which enables it to "bridge" between cells by binding to cell surface glycolipids or glycoproteins, may also be used for cross-linking of glycoconjugates in the extracellular matrix or in solution.

As has already been mentioned, CLL-I in embryonic muscle, being primarily intracellular, is externalized with differentiation, ultimately becoming localized on the surface of myotubes and in extracellular spaces (56). This translocation seems to be a fairly general phenomenon, and may be triggered by specific external signals (60).

Both CLL-I and CLL-II have been localized in secretory vesicles of the goblet cells of the intestinal mucosa, with CLL-I being the minor component (15, 61). Both lectins apparently bind to purified chicken intestinal mucin, which is a very potent inhibitor of their hemagglutinating activities. Immunohistochemical evidence demonstrates that CLL-II is associated with the intestinal mucosal surface, presumably with mucin. The lectin is secreted upon cholinergic stimulation (61). These results suggest that the lectin may play a role in organizing the mucin in the secretory vesicles and on the luminal surface and/or in cross-linking mucin molecules with glycoproteins of other types on the intestinal epithelial surface.

The third class of soluble lectins, the chicken heparin lectin, was also shown to be externalized. CHL from embryonic muscle was shown to be developmentally regulated in vivo (21). Studies in embryonic muscle cultures showed that the lectin was secreted into the medium, and the secretion started only after substantial differentiation of the muscle cells had taken place (62). The hemagglutinating activity, mediated by CHL, is markedly inhibited by specific glycosaminoglycans, as well as simple saccharides such as N-acetylgalactosamine and cellobiose (22, 23). Hapten inhibition studies suggested that the lectin interacts with glycosaminoglycans containing iduronic acid, most likely to be heparin, but possibly dermatan sulfate or heparan sulfate. Based on these inhibition studies, it may be envisaged that the functional receptors for CHL will vary in different tissues and will be determined by the availability of a specific glycosaminoglycan in the given tissue. Extracts prepared from the SAM from myoblast cultures, as well as crude glycosaminoglycans prepared from embryonic muscle tissue, were extremely potent inhibitors of CHL hemagglutinating activity. The active material in the extracts of SAM was shown to contain a glycosaminoglycan with enzymatic cleavage characteristics resembling dermatan sulfate (62).

One possible interpretation of these findings is that CHL may interact with basal membrane or cell surface glycosaminoglycan receptors to mediate the organization of cell matrix or cell-cell interactions. Consistent with this interpretation is the finding of a CHL inhibitor, heparan sulfate, that is associated with cell surfaces (63, 64), and it has been implicated in cell adhesive responses to extracellular matrices (65, 66).

MITOGENIC ACTIVITY OF THE LECTINS

The chicken lectins, CLL-I and CLL-II, exhibit mitogenic activity for murine lymphocytes (67, 68). Since CLL-I has been detected in chicken spleen (15), and since the intracellular lectins can be externalized, a possible role of these lectins in regulating lymphocyte proliferation should be considered. A possible physiological role of the mitogenic effect was implicated in an experimental system involving the differentiation of the mesenchymal cells, which had been dissociated from the limb buds of chic embryos (69). The cells incubated with the lectin showed activation of cell proliferation, an increased number of cartilage nodules, and higher incorporation of ^{35}S-sulfate per cell. The results demonstrate that chick lectin promotes cartilage differentiation in vitro and imply that the endogenous lectin of this tissue (40) may be involved in chondriogenesis in vivo.

CONCLUDING REMARKS

The study of developmentally regulated soluble lectins has been stimulated by the growing interest in the molecular mechanisms underlying cell-cell interactions. The saccharide specificity and the ability to agglutinate cells make the lectins an adequate candidate to function as a bridging molecule in cell recognition and adhesion. Evidence supporting this hypothesis has been presented in a few developmental systems, and although this function has not been proved conclusively, it has yielded some exciting speculations. β-D-galactoside-binding lectins have been found on the cell surfaces of malignant cells, and their possible influence on the pathogenesis of cancer is being considered. Metastasis, the most devastating aspect of neoplasia, may be mediated by tumor cell lectins, as they can bind to the cell surface glycoconjugates of the host tissues (70).

Several studies, undertaken to prove the role of lectins in cell-cell interactions, have led to the demonstration of the plausibility of other potential functions of these lectins. Localization studies have shown that lectins seem to play different roles in different tissues. One such role that has been considered is the intracellular organization of glycoconjugates and their externalization from the cells. Other possibilities, such as the interaction of the lectin with the extracellular matrix and its activity as a proliferative or differentiating factor, are functions that merit detailed consideration. Additional studies are needed to assess defined functions for these lectins and to elucidate their mechanisms of action.

REFERENCES

1. G. Ashwell, and J. Harford, 1982, Carbohydrate-specific receptors of the liver, *Annu. Rev. Biochem.* **51:**531-554.
2. S. H. Barondes, 1981, Lectins: Their multiple endogenous cellular functions, *Annu. Rev. Biochem.* **50:**207-231.
3. S. H. Barondes, 1985, Vertebrate lectins: Properties and functions, in *The Chemistry and Biology of Lectins,* I. Liener, I. Goldstein, and N. Sharon, eds., Academic Press, New York.
4. V. I. Teichberg, I. Silman, D. D. Beitsch, and G. Resheff, 1975, A β-D-Galactoside binding protein from electric organ tissue of Electrophorus electricus, *Proc. Natl. Acad. Sci. (USA)* **72:**1383-1387.
5. F. L. Harrison, and C. J. Chesterton, 1980, Factors mediating cell-cell recognition and adhesion: Galaptins, a recently discovered class of bridging molecules, *FEBS Lett.* **122:**157-165.
6. A. De Waard, S. Hickman, and S. Kornfeld, 1976, Isolation and properties of β-galactoside binding lectins of calf heart and lung, *J. Biol. Chem.* **251:**7581-7587.
7. H. Den, and D. A. Malinzak, 1977, Isolation and properties of a β-D-galactoside specific lectin from chick embryo thigh muscle, *J. Biol. Chem.* **252:**5444-5448.
8. T. P. Nowak, D. Kobiler, L. E. Roel, and S. H. Barondes, 1977, Developmentally regulated lectin from embryonic chick pectoral muscle: Purification by affinity chromatography, *J. Biol. Chem.* **252:**6026-6030.
9. J. T. Powell, 1980, Purification and properties of lung lectin: Rat lung and human lung β-galactoside binding proteins, *Biochem. J.* **187:**123-129.
10. G. T. Montellione, S. Callaham, and T. R. Podleski, 1981, Physical and chemical characterization of the major lactose blockable lectin activity from fetal calf skeletal muscle, *Biochem. Biophys. Acta* **670:**110-123.
11. G. Levi, and V. I. Teichberg, 1981, Isolation and physicochemical characterization of electrolectin, a β-D-galactoside binding lectin from the electric organ of Electrophorus electricus, *J. Biol. Chem.* **256:**5735-5740.
12. E. B. Briles, W. Gregory, P. Fletcher, and S. Kornfeld, 1979, Vertebrate lectins: Comparison of properties of β-galactoside-binding lectins from tissues of calf and chicken, *J. Cell Biol.* **81:**528-537.
13. R. A. Childs, and T. Feizi, 1979, β-galactoside-binding muscle lectins of man and monkey show antigenic crossreactions with those of bovine origin, *Biochem. J.* **183:**755-758.
14. D. Kobiler, E. C. Beyer, and S. H. Barondes, 1978, Developmentally regulated lectins from chick muscle, brain and liver have similar chemical and immunological properties. *Develop. Biol.* **64:**265-272.
15. E. C. Beyer, K. T. Tokuyasu, and S. H. Barondes, 1979, Localization of an endogenous lectin in chicken liver, intestine and pancreas, *J. Cell Biol.* **82:**565-571.
16. E. C. Beyer, S. E. Zweig, and S. H. Barondes, 1980, Two lactose binding lectins from chicken tissues. Purified lectin from intestine is different from those in liver and muscle, *J. Biol. Chem.* **255:**4236-4239.
17. E. C. Beyer, and S. H. Barondes, 1982, Quantitation of two endogenous lectins in embryonic and chicken tissues, *J. Cell Biol.* **92:**23-27.
18. G. Levi, and V. I. Teichberg, 1982, Antigenic crossreactivity between electrolectins, *FEBS Lett.* **148:**145-148.
19. F. L. Harrison, and C. J. Chesterton, 1980, Erythroid developmental agglutinin is a protein lectin mediating specific cell-cell adhesion between differentiating rabbit erythroblasts, *Nature* **286:**502-504.

20. S. F. Godsave, F. L. Harrison, and C. J. Chesterton, 1981, Molecular properties of a galaptin and its activity in normal and leukemic erythroid tissue, *J. Supramol. Struc.* **17:**223-230.
21. F. J. Mir-Lechaire, and S. H. Barondes, 1978, Two distinct developmentally regulated lectins in chick embryo muscle, *Nature* **272:**256-258.
22. D. Kobiler, and S. H. Barondes, 1979, Lectin from embryonic chick muscle that interacts with glycosaminoglycans, *FEBS Lett.* **101:**257-261.
23. H. Ceri, D. Kobiler, and S. H. Barondes, 1981, Heparin-inhibitable lectin: Purification from chicken liver and embryonic chicken muscle, *J. Biol. Chem* **256:**390-394.
24. M. M. Roberson, H. Ceri, P. J. Shadle, and S. H. Barondes, 1981, Heparin-inhibitable lectins: Marked similarities in chicken and rat, *J. Supramol. Struct.* **15:**395-402.
25. G. I. Franklin, F. S. Walsh, and E. J. Thompson, 1980, Endogenous lectins of human muscle, *FEBS Lett.* **118:**200-204.
26. L. B. Grabel, C. G. Glabe, M. S. Singer, G. R. Martin, and S. D. Rosen, 1981, A fucan specific lectin on teratocarcinoma stem cells, *Biochem. Biophys. Res. Comm.* **102:**1165-1171.
27. L. B. Grabel, M. S. Singer, G. R. Martin, and S. D. Rosen, 1983, Teratocarcinoma stem cell adhesion: The role of divalent cations and a cell surface lectin, *J. Cell. Biol.* **96:**1532-1537.
28. T. K. Gartner, K. Stocker, and D. C. Williams, 1980, Thrombolectin. A lectin isolated from Bothrops atrox venom, *FEBS Lett.* **117:**13-16.
29. M. M. Roberson, and S. H. Barondes, 1982, Lectin from embryos and oocytes of Xenopus laevis, *J. Biol. Chem.* **257:**7520-7524.
30. K. M. Yamada, and K. Olden, 1978, Fibronectins: Adhesive glycoproteins of cell surface and blood, *Nature* **275:**179-184.
31. D. L. Simpson, D. R. Thorne, and H. H. Loh, 1977, Developmentally regulated lectin in neonatal rat brain, *Nature* **266:**367-369.
32. M. J. Pitts, and D. C. H. Yang, 1980, Isolation of a developmentally regulated lectin from chick embryo, *Biochem. Biophys. Res. Comm.* **95:**750-757.
33. H. Den, D. A. Malinzak, and A. Rosenberg, 1976, Lack of evidence for the involvement of a β-D-galactosyl-specific lectin in the fusion of chick myoblasts, *Biochem. Biophys. Res. Comm.* **69:**621-627.
34. T. P. Nowak, P. L. Haywood, and S. H. Barondes, 1976, Developmentally regulated lectin in embryonic chick muscle and a myogenic cell line, *Biochem. Biophys. Res. Comm.* **68:**650-657.
35. D. Kobiler, and S. H. Barondes, 1977, Lectin activity from embryonic chick brain, heart and liver. Changes with development, *Develop. Biol.* **60:**326-330.
36. T. K. Gartner, and T. R. Podleski, 1976, Evidence that the types and specific activity of lectins control fusion of L_6 myoblasts, *Biochem. Biophys. Res. Comm.* **70:**1142-1149.
37. G. M. W. Cook, S. E. Zalik, N. Milos, and V. Scott, 1979, A lectin which binds specifically to β-D-galactoside groups is present at the earliest stages of chick embryo development, *J. Cell Sci.* **38:**293-304.
38. S. E. Zalik, N. Milos, and I. Ledsham, 1983, Distribution of 2-β-D-galactoside-binding lectins in the gastrulating chick embryo, *Cell Differentiation* **12:**121-128.
39. K. Kitamura, 1980, The changes in lectin activity during the development of embryonic chick skin, *J. Embryol. Exp. Morph.* **59:**59-69.
40. T. Yamagata, and T. Nishiwaki, 1979, Chick embryonic cartilage has a developmentally regulated lectin specific for galactose containing saccharides, *Proc. Japan Acad.* **55:**393-397.
41. F. Gremo, D. Kobiler, and S. H. Barondes, 1978, Distribution of an endogenous lectin in the developing chick optic tectum. *J. Cell Biol.* **79:**491-499.
42. G. S. Eisenbarth, R. R. Ruffalo, F. S. Walsh, and M. Nirenberg, 1978, Lactose sensitive lectin of chick retina and spinal cord, *Biochem. Biophys. Res. Comm.* **83:**1246-1252.
43. J. T. Powell, and P. L. Whitney, 1980, Postnatal development of rat lung. Changes in lung lectin, elastin, acetylcholinesterase and other enzymes, *Biochem. J.* **188:**1-8.

44. E. C. Beyer, and S. H. Barondes, 1980, Chicken tissue binding sites for a purified chicken lectin, *J. Supramol. Struct.* **13:**219-227.
45. K. A. Knudsen, and A. F. Horowitz, 1978, Differential inhibition of myoblast fusion, *Develop. Biol.* **66:**294-307.
46. T. K. Gartner, and T. R. Podleski, 1975, Evidence that a membrane bound lectin mediates fusion of L_6 myoblasts, *Biochem. Biophys. Res. Comm.* **67:**972-978.
47. S. J. Kaufman, and M. L. Lawless, 1980, Thiodigalactoside binding lectin and skeletal myogenesis, *Differentiation* **16:**41-48.
48. H. Den, and J. H. Chin, 1981, Endogenous lectin from chick embryo skeletal muscle is not involved in myotube formation in vitro, *J. Biol. Chem.* **256:**8069-8073.
49. R. G. MacBride, and R. J. Przybylski, 1980, Purified lectin from skeletal muscle inhibits myotube in vitro, *J. Cell Biol.* **85:**617-625.
50. H. Den, D. A. Malinzak, H. J. Keating, and A. Rozenberg, 1975, Influence of Con-A, wheat germ agglutinin and soybean agglutinin on the fusion of myoblasts in vitro, *J. Cell Biol.* **67:**826-834.
51. A. Sandra, M. A. Leon, and R. J. Przybylski, 1977, Suppression of myoblast fusion by concanavalin A: Possible involvement of membrane fluidity, *J. Cell Sci.* **28:**251-271.
52. K. Olden, J. Law, V. A. Hunter, R. Romain, and J. B. Parent, 1981, Inhibition of fusion of embryonic muscle cells in culture by tunicamycin is prevented by leupeptin, *J. Cell Biol.* **88:**199-204.
53. K. Olden, B. A. Bernard, M. J. Humphries, T.-K. Yeo, K.-T. Yeo, S. L. White, S. A. Newton, H. C. Bauer, and J. B. Parent, 1985, Function of glycoprotein glycans, *Trends Biochem. Sci.* **10:**78-82.
54. K. Olden, S. L. White, and S. A. Newton, 1985, *Critical Reviews in Chemistry,* CRC Press (in press).
55. T. R. Podleski, and I. Greenberg, 1980, Distribution and activity of endogenous lectin during myogenesis as measured with antilectin antibody, *Proc. Natl. Acad. Sci. (USA)* **77:**1054-1058.
56. V. I. Teichberg, 1978, Electrolectins: β-D-galactoside binding proteins, *Methods Enzymol.* **50:**291-302.
57. N. Milos, and S. E. Zalik, 1981, Effect of the β-D-Galactoside-binding lectin on cell to substratum and cell to cell adhesion of cells from the extraembryonic endoderm of the early chick blastoderm, *Wil. Roux's Arch.* **190:**259-266.
58. N. Milos, and S. E. Zalik, 1982, Mechanisms of adhesion among cells of the early chick blastoderm: Role of the β-galactoside binding lectin in the adhesion of extraembryonic endoderm cells, *Differentiation* **21:**175-182.
59. S. E. Zalik, N. Milos, and I. Ledsham, 1982, Distribution of a galactose-specific lectin in endoderm cells from early chick embryos, *Cell Tissue Res.* **225:**223-228.
60. S. H. Barondes, E. C. Beyer, W. R. Springer, and D. N. Cooper, 1981, Endogenous lectins in chickens and slime molds: Transfer from intracellular to extracellular sites, *J. Supramol. Struct.* **16:**233-242.
61. E. C. Beyer, and S. H. Barondes, 1982, Secretion of endogenous lectin by chicken intestinal goblet cells, *J. Cell Biol.* **92:**28-33.
62. H. Ceri, P. J. Shadle, D. Kobiler, and S. H. Barondes, 1979, Extracellular lectin and its glycosaminoglycan inhibitor in chick muscle cultures, *J. Supramol. Struct.* **11:**61-67.
63. P. M. Kraemer, 1977, Heparin releases heparan sulfate from cell surface, *Biochem. Biophys. Res. Comm.* **78:**1334-1340.
64. L. Kjellen, I. Petterson, and M. Hook, 1981, Cell surface heparan sulfate: An intercalated membrane protoglycan, *Proc. Natl. Acad. Sci. (USA)* **78:**5371-5375.
65. J. Laterra, J. E. Silbert, and L. A. Culp, 1983, Cell surface heparan sulfate mediates some

adhesive responses to glycosaminoglycan-binding matrices, including fibronectins, *J. Cell Biol.* **96:**112-123.
66. S. C. Stamatoglou, and J. M. Keller, 1983, Correlation between cell substrate attachment in vitro and cell surface heparan-sulfate affinity for fibronectin and collagen, *J. Biol. Chem.* **96:**1820-1823.
67. J. S. Lipstick, E. C. Beyer, S. H. Barondes, and N. O. Kaplan, 1980, Lectins from chicken tissues are mitogenic for Thy-1 negative murine spleen cells, *Biochem. Biophys. Res. Comm.* **97:**56-61.
68. M. J. Pitts, and D. C. H. Yang, 1981, Mitogenicity and binding properties of β-galactoside-binding lectin from chick embryo kidney, *Biochem. J.* **195:**435-439.
69. E. Matsutani, and T. Yamagata, 1982, Chick endogenous lectin enhances chondriogenesis of cultured chick limb bud cells, *Develop. Biol.* **92:**544-548.
70. A. Raz, and R. Lotan, 1981, Lectin-like activities associated with human and murine neoplastic cells, *Cancer Res.* **41:**3642-3647.

11

Possible Involvement of Intracellular Lectins in Intracellular Transport of Glycoproteins

Sheila A. Newton, Kiang-Teck Yeo, Tet-Kin Yeo, James Brian Parent, and Kenneth Olden*

Howard University Cancer Center

Cell-associated carbohydrate recognition proteins, or lectins, have been described in numerous animal systems. Documented specificities include galactose, fucose, N-acetylglucosamine, mannose, and mannose derivatives (see chaps. 1-10). With few exceptions, no specific endogenous ligands have been found for most of these lectins; however, the specificity of the binding sites of these lectins inferred from hapten inhibition studies suggests the existence of endogenous glycoconjugate ligands in the tissues from which they are derived.

The general intracellular pathway for transport and externalization of membrane, lysosomal, and secretory proteins in animal cells is well understood at the morphological level (1, 2). They are synthesized on the rough endoplasmic reticulum (RER), transported through smooth ER to the Golgi apparatus, and from the Golgi to their final destinations. Whether all proteins traverse the same pathway from the RER to the Golgi and why they travel at the various rates are not known. Also, it is not known how segregation of glycoproteins into the two post-Golgi secretory pathways (constitutive and regulated involving intermediate storage vesicles) is accomplished. Do proteins processed through the different post-Golgi pathways share some common structural feature that mediates their segregation; that is, are they

*K. Olden is also affiliated with the Membrane Biochemistry Section, Laboratory of Molecular Biology, National Cancer Institute, National Institutes of Health, Bethesda, Md.

encoded with information that regulates their rate of intracellular transport or directs them to their respective destinations? The molecular interactions responsible for the transport of proteins from the RER to their final destination are not known. Is it achieved by a nonspecific bulk flow process or by a specific carrier-mediated process?

Intracellular transport and selective assortment of lysosomal, secretory (constitutive and regulated), and membrane proteins must require very sophisticated intracellular recognition systems. The recognition systems are presumably composed of discrete molecular entities (receptors) within the structural matrix of the ER/Golgi complex that recognize specific molecular entities (ligands) on the exported products.

During the last decade, considerable effort has been expended to elucidate the molecular mechanisms by which proteins are transported to a variety of noncytoplasmic locations (2). Since most of the products processed by the intracellular transport system of eukaryotic cells are glycoproteins (3), the focus of much research has been on the oligosaccharide moiety (4-11). Although the biological function of oligosaccharide chains on secretory glycoproteins is not known, it has been proposed that the attachment of carbohydrates to these proteins is important in intracellular transport and assortment (3, 6, 8, 11, 12, 13) and in receptor-mediated uptake of lysosomal enzymes (14, 15, 16). Complex carbohydrates have ample structural diversity to serve as ligands in recognition phenomena, and hence are excellent candidates for encoding biological information.

The covalent attachment of oligosaccharide chains represents a significant modification in the structure of a protein. However, the relationship, if any, between these complex series of co-translational and post-translation events and intracellular transport and compartmentation is largely unknown. Why do some oligosaccharides remain as high-mannose, whereas others become complex, and why are some complex chains biantennary, triantennary, or tetraantennary? Heterogeneity in carbohydrate structure is often dismissed as random or trivial. However, heterogeneity may be the key to their specialized function in intracellular transport.

In a number of instances, inhibitors of glycosylation block the secretion of glycoproteins (5-11). For example, it has been found that the secretion of IgA, IgG, and IgE by myeloma tumor cells is prevented by inhibiting the glycosylation of the immunoglobin heavy chain (17, 18). Recently, Firestone (19) has shown that glycosylation of the mammary tumor virus (MTV) precursor polyprotein (Mr 74,000) is required for its compartmentation and specific cleavage. Datema and co-workers (20) reported that bromoconduritol, the α-glucosidase inhibitor that prevents trimming of the innermost glucose residue from the Glc_3Man_9-$GlcNAc_2$ precursor of high-mannose oligosaccharide, inhibited the release of infectious virus. Also, deglycosylated gp90 of Rauscher murine leukemia virus is neither proteolytically processed nor transported to the surface (21). Also, patients homozygous for the Z-allele of α_1-antitrypsin have decreased amounts of this protein in the plasma due to accumulation in the RER (22). Structural analysis of α_1-antitrypsin of the Z-variant has shown that there are major alterations in the N-linked carbohydrate moiety of this glycoprotein; the high-mannose oligosaccharide precursor is not completely processed to the complex structure. Specific deletions in carbohydrate residues of the mμ-chain of IgM have been shown to influence its rate of secretion (23). Moreover, treatment with 1-deoxynojirimycin (DNJ), an inhibitor of glucosidase I and II of the RER, inhibited the secretion of α_1-antitrypsin from the primary cultures of rat hepatocytes (24), α_1-antitrypsin and α_1-antichymotrypsin from human hepatoma cells (25), and the delivery of newly synthesized cathepsin D and β-hexosaminidase to lysosomes (26). While carbohydrate depletion impairs the intracellular transport of invertase and acid phosphatase in yeast (27) and glycoprotein G in cells infected with vesicular stomatitis virus (VSV) (28, 29, 30), their inhibition is due to alterations in physical properties that result in decreased solubility and aggregation within the ER/Golgi. However, there is considerable evidence that mannose-6-phosphate residues are involved in the intracellular transport of lysosomal glycoproteins (12, 13, 14, 31), and it was recently shown that treatment of macrophages with swainsonine, an inhibitor of α-mannosidase II, enhanced the secretion of lysosomal hydrolases that react with the mannose receptor (32).

In contrast, most studies with inhibitors of protein glycosylation, such as tunicamycin, have failed to reveal a requirement for oligosaccharide attachment and intracellular transport and compartmentalization. For example, we have shown that glycosylation is not directly required for the export of fibronectin (33) or the acetylcholine receptor of muscle (34). Similarly, Struck and co-workers (35) demonstrated that tunicamycin, an inhibitor of the synthesis of N-glycosidically linked oligosaccharides, did not affect the secretion of transferrin or low-density lipoprotein from chicken hepatocytes. Moreover, the deglycosylated forms of K-46 immunoglobulin light chain, bovine pancreatic RNAase, and α-lactalbumin are secreted normally. Additionally, not all secretory proteins are glycoproteins. For example, serum albumin, the major secretory product of the liver, contains no bound carbohydrate at any time during its life history (36).

It is apparent that considerable variation exists, and one cannot conclude from the above results that carbohydrates do or do not play a role in intracellular transport. Part of the confusion may originate from the use of various proteins in many different systems. For this reason, we have investigated this problem in considerable detail in the liver system that secretes a large number of well-characterized glycoproteins. The studies to be summarized here address two of the most important problems in cell biology—the function of the oligosaccharide moieties of glycoproteins and elucidation of the molecular events involved in intracellular transport and secretion.

The four approaches most frequently used to investigate the functions of the oligosaccharide moieties include: (a) destruction or elimination of the sugar moiety by the use of reagents such as periodate or glycosidases (37); (b) prevention of the attachment of the sugar to the protein during synthesis using inhibitors (5-11, 38); (c) isolation of cell mutants having defects in the biosynthesis of various intermediate oligosaccharide structures (5, 38); and, more recently, (d) utilization of inhibitors of processing to modify oligosaccharide structure (5-11, 39). The latter three approaches have been particularly useful in that they provide information about the role of the oligosaccharide moiety in intracellular transport or processing of glycoproteins.

The availability of specific inhibitors acting at different stages in the glycosylation process has permitted the synthesis of protein without oligosaccharide moieties, or protein with altered carbohydrate structures. Tunicamycin, swainsonine, DNJ, and bromoconduritol are the best agents to inhibit protein glycosylation specifically, and have become widely used since their discovery and characterization (5-11, 40, 41). Tunicamycin is a structural analog of UDP-2-acetamide-2-deoxyglucose, and blocks the initial step in the glycosylation of dolichol phosphate, an intermediate in the assembly of core oligosaccharides linked N-glycosidically to protein (42-45). Swainsonine, an inhibitor of Golgi α-mannosidase II (46-52), and DNJ and bromoconduritol, inhibitors of endoplasmic reticulum glucosidase I and II (53, 54) or glucosidase II (55, 56), respectively, are the most useful for specifically modifying the structure of N-linked oligosaccharides (9, 11). Unfortunately, no comparable inhibitors are available for perturbing the biosynthesis of 0-linked oligosaccharides and subsequently studying the deglycosylated protein.

EVIDENCE FOR SELECTIVITY IN TRANSPORT OF SECRETORY GLYCOPROTEINS

Our choice of liver secretory glycoproteins for these studies is based on several advantageous aspects of this system. First, transformants of human hepatocytes have been isolated that are easily propagated in culture (57). Second, the transformant (Hep G2) used in the present study continues to synthesize, in large quantities, and secrete more than 20 of the serum glycoproteins. Third, antibodies directed against the various glycoproteins are commercially available. Fourth, the primary structure of the carbohydrate moieties of the glycoproteins of the normal liver are generally well characterized (58-70). Hep G2 was isolated by Dr. Barbara Knowles and associates from liver biopsies of patients with primary hepatoblastoma and hepatocellular carcinoma. The hepatoma cell line was derived from a single cell by cloning in soft agar. The fact that they retain the biosynthetic capabilities of normal liver parenchymal cells makes them very useful for investigating the control of secretion and intracellular processing.

To examine the regulation of intracellular transport of secretory proteins in Hep G2 cells, we have studied the synthesis, processing, and secretion of nine proteins: fibronectin, α-fetoprotein, α_1-protease inhibitor, transferrin, plasminogen, α_2-macroglobulin, ceruloplasmin, fibrinogen, and albumin (68). With the exception of albumin, all of the proteins are glycoproteins with N-linked complex oligosaccharides. Plasminogen from normal plasma has one O-linked oligosaccharide per chain (69). The secreted form of all of these proteins, with the exception of fibrinogen, migrated as single bands of SDS-polyacrylamide gels under reducing conditions. Fibrinogen is composed of three pairs of identical subunits, α, β, and γ, with apparent molecular weights (M_r) in SDS-polyacrylamide gel electrophoresis (SDS-PAGE) of 63,000, 55,000, and 47,000 KDa, respectively (70). Hep G2 cells have been shown to secrete the cellular form of fibronectin (S. K. Akiyama and K. M. Yamada, personal communication).

To examine the kinetics of secretion of the nine proteins mentioned above, Hep G2 cultures were incubated for 10 min at 37°C in medium containing [^{35}S]methionine, washed with phosphate-buffered saline containing an excess of unlabeled methionine and chased. Specific proteins were quantitatively isolated from both the medium and cell lysate by immunoisolation with monospecific antibodies. We found that all of the proteins studied are secreted with one of three discrete rates, which we termed rapid, intermediate, and slow. Rapidly secreted proteins include albumin, fibronectin, α-fetoprotein, and α_1-protease inhibitor, with half of the newly synthesized protein appearing in the medium in 30-40 min. Proteins secreted at an intermediate rate include ceruloplasmin, α_2-macroglobulin, and plasminogen, with half of each protein being secreted in 75-80 min after synthesis. Slowly secreted proteins include transferrin and fibrinogen, with an intracellular retention half-time of 110-120 min; all three subunits of fibrinogen are secreted with similar kinetics. Similar results were obtained by using data either on the loss of proteins from the cell or the accumulation of proteins in the medium, and the rates were highly reproducible in individual separate experiments.

The mechanisms regulating the intracellular transport of proteins destined for export are unknown. A simple model for secretion in a constitutively secreting cell type is that the luminal volume of the RER is continually being packaged into transport vesicles for shipment first to the Golgi and then to the cell surface. However, this nonspecific "bulk-flow," or "conveyor-belt," model for secretion predicts that the kinetics of secretion of all proteins in a single cell should be identical. Therefore, the finding that different secretory proteins of hepatoma cells are secreted at markedly different rates is not consistent with a bulk-flow model for secretion (71).

Our results imply that intracellular transport is a selective process, and that proteins within a secretory class may share structural determinants that specify their rate of export. The structural determinants of the protein could function as transport signals (ligands) that are recognized by specific intracellular transport receptors, and proteins in the same secretory class would presumably share these determinants. These structural determinants might include the oligosaccharide structure on glycoproteins or some characteristic feature of their amino acid sequence or conformation. Specific transport receptors in the ER or Golgi could either sort different classes of secretory proteins into different secretory pathways or differentially regulate the rate of transport of different classes of secretory proteins along one common pathway. If binding to the receptor is due to the random collision of ligand and receptor, then glycoproteins with a strong binding affinity would have a small intracellular pool and newly synthesized protein would be exported rapidly. Alternatively, if the binding is weak, then the intracellular pool would be large and would be secreted slowly over a long time span. An alternative model to explain our finding is that there could be topological sorting of the different classes of secretory proteins at the level of RER and co-transport to the cell surface.

We cannot comment on correlation between carbohydrate structure and rates of secretion, since the carbohydrate structures on glycoproteins from transformed cells are usually altered (72, 73), and preliminary studies suggest that indeed the oligosaccharide units are larger and more highly branched.

OLIGOSACCHARIDE-DEPENDENT SECRETION OF SOME GLYCOPROTEINS

To determine whether inhibition of glycosylation would affect the kinetics of secretion of the liver proteins, Hep G2 cultures were treated for 3 hr at 37°C with tunicamycin prior to addition of [^{35}S]methionine. Labeling was terminated after a 10-min pulse, and the cultures were washed with phosphate-buffered saline and chased for various periods in tunicamycin-free medium containing excess unlabeled methionine. Specific proteins were quantitatively isolated from both the medium and cell lysate by immunoisolation with monospecific antibodies. We found that tunicamycin treatment markedly inhibited the secretion of α_2-macroglobulin, α_1-protease inhibitor, and ceruloplasmin, but had no effect on the intracellular transport of albumin nor on the glycoproteins transferrin, fibrinogen, α-fetoprotein, fibronectin, and plasminogen (74).

The most dramatic effect of tunicamycin treatment was on the secretion of α_2-macroglobulin, with more than 90% of the protein still present in the cell fraction in an undegraded form almost 7 hr after synthesis; in control cells, 50% of newly synthesized α_2-macroglobulin is secreted by 75-80 min. A small fraction of α_2-macroglobulin escapes inhibition of glycosylation by tunicamycin and migrates more slowly on the SDS-polyacrylamide gel than the major nonglycosylated form, and the glycosylated form is secreted quantitatively during the chase period. Only 40-45% of nonglycosylated α_1-protease inhibitor is secreted during a chase period of almost 7 hr, whereas 50% of the glycosylated form is secreted in 30-40 min in control cells. Half of newly synthesized ceruloplasmin is secreted in 165 min by tunicamycin-treated cells, compared with 75-80 min in untreated cells.

There is no apparent correlation between the kinetic class of the secretory protein and sensitivity of secretion to tunicamycin treatment. For example, of the three glycoproteins that are exported with rapid kinetics, secretion of only one of these, α_1-protease inhibitor, is inhibited by tunicamycin treatment. Similarly, of the three glycoproteins secreted with intermediate kinetics, tunicamycin treatment inhibits the secretion of only two of these proteins, α_2-macroglobulin and ceruloplasmin, but not the third, plasminogen. These results agree with previous reports that tunicamycin treatment of hepatocytes and hepatoma cells inhibits secretion of α_1-protease inhibitor but not albumin or transferrin (25, 70, 75).

It is presently unclear why tunicamycin treatment inhibits secretion of specific glycoproteins. Removal of carbohydrate by tunicamycin treatment has been shown to alter the conformation of some glycoproteins and induce the formation of insoluble aggregates (5, 11), which could inhibit intracellular transport of proteins. While we cannot exclude the possibility of alteration in tertiary conformation, we do know that the carbohydrate-depleted proteins, whose secretion is disrupted by tunicamycin treatment of Hep G2 cultures, are released in a soluble form after cell disruption (T. -K. Yeo et al., manuscript in preparation).

To determine whether the deglycosylated protein accumulated in the RER, the Golgi, or both, pulse-chase experiments were performed and proteins were immunoprecipitated from the RER and Golgi fractions following isolation by Percoll density gradient centrifugation. Tunicamycin treatment markedly inhibited the transport of α_1-protease inhibitor, ceruloplasmin, and α_2-macroglobulin from the RER to the Golgi, but had only a marginal effect on transport from the Golgi to the cell surface. The retention of these deglycosylated proteins in the RER was not due to changes in their solubility properties, since they, like their glycosylated counterparts, are associated with the supernatant fraction of cells disrupted by freeze-thawing or lysed with 0.2% Triton X-100. The three inhibited proteins were also shown to be soluble by gel chromatography and SDS-PAGE under nonreducing conditions (T. -K. Yeo et al., manuscript in preparation).

These results indicate that carbohydrate is required for the transport of three glycoproteins whose solubility is apparently not affected by deglycosylation (76). We have suggested that carbohydrate-specific receptors are present for this subgroup of glycoproteins that mediate their transport from the RER to the Golgi (74). However, we have not excluded the possibility that deglycosylation leads to a change in the tertiary conformation of these glycoproteins such that they can no longer interact with specific transport receptors that recognize a structural feature other than the carbohydrate moiety.

To further investigate the potential role of oligosaccharide moieties in the intracellular transport of liver secretory glycoproteins, we have used specific inhibitors that act at different stages in the processing of the oligosaccharide chain, resulting in the biosynthesis of glycoprotein with altered oligosaccharide structures. To date we have used swainsonine, an inhibitor of Golgi mannosidase II (46–52), and DNJ, an inhibitor of ER glucosidase I and II (53, 54).

To examine the effect of swainsonine on the kinetics of glycoprotein secretion, control and swainsonine-treated cells were pulsed with [^{35}S] methionine for 10 min in the presence or absence of the inhibitor, and after various chase times the culture media and cell lysates were used for the isolation of the various secretory proteins (39). We found that all of the secretory glycoproteins appeared in the medium faster when the cells were treated with swainsonine than in untreated cultures.

Since the different forms of α_1-antitrypsin, that is, intracellular high-mannose precursor form (M_r 51,000), swainsonine-induced hybrid form (M_r 55,000), and fully processed complex form (M_r 59,000), are easily detectable by SDS-PAGE, immunoprecipitation of the cell lysates can provide valuable information regarding the time required for conversion of one form to the other. We found that the conversion of the precursor to the hybrid in the presence of swainsonine occurred faster than the conversion of the precursor to the complex form in control cells. Upon quantitation of the radioactivity present in the respective bands, the conversion to the hybrid form is accelerated by 10 min as compared with its normal counterpart. The glycoproteins secreted by the swainsonine-treated cells were sensitive to digestion with endo-β-N-acetylglucosaminidase H (endo H), consistent with expectations for the hybrid oligosaccharide structure. Also, the M_r of the hybrid glycoprotein was consistently smaller than that of the fully processed complex form.

It seems unlikely that the effect of swainsonine is on the first portion (from

RER to *cis*-Golgi) of the intracellular transport pathway, since the enzyme it inhibits is localized in the cis-Golgi (77, 78, 79). Therefore, the swainsonine-induced structural alteration in the carbohydrate moiety of glycoproteins apparently enhances their transport within the Golgi compartment. The finding that glycoproteins with hybrid-type oligosaccharide structures are secreted into the medium indicates that trimming of α,1,3- and α,1,6-mannose residues by α-mannosidase II is not obligatory for the intracellular transport and ultimate secretion of glycoproteins.

While the molecular basis for the earlier appearance of hybrid-containing glycoprotein in the culture medium is currently unknown, it is possible that its interaction with an endogenous carbohydrate-binding protein (lectin) is altered. For example, a carbohydrate-binding protein may be present that mediates transport of mannose-terminated oligosaccharides across the Golgi stacks faster than glycoproteins with complex oligosaccharides and possibly via a different pathway.

Finally, we have used DNJ treatment of human hepatoma cells to examine the effect of glucose removal from precursor N-linked oligosaccharides on the intracellular transport of ceruloplasmin, α-fetoprotein, fibronectin, α_2-macroglobulin, α_1-protease inhibitor, and transferrin (J. B. Parent et al., in preparation). We found that DNJ inhibits export of specific glycoproteins (α_1-protease inhibitor, ceruloplasmin, and α_2-macroglobulin) but has no effect on the transport of any of the others (α-fetoprotein, fibronectin, and transferrin). In fact, DNJ inhibits the transport of the same three glycoproteins inhibited by tunicamycin, and the glycoproteins also accumulated in the same organelle (the RER) as in tunicamycin-treated cells. We have suggested that $Man_{8-9}GlcNAc_2$ may be part of the binding site for a receptor that mediates rapid transport of α_1-protease inhibitor, ceruloplasmin, and α_2-macroglobulin (11). Why the intracellular transport of other glycoproteins is not inhibited is subject of future investigations.

In all of these studies with the various pharmacologic agents, it is important that the secretion of albumin is not affected. This indicates that the effects of the drugs (which are in contact with the cells for only 3-5 hr) are specific for glycoproteins, and the intracellular transport machinery is not generally impaired.

DIFFERENCES IN OVERALL RATES OF PROTEIN SECRETION VERSUS INTRACELLULAR TRANSPORT

It was not known if differences in protein secretion rates primarily represent the time required for the protein to be transported from the RER to the Golgi; to traverse the Golgi stacks; to be transported from the Golgi outwards in secretory vesicles; or some combination of the above. To answer this question, we have employed three independent experimental approaches; that is, cellular fractionation, acquisition of endo H resistance, and binding to immobilized plant lectins to measure the rate of glycoprotein transport within the RER and Golgi.

To determine if the differences in rates of protein transport in the RER are responsible for the variability in the overall rates of export observed for three glycoproteins with markedly different secretion rates, we measured the rates of transfer from the RER to the Golgi, and from the RER to the cell surface (80). Multiple cultures of Hep G2 cells were pulse labeled for 10 min with [^{35}S]methionine and chased for varying intervals, and cell lysates were prepared. The lysates were used directly for immunoprecipitation of proteins (transferrin, ceruloplasmin, and α_1-protease inhibitor) representative of three secretory classes (slow, intermediate, and rapid, respectively). In a parallel set of pulse-labeled experiments, the three secretory glycoproteins were immunoisolated from RER fractions obtained by using a modified Percoll gradient fractionation protocol. We found that there is variability in the rates of glycoprotein transport within the RER; however, it is also apparent that, on the average, the three glycoproteins reside in the RER for no more than two-fifths of their total intracellular residency times. This indicates that the marked variability in rates of protein secretion cannot be explained solely on the basis of differences in rates of export from the RER.

To utilize another approach to determine how long it takes (following synthesis on the RER) for α_1-protease inhibitor, ceruloplasmin, and transferrin to enter specific Golgi compartments, we have measured the kinetics of acquisition of binding sites on these secretory glycoproteins to immobilized lectins (80). This is a valid approach, since specific sugar residues are removed or added to N-linked oligosaccharides as glycoproteins traverse the *cis, medial,* and *trans* compartments of the Golgi. For this study, nearly confluent cultures were labeled for 5 min with [^{35}S]methionine and chased for various intervals prior to preparation of cell lysates. The labeled glycoproteins were isolated from the cell lysates by immunoprecipitation and incubated with the appropriate agarose-lectin, and bound glycoprotein was extracted and analyzed by SDS-PAGE fluorography. We found that maximal binding of transferrin to wheat germ agglutinin (WGA) and *Ricinus communis* agglutinin (RCA_I) occurred at approximately 75 and 90 min, respectively, maximal binding of ceruloplasmin to both lectins occurred in approximately 30 min, and maximal binding of α_1-protease inhibitor to WGA and RCA_I required approximately 15 and 30 min, respectively.

As a third approach to monitoring glycoprotein movement within the various compartments of the ER/Golgi pathway, we used the time course of conversion of the oligosaccharide moiety of N-linked glycoproteins from the endo H-sensitive (high-mannose structure) forms to the endo H-resistant (complex structure) forms (80). In good agreement with the lectin binding results, we find that half of pulse-labeled transferrin, ceruloplasmin, and α_1-protease inhibitor becomes resistant to endo H by approximately 65, 30, and 18 min, respectively. Recent studies have shown that glycoproteins become endo H-resistant at a stage after the action of GlcNAc transferase and mannosidase II (81), which suggests that endo H-resistance is conferred between the time that the glycoproteins traverses the medial Golgi and the addition of terminal galactose in the *trans* Golgi (82, 83). Therefore, it is reasonable to consider that the time required for newly synthetized glycoprotein to become endo H-resistant approximates the average time required for transport from the RER to the medial Golgi compartment. However, this approach is valid only when the endo H-resistant fraction found in the medium is also taken into account in computing the average time required for endo H-resistance on total newly synthesized protein. Exclusion of this fraction leads to a serious underestimation of the rate of transport from the RER.

The data presented here are consistent with the idea that the variable secretory rates of glycoproteins in Hep G2 cells reflect heterogeneous rates of glycoprotein transport within both the RER and Golgi. We suggest that the variability in rates of transport reflects differential regulation by interaction with specific transport receptors within the ER and Golgi (8).

It is not known why DNJ and tunicamycin impair the export of the same three glycoproteins. Similarly, it is unclear why removal of glucose from $Glc_3Man_9GlcNAc_2$ is required for the intracellular transport of some glycoproteins but not others.

The above results clearly indicate that glycosylation and specific carbohydrate structures are required for the intracellular transport of several liver secretory glycoproteins. While not conclusive, these findings strongly suggest the involvements of intracellular lectins in the secretion of liver glycoproteins as carbohydrate alteration or depletion affects their transport kinetics.

ACKNOWLEDGMENT

The work reviewed above was partially supported by NIH Grant R01-GM-29804 to K. Olden.

REFERENCES

1. G. E. Palade, 1975, Intracellular aspects of the process of protein synthesis, *Science* **189:**347-358.
2. M. G. Farquhar, and G. E. Palade, 1981, The Golgi apparatus (complex)—(1954-1981)—From artifact to center stage, *J. Cell Biol.* **91:**77s-103s.

3. E. H. Eylar, 1965, On the biological role of glycoproteins, *J. Theor. Biol.* **10:**89-113.
4. D. K. Struck, and W. J. Lennarz, 1980, The function of saccharide-lipids in synthesis of glycoproteins, in *The Biochemistry of Glycoproteins and Proteoglycans,* W. J. Lennarz, ed., Plenum Press, New York, pp. 35-83.
5. R. Gibson, S. Kornfeld, and S. Schlesinger, 1980, A role for oligosaccharides in glycoprotein biosynthesis, *Trends Biochem. Sci.* **5:**290-293.
6. K. M. Yamada, and K. Olden, 1982, Actions of tunicamycin on vertebrate cells, in *Tunicamycin,* G. Tamura, ed., Japan Scientific Societies Press, Tokyo, pp. 119-144.
7. R. T. Schwarz, and R. Datema, 1982, The lipid pathway of protein glycosylation and its inhibitors: The biological significance of protein-bound carbohydrates, *Adv. Carbohydr. Chem. Biochem.* **40:**287-379.
8. K. Olden, J. B. Parent, and S. L. White, 1982, Carbohydrate moieties of glycoproteins: A re-evaluation of their function, *Biochim. Biophys. Acta* **650:**209-232.
9. K. von Figura, and A. Hasilik, 1984, Genesis of lysosomal enzyme deficiencies, *Trends Biochem. Sci.* **9:**29-31.
10. R. T. Schwarz, and R. Datema, 1984, Inhibitors of trimming: New tools in glycoprotein research, *Trends Biochem. Sci.* **9:**32-34.
11. K. Olden, B. A. Bernard, M. J. Humphries, T.-K. Yeo, K.-T. Yeo, S. L. White, S. A. Newton, H. C. Bauer, and J. B. Parent, 1985, Function of glycoprotein glycans, *Trends Biochem. Sci.* **10:**78-82.
12. A. Kaplan, D. T. Achord, and W. S. Sly, 1977, Phosphohexosyl components of a lysosomal enzyme are recognized by pinocytosis receptors on human fibroblasts, *Proc. Natl. Acad. Sci. (USA)* **74:**2026-2030.
13. A. Gonzalez-Noriega, J. H. Grubb, V. Talkad, and W. S. Sly, 1980, Chloroquine inhibits lysosomal enzyme pinocytosis and enhances lysosomal enzyme secretion by impairing receptor recycling, *J. Cell Biol.* **85:**839-852.
14. A. Hasilik, and E. F. Neufeld, 1980, Biosynthesis of lysosomal enzymes in fibroblasts: Phosphorylation of mannose residues, *J. Biol. Chem.* **255:**4946-4950.
15. P. D. Stahl, and P. Schlesinger, 1980, Receptor-mediated pinocytosis of mannose/N-acetylglucosamine-terminated glycoproteins and lysosomal enzymes by macrophages, *Trends Biochem. Sci.* **5:**194-196.
16. E. F. Neufeld, and G. Ashwell, 1980, Carbohydrate recognition systems for receptor-mediated pinocytosis, in *The Biochemistry of Glycoproteins and Proteoglycans,* W. J. Lennarz, ed., Plenum Press, New York, pp. 241-266.
17. F. Melchers, 1973, Biosynthesis, intracellular transport, and secretion of immunoglobulins. Effect of 2-deoxy-D-glucose in tumor plasma cells producing and secreting immunoglobulin G_1, *Biochemistry* **12:**1471-1476.
18. S. Hickman, A. Kulczycki, R. G. Lynch, and S. Kornfeld, 1977, Studies of the mechanism of tunicamycin inhibition of IgA and IgE secretion by plasma cells, *J. Biol. Chem.* **252:**4402-4408.
19. G. Firestone, 1983, The role of protein glycosylation in the compartmentalization and processing of mouse mammary tumor virus glycoproteins in mouse mammary tumor virus-infected rat hepatoma cells, *J. Biol. Chem.* **258:**6155-6161.
20. R. Datema, P. A. Romero, G. Legler, and R. T. Schwarz, 1982, Inhibition of formation of complex oligosaccharides by the glucosidase inhibitor bromoconduritol, *Proc. Natl. Acad. Sci. (USA)* **79:**6787-6791.
21. E. Polonoff, C. A. Machida, and D. Kabat, 1982, Glycosylation and intracellular transport of membrane glycoproteins encoded by murine leukemia viruses, *J. Biol. Chem.* **257:**14023-14028.
22. A. Hercz, E. Katona, E. Cutz, J. R. Wilson, and M. Barton, 1978, α_1-Anti-trypsin: The presence of excess mannose in the z variant isolated from liver, *Science* **201:**1229-1232.

23. C. Sidman, M. J. Potash, and G. Köhler, 1981, Roles of protein and carbohydrate in glycoprotein processing and secretion. Studies using mutants expressing altered IgM mu chains, *J. Biol. Chem.* **256:**13180-13187.
24. V. Gross, T. Andus, T. A. Tran-Thi, R. T. Schwarz, K. Decker, and P. C. Heinrich, 1983, 1-Deoxynojirimycin impairs oligosaccharide processing of alpha-1-proteinase inhibitor and inhibits its secretion in primary cultures of rat hepatocytes, *J. Biol. Chem.* **258:**12203-12209.
25. H. F. Lodish, and N. Kong, 1984, Glucose removal from N-linked oligosaccharides is required for efficient maturation of certain secretory glycoproteins from the rough endoplasmic reticulum to the Golgi complex, *J. Cell Biol.* **98:**1720-1729.
26. P. Lemansky, V. Gieselman, A. Hasilik, and K. von Figura, 1984, Cathepsin D and beta-hexosaminidase synthesized in the presence of 1-deoxynojirimycin accumulate in the endoplasmic reticulum, *J. Biol. Chem.* **259:**10129-10135.
27. H. R. Onishi, J. S. Tkacz, and J. O. Lampen, 1979, Glycoprotein nature of yeast alkaline phosphatase. Formation of active enzyme in the presence of tunicamycin, *J. Biol. Chem.* **254:**11943-11952.
28. R. Leavitt, S. Schlesinger, and S. Kornfeld, 1977, Impaired intracellular migration and altered solubility of nonglycosylated glycoproteins of vesicular stomatitis virus and Sindbis virus, *J. Biol. Chem.* **252:**9018-9023.
29. R. Gibson, R. Leavitt, S. Kornfeld, and S. Schlesinger, 1978, Synthesis and infectivity of vesicular stomatitis virus containing nonglycosylated G protein, *Cell* **13:**671-679.
30. R. Gibson, S. Schlesinger, and S. Kornfeld, 1979, The nonglycosylated glycoprotein of VSV is temperature-sensitive and undergoes intracellular aggregation at elevated temperatures, *J. Biol. Chem.* **254:**3600-3607.
31. I. Tabas, and S. Kornfeld, 1980, Biosynthetic intermediates of beta-glucuronidase contain high mannose oligosaccharides with blocked phosphate residues, *J. Biol. Chem.* **255:**6633-6639.
32. K. -N. Chung, V. L. Shepherd, and P. D. Stahl, 1984, Swainsonine and castanospermine blockade of mannose glycoprotein uptake by macrophages. Apparent inhibition of receptor-mediated endocytosis by endogenous ligands, *J. Biol. Chem.* **259:**14637-14641.
33. K. Olden, R. M. Pratt, and K. M. Yamada, 1978, Role of carbohydrates in protein secretion and turnover: Effects of tunicamycin on the major cell surface glycoprotein of chick embryo fibroblasts, *Cell* **13:**461-473.
34. J. M. Prives, and K. Olden, 1980, Carbohydrate requirement for expression and stability of acetylcholine receptor on the surface of embryonic muscle cells in culture, *Proc. Natl. Acad. Sci. (USA)* **77:**5263-5267.
35. D. K. Struck, P. B. Siuta, M. D. Lane, and W. J. Lennarz, 1978, Effect of tunicamycin on the secretion of serum proteins by primary cultures of rat and chick hepatocytes. Studies on transferrin, very-low-density lipoprotein, and serum albumin, *J. Biol. Chem.* **253:**5332-5337.
36. T. Peters, B. Fleisher, and S. Fleisher, 1971, The biosynthesis of rat serum albumin, *J. Biol. Chem.* **246:**240-244.
37. P. Atkinson, and J. Hakimi, 1980, Alterations in glycoproteins of the cell surface, in *The Biochemistry of Glycoproteins and Proteoglycans,* W. J. Lennarz, ed., Plenum Press, New York, pp. 191-239.
38. I. S. Trowbridge, R. Hyman, and C. Mazauskas, 1978, The synthesis and properties of T25 glycoprotein in Thy-1-negative mutant lymphoma cells, *Cell* **14:**21-29.
39. T. -K. Yeo, K. -T. Yeo, J. B. Parent, and K. Olden, 1985, Swainsonine treatment accelerates intracellular transport and secretion of glycoproteins in human hepatoma cells, *J. Biol Chem.* **260:**2565-2569.
40. A. Takatsuki, K. Arima, and G. Tamura, 1971, Tunicamycin, a new antibiotic. I. Isolation and characterization of tunicamycin, *J. Antibiot.* **24:**215-223.
41. A. Takatsuki, and G. Tamura, 1971, Effect of tunicamycin on the synthesis of macromole-

cules in cultures of chick embryo fibroblasts infected with Newcastle disease virus, *J. Antibiot.* **24:**785-794.

42. S. C. Kuo, and J. O. Lampen, 1976, Tunicamycin inhibition of [^{3}H]glucosamine incorporation into yeast glycoproteins: Binding of tunicamycin and interaction with phospholipids, *Arch. Biochem. Biophys.* **172:**574-581.
43. A. Takatsuki, K. Kohno, and G. Tamura, 1975, Inhibition of biosynthesis of polyisoprenol sugars in chick-embryo microsomes by tunicamycin, *Agric. Biol. Chem.* **39:**2089-2091.
44. J. S. Tkacz, and J. O. Lampen, 1975, Tunicamycin inhibition of polyisoprenyl N-acetylglucosaminyl pyrophosphate formation in calf-liver microsomes, *Biochem. Biophys. Res. Commun.* **65:**248-257.
45. D. K. Struck, and W. J. Lennarz, 1977, Evidence for the participation of saccharide-lipids in the synthesis of the oligosaccharide chain of ovalbumin, *J. Biol. Chem.* **252:**1007-1013.
46. S. M. Colegate, P. R. Dorling, and C. R. Huxtable, 1979, A spectroscopic investigation of swainsonine: An α-mannosidase inhibitor isolated from *Swainsona canescens, Aust. J. Chem.* **32:**2257-2264.
47. D. R. P. Tulsiani, T. M. Harris, and O. Touster, 1982, Swainsonine inhibits the biosynthesis of complex glycoproteins by inhibition of Golgi mannosidase II, *J. Biol. Chem.* **257:**7936-7939.
48. P. R. Dorling, C. R. Huxtable, and S. M. Colegate, 1980, Inhibition of lysosomal alpha-mannosidase by swainsonine, an indolizidine alkaloid isolated from *Swainsona canescens, Biochem. J.* **191:**649-651.
49. A. D. Elbein, P. R. Dorling, K. Vosbeck, and M. Horisberger, 1981, Swainsonine prevents the processing of the oligosaccharide chains of influenza virus hemagglutinin, *J. Biol. Chem.* **257:**1573-1576.
50. V. Gross, T. A. Tran-Thi, K. Vosbeck, and P. C. Heinrich, 1983, Effect of swainsonine on the processing of the asparagine-linked carbohydrate chains of α_1-antitrypsin in rat hepatocytes. Evidence for the formation of hybrid oligosaccharides, *J. Biol. Chem.* **258:**4032-4036.
51. R. G. Arumugham, and M. L. Tanzer, 1983, Abnormal glycosylation of human cellular fibronectin in the presence of swainsonine, *J. Biol. Chem.* **258:**11883-11889.
52. A. D. Elbein, R. Solf, P. R. Dorling, and K. Vosbeck, 1981, Swainsonine: An inhibitor of glycoprotein processing, *Proc. Natl. Acad. Sci. (USA)* **78:**7393-7397.
53. P. A. Romero, R. Datema, and R. T. Schwarz, 1983, N-methyl-1-deoxynojirimycin, a novel inhibitor of glycoprotein processing, and its effect on fowl plague virus maturation, *Virology* **130:**238-242.
54. A. D. Elbein, 1984, Inhibitors of the biosynthesis and processing of N-linked oligosaccharides, *CRC Crit. Rev. Biochem.* **16:**21-49.
55. G. Legler, 1977, Glucosidases, *Methods Enzymol.* **46:**368-381.
56. G. Legler, and W. Lotz, 1973, Investigations on the mechanism of action of glycoside splitting enzymes. VII. Functional groups at the active site of an α-glucosidase from *Saccharomyces cerevisiae, Hoppe-Seyler's Z. Physiol. Chem.* **354:**243-254.
57. B. B. Knowles, C. C. Howe, and D. P. Aden, 1980, Human hepatocellular carcinoma cell lines secrete the major plasma proteins and hepatitis B surface antigen, *Science* **209:**497-499.
58. B. Fournet, J. Montreuil, G. Strecker, L. Dorland, J. Haverkamp, J. F. G. Vliegenhart, J. P. Binette, and K. Schmid, 1978, Determination of the primary structures of 16 asialo-carbohydrate units derived from human plasma alpha$_1$-acid glycoprotein by 360-MHZ ^{1}H-NMR spectroscopy and permethylation analysis, *Biochemistry* **17:**5206-5214.
59. G. O. Schwarzmann, V. B. Hatcher, and R. W. Jeanloz, 1978, Purification and structural elucidation of several carbohydrate side chains from alpha 1-acid glycoprotein, *J. Biol. Chem.* **253:**6983-6987.
60. M. L. Hayes, and F. J. Castellino, 1979, Carbohydrate of the human plasminogen variants. II. Structure of the asparagine-linked oligosaccharide unit, *J. Biol. Chem.* **254:**8772-8776.

61. M. L. Hayes, and F. J. Castellino, 1979, Carbohydrate of the human plasminogen variants. III. Structure of the O-glycosidically linked oligosaccharide unit, *J. Biol. Chem.* **254:**8777-8780.
62. L. C. Hodges, R. Laine, and S. K. Chan, 1979, Structure of the oligosaccharide chains in human alpha$_1$-protease inhibitor, *J. Biol. Chem.* **254:**8208-8212.
63. T. Mega, E. Lujan, and A. Yoshida, 1980, Studies on the oligosaccharide chains of human alpha$_1$-protease inhibitor. II. Structure of oligosaccharides, *J. Biol. Chem.* **255:**4057-4061.
64. E. Regoeczi, P. Taylor, M. T. Debanne, L. Marz, and M. W. Hatton, 1979, Three types of human asialo-transferrin and their interactions with the rat liver, *Biochem. J.* **184:**399-407.
65. G. Spik, B. Bayard, B. Fournet, G. Strecker, S. Bouguelet, and J. Montreuil, 1975, Studies on glycoconjugates. LXIV. Complete structure of two carbohydrate units of human serotransferrin, *FEBS Lett.* **50:**296-299.
66. K. Yamashita, C. J. Liang, S. Funakoshi, and A. Kobata, 1981, Structural studies of asparagine-linked sugar chains of human ceruloplasmin. Structural characteristics of the triantennary complex type sugar chains of human plasma glycoproteins, *J. Biol. Chem.* **256:**1283-1289.
67. A. B. Zinn, J. S. Marshall, and D. M. Carlson, 1978, Preparation of glycopeptides and oligosaccharides from thyroxine-binding globulin, *J. Biol. Chem.* **253:**6761-6767.
68. J. B. Parent, H. C. Bauer, and K. Olden, 1985, Three secretory rates in human hepatoma cells, *Biochim. Biophys. Acta* **846:**44-50.
69. M. L. Hayes, and F. J. Castellino, 1979, Carbohydrate of the human plasminogen variants. I. Carbohydrate composition, glycopeptide isolation, and characterization, *J. Biol. Chem.* **254:**8768-8771.
70. R. R. Townsend, E. Hilliker, Y. T. Li, R. A. Laine, W. R. Bell, and Y. C. Lee, 1982, Carbohydrate structure of human fibrinogen. Use of 300-MHz ^{1}H-NMR to characterize glycosidase-treated glycopeptides, *J. Biol. Chem.* **257:**9704-9710.
71. G. J. A. M. Strous, and H. F. Lodish, 1980, Intracellular transport of secretory and membrane proteins in hepatoma cells infected by vesicular stomatitis virus, *Cell* **22:**709-717.
72. J. P. Quigley, R. H. Goldfarb, C. Scheiner, J. O'Donnell-Tormey, and T. -K. Yeo, 1980, Plasminogen activator and the membrane of transformed cells, in *Tumor Cell Surfaces and Malignancy,* R. O. Hynes and C. F. Fox, eds., Alan R. Liss, New York, pp. 773-796.
73. D. A. Wagner, R. Ivatt, A. T. Destree, and R. O. Hynes, 1981, Similarities and differences between the fibronectins of normal and transformed hamster cells, *J. Biol. Chem.* **256:**11708-11715.
74. H. C. Bauer, J. B. Parent, and K. Olden, 1985, Role of carbohydrate in glycoprotein secretion by human hepatoma cells, *Biochem. Biophys. Res. Commun.* **128:**368-375.
75. B. E. Ledford, and D. F. Davis, 1983, Kinetics of serum protein secretion by cultured hepatoma cells. Evidence for multiple secretory pathways, *J. Biol. Chem.* **258:**3304-3308.
76. K. Olden, S. L. White, and S. A. Newton, 1986, Glycoproteins: Potential role of glycan moiety in regulation of physicochemical stability and proteolysis, in *Thermotolerance and Thermophily: Observations and Mechanisms,* Critical Reviews in Chemistry, I. Henle, ed., in press.
77. D. E. Goldberg, and S. Kornfeld, 1983, Evidence for extensive subcellular organization of asparagine-linked oligosaccharide processing and lysosomal enzyme phosphorylation, *J. Biol. Chem.* **258:**3159-3165.
78. W. G. Dunphy, E. Fries, L. J. Urbani, and J. E. Rothman, 1981, Early and late functions associated with the Golgi apparatus reside in distinct compartments, *Proc. Natl. Acad. Sci. (USA)* **78:**7453-7457.
79. R. Polhmann, A. Waheed, A. Hasilik, and K. von Figura, 1982, Synthesis of phosphorylated recognition marker in lysosomal enzymes is located in the cis part of Golgi apparatus, *J. Biol. Chem.* **257:**5323-5325.

80. K. -T. Yeo, J. B. Parent, T. -K. Yeo, and S. Olden, 1985, Variability in transport rates of secretory glycoproteins through the endoplasmic reticulum and Golgi in human hepatoma cells, *J. Biol. Chem.* **260:**7896-7902.
81. D. R. P. Tulsiani, and O. Touster, 1983, Swainsonine causes the production of hybrid glycoproteins by human skin fibroblasts and rat liver Golgi preparations, *J. Biol. Chem.* **258:**7578-7585.
82. W. G. Dunphy, R. Brands, and J. E. Rothman, 1985, Attachment of terminal N-acetylglucosamine to asparagine-linked oligosaccharides occurs in central cisternae of the Golgi stack, *Cell* **40:**463-472.
83. J. Roth, and E. F. Berger, 1982, Immunocytochemical localization of galactosyltransferase in HeLa cells: Codistribution with thiamine pyrophosphatase in trans-Golgi cisternae, *J. Cell Biol.* **93:**223-229.

12

Use of Chimeric Toxins in Elucidating the Function of Endogenous Animal Lectins

David L. Simpson
University of Pennsylvania

ABSTRACT

The first mammalian lectin, or carbohydrate-binding protein, identified was the asialoglycoprotein receptor (1, 2, 3). Since that time, a large number of endogenous lectins have been identified (4, 5a, 5b, 6) in a wide variety of vertebrate tissues and cultured cells. These lectins, like those of plants (7), are capable of recognizing and interacting with a wide variety of glycoconjugates. This specificity and the fact that many of these lectins are developmentally regulated and are present on the surface of the cell (4) have stimulated an intense interest in their biological function(s). There is some information regarding the function of these interesting proteins in plants (5a, 5b), and what is known about the function of animal lectins has been discussed in previous chapters. For example, there is considerable evidence that some animal lectins mediate the receptor-mediated endocytosis of certain serum glycoproteins (3, 8) and the uptake of lysosomal hydrolases in fibroblasts (9, 10). Other interesting speculations (4, 5a, 5b) include their putative function(s) as aggregation or recognition factors in the organization of tissues during histogenesis, their involvement as specific adhesion factors, and their roles as intracellular and intercellular recognition factors. To a large extent, our lack of knowledge regarding lectin function is due to limitations in the experimental approaches (4, 5a). The purpose of this chapter is to demonstrate the usefulness of an exciting new methodology termed *chimeric toxin technology* (11), the design and synthesis of chimeric, or hybrid-protein toxins, which are tools or reagents useful in studying animal lectin function.

INTRODUCTION

Endogenous lectins, or carbohydrate-binding proteins, have recently been identified in a wide variety of vertebrate tissues (4, 5a, 5b). Striking changes in the concentrations of many vertebrate lectins during cell differentiation and development, in conjunction with their ability to selectively interact with complementary glycoconjugate receptors, suggest that these carbohydrate-binding proteins may participate in glycoconjugate-mediated recognition phenomena during development (4).

It is not the intent of this chapter to review the field of vertebrate lectins or their implicated or speculative function(s). The first review article dedicated to animal lectins (4) was published in 1978; many more lectins have since been identified and characterized, and considerable progress has been made in determining their function in lower organisms (12) and in plants (5, 13). Since the original review a number of excellent articles have appeared that have included discussions of animal lectins and their functions. Those lectins thought to be involved in receptor-mediated uptake of glycoconjugates have been reviewed extensively by Ashwell and Harford (3), Neufeld and Ashwell (8), Sly and Stahl (10), and Neufeld and associates (9, 14). An informative review by Barondes (5a) deals primarily with the cellular functions of lectins in general. Other comprehensive reviews dealing with various aspects of this subject have been published by Brown and Hunt (13), Olden and co-workers (6), and Levi (15). However, this publication represents the first comprehensive treatise on animal lectins.

The purpose of this chapter is to outline an experimental strategy that encompasses a biochemical, cell biological, and genetic approach that takes advantage of chimeric toxin technology. This new and promising methodology (11), which can be used for the selection of lectin or glycoconjugate-receptor variant cell lines, is useful in establishing lectin function.

CURRENT EXPERIMENTAL APPROACHES TO DETERMINING LECTIN FUNCTION

The definitive determination of lectin function, like that of any protein, requires a multitude of approaches. These approaches have been outlined by Barondes (5a) in his review of lectin function. In this article he discusses the value and the limitations of each individual approach used in the past, and emphasizes the importance of evaluating the problem in a number of different ways. Since approaches other than the use of chimeric toxins (11) have been reviewed, no attempt will be made here to evaluate them, and readers are instead referred to Barondes's article (5a).

CHIMERIC, OR HYBRID-PROTEIN, TOXINS

Nature has designed and provided us with a number of plant and bacterial toxins (16) that kill sensitive cells by inhibiting protein synthesis. While the precise mechanisms by which these native toxins interfere with protein synthesis may differ, the basic design and structure of the toxins is similar in several respects as seen in Figure 12-1. They consist of two polypeptide chains (17), a binding protein, and a protein with catalytic activity.

Ricin

Perhaps the best known example of such native toxins is ricin. Ricin is a toxic glycoprotein from castor beans, the seeds of *Ricinus communis.* It is composed of two polypeptide chains held together by a single disulfide bond (17). The B subunit, RTB, is a galactose-specific carbohydrate-binding protein, a lectin. This nontoxic B chain is able to recognize and bind galactose-containing glycoconjugates (17) on the cell surface. The A chain, RTA, is an enzyme that acts catalytically on the 60S subunit of ribosomes (18, 19) to inhibit protein synthesis. These chains can be separated and isolated in pure

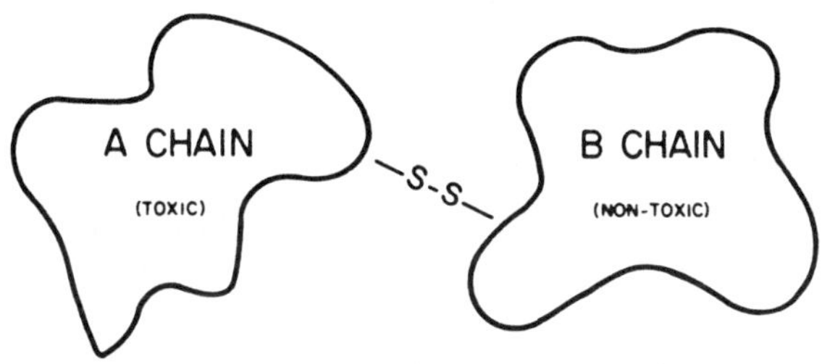

NATIVE TOXINS

A CHAIN PROPERTIES	B CHAIN PROPERTIES
•TOXIC	•NON-TOXIC
ENZYME THAT INHIBITS PROTEIN SYNTHESIS	BINDS TO CELL-SURFACE RECEPTOR(S)
ACTS CATALYTICALLY	ACT STOICHIOMETRICALLY
NO CELL-SURFACE BINDING ACTIVTY	NO ENZYME ACTIVITY
REQUIRES B CHAIN TO KILL CELLS	REQUIRES A CHAIN TO KILL CELLS
FREE A CHAIN NON-TOXIC TO CELLS	FREE B CHAIN NON-TOXIC TO CELLS

Figure 12-1. Native toxins. The basic structure and design of a number of plant and bacterial toxins and the respective properties of their A chains and B chains.

form (20). Neither chain alone is a potent toxin. The B chain is able to bind to cells but has no appreciable toxicity. The isolated A chain of ricin, while able to inactivate protein synthesis in vitro, is not toxic to cells because it depends on its association with the B chain to bind to the cell surface, traverse the membrane, and gain entry to the cytosol. Thus, the free A chain, RTA, is four to five orders of magnitude less toxic to cells than the intact native ricin molecule. Ricin exhibits its extremely potent toxicity only when RTA and RTB are linked by a disulfide bond.

Diphtheria Toxin

Diphtheria toxin (21, 22), a bacterial toxin, works in a manner similar to ricin. The B chain, DTB, binds to target cell surfaces but is in itself nontoxic. The toxic A chain, DTA, is an enzyme that catalyzes the transfer of adenosine diphosphate (ADP) ribose from nicotinamide adenine nucleotide (NAD^+) to elongation factor 2, thus inhibiting protein synthesis. The isolated DTA chain is toxic in a cell-free system, but like ricin, is a poor toxin on cells normally sensitive to the intact native toxin. Isolated DTB binds to the surface of the cells but is not toxic to cells, even when internalized.

In addition to ricin and diphtheria toxin, other native plant and bacterial toxins are available (11), in addition to a number of interesting A-chain-like proteins from plants such as gelonin (23), which can be used in the design and synthesis of chimeric toxins.

Chimeric Toxins

The paramount objective in the design and synthesis of a chimeric toxin is to improve on the specificity of targeting while maintaining the extremely potent catalytic, yet nonspecific, killing potential. While ricin, for example, does exhibit some selectivity in its binding in that it binds only to galactose (16) and not to mannose-, fucose-, or glucosamine-containing receptors, galactose-containing glycoconjugates are present on the surfaces of most animal cells. The effect is that the native toxin shows little cell-type specificity.

The primary goal of the design and synthesis of chimeric toxins is to replace the binding subunit of the native toxin, the B chain, with another ligand that confers a different and more precise and selective targeting specificity to the hybrid toxin, while at the same time conserving the potent catalytic cytotoxicity of the native A chain. This concept of B chain replacement or substitution by a number of potential ligands is illustrated in Figure 12-2.

The objective of hybrid protein toxin synthesis is to specifically direct the entry of proteins, usually toxins, into cells via alternative cell surface receptors or antigens. If possible, these chimeric toxins are usually constructed to take advantage of the cell's receptor-mediated ligand transport systems. A number of these receptors for protein uptake systems are known to be carbohydrate-mediated in that they are endogenous carbohydrate-binding proteins, or lectins (3, 8, 9, 10). Again, it is not possible within the scope of this chapter to review what is presently known about chimeric toxins. The field of chimeric toxin or hybrid-protein conjugates has expanded rapidly in the past three years. Much is known about the chemistry and the biology of a wide variety of synthetic protein toxins, and this information is detailed in a number of excellent review articles (24, 25), the most comprehensive being that of Olsnes and Pihl (11).

The interest, excitement, and rapid growth in this area are evidenced by a recent volume of *Immunological Reviews* entitled "Antibody Carriers of Drugs in Tumor Therapy" (26). This volume is devoted to the use of monoclonal and polyclonal antisera in the design and synthesis of chimeric toxins whose toxic selectivity is directed toward cancer cells. These reagents have also been called *immunotoxins* (27, 28) and the methodology has also been referred to as the *retiarian* approach to cancer chemotherapy (29).

Monoclonal antibodies are among the most specific, selective, and powerful reagents that are now available to us. If the B chain of a native toxin is replaced with a monoclonal antibody, these chimeric, or immunotoxin, conjugates have all the potential and characteristic properties of Paul Ehrlich's

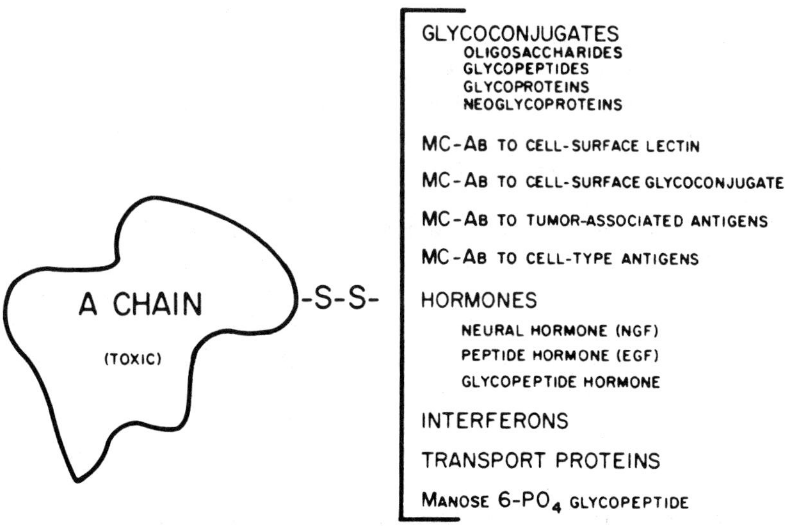

Figure 12-2. Chimeric toxins. The basic design and structure of chimeric toxins, illustrating the variety of ligands that can be substituted in place of the B chain of native toxins.

zauberklugen, or "magic bullets" (30). These reagents, proposed by Ehrlich, are antibody-like and are able to seek and destroy target cells or tissues in an organism that give rise to a disease state.

No attempt will be made here to review the field of chimeric toxins, but for the purpose of this chapter it is pertinent to stress that a large number of biologically active, highly toxic, and extremely selective conjugates have been made by covalently linking glycoconjugates (11, 24, 25, 31, 32), lectins (11, 23, 33, 35) and monoclonal or polyclonal antibodies (11, 28) to the purified toxic A chain of ricin, diphtheria toxin, and to gelonin.

Peptide and glycopeptide hormone conjugates have also been constructed with epidermal growth factor (24, 25, 32, 36), insulin (11, 37, 38), human chorionic gonadotropin (39, 40), human placental lactogen (41), lysosomal hydrolase receptors (42), and anti-Thy 1.2 antigen present on certain cells (43, 44). Presently, in this laboratory, both A chain and B chain hybrid conjugates of nerve growth factor (NGF) are being constructed.

RATIONALE FOR THE USE OF CHIMERIC TOXINS

Lectin Function in the Cellular Slime Molds

The function of lectins in the cellular slime molds, in that they appear to mediate species-specific cellular recognition and adhesion (4), is perhaps one of the best examples of how the recognition properties of these protein, as measured in vitro, are utilized in vivo to mediate a vital phase in the development of these simple eukaryotic organisms. This work will be summarized only briefly since the rationale for the use of chimeric toxins stems, in part, from these studies and since the work has been reviewed in detail elsewhere (4, 5, 12, 45).

Cellular slime molds have served as an excellent model for studies of intercellular recognition and adhesion at the molecular level. The main advantage is the unique life cycle of these organisms (46) that exist first in a vegetative or nonsocial stage as unicellular amoebae, feeding on bacteria and dividing every few hours. When the food supply is depleted or removed, the unicellular amoebae differentiate into an aggregation-competent state over a period of 9 to 11 hours. The particular advantage in using these organisms to study cell-cell recognition is the large number of species (45) shown to exhibit selective sorting out of amoeba as they form multicellular aggregates, which is presumably due to species-specific intercellular affinities.

The critical observation made by Steve Rosen was that extracts of cohesive *D. discoideum* contain an agglutinin of erythrocytes (47), whereas nonsocial cells do not. Agglutinin activity was identified in soluble extracts of two species of cellular slime molds, *Dictyostelium discoideum* (48) and *Polysphondylium pallidum* (49), as these cells developed from a single-cell vegetative state to a cohesive aggregation-competent state. These developmentally regulated agglutination activities were specifically inhibited by certain monosaccharides, but not by others at the same concentration. This suggested that the molecules responsible for the agglutination in developing slime molds were multivalent carbohydrate-binding protein, referred to as lectins. The developmentally regulated lectin from *D. discoideum* called *discoidin* (50, 51) and *P. pallidum* called *pallidin* (52) were purified employing affinity techniques based on their respective carbohydrate recognition specificities.

The purified lectins, discoidin and pallidin, could be distinguished on the basis of their hapten inhibition specificity (50, 51) as well as by their subunit molecular weights, behavior in the ultracentrifuge, amino acid compositions, and isoelectric points (50, 52).

These lectins were also shown to be present on the surface of cohesive slime molds cells (48, 49, 53, 54, 55), an advantageous localization if these molecules are involved in cellular recognition and adhesion. Cohesive slime mold cells formed rosettes with erythrocytes and this interaction was sensitive to the same sugars that selectively inhibited agglutination by purified lectin. Antibody to purified discoidin and pallidin bound to the surface of cohesive cells, and in reduced amounts or not at all to vegetative cells as shown by immunoferritin and immunofluorescent techniques (53).

It was also possible to demonstrate the presence of developmentally regulated species-specific receptors, presumably glycoconjugates, for these cell surface lectins in a refined series of binding studies (56, 57, 58). These lectin-specific high-affinity receptors also appear on the surface of slime mold cells (56) as they become progressively more cohesive. They have a significantly higher affinity for the lectins purified from the same species as they do lectins from other species.

Simple sugars and desialylated glycoproteins (48, 49, 50), which specifically blocked the agglutination of erythrocytes, also blocked the agglutination of slime mold cells in a gyrated suspension. This suggests that in vivo aggregation can be competitively inhibited by glycoproteins whose oligosaccharide structure is similar to that of the native receptor for the slime mold lectins. The direct involvement of pallidin in the developing cohesiveness of these cells was also suggested when Fab fragments derived from antisera raised to purified pallidin inhibited both the agglutination of erythrocytes and the endogenous cohesiveness of differentiated *P. pallidum* cells (59).

In summary, these studies (12, 45) indicate that species-specific lectins, carbohydrate-binding protein, appear on the cell surface of slime mold cells as they differentiate from a vegetative single-cell state to a social aggregation-competent state. The presence of lectin parallels the development of cell-cell cohesiveness. Appearance of lectin is also correlated with the appearance of high-affinity species-specific glycoconjugate receptors on the cell surface (56, 62). Both the agglutination of erythrocytes and the developing cohesiveness of slime mold cells is inhibited by certain simple saccharides and glycoproteins and by Fab fragments derived from antisera raised to the purified lectins. As more recent evaluation of the functional role of these interesting developmentally regulated lectins in cellular slime molds is found in a series of review articles (12, 45).

Our experimental approach in establishing the function of these lectins that we believed to be involved in species-specific intercellular recognition and adhesion was to first establish what criteria must be met to implicate a macromolecule, in this case a lectin, in these processes.

First, the macromolecule should be present on the cell surface. It should be present when cells are in a cohesive state but absent when cells are not cohesive. Addition of purified macromolecule to cells should promote their cohesiveness. Reagents that block the active binding site of the macromolecule should also block cell cohesion produced either by the added purified macromolecule or by the endogenous macromolecule present on the surface of cohesive cells. Most importantly, the precise recognition specificity of the macromolecule (lectin) should be different in species of cellular slime molds that show selective intercellular affinities with respect to one another. In an extensive series of experiments (49, 53, 56, 59) we were able to show that both lectins, discoidin and pallidin, satisfy all of the established criteria for a cell recognition and adhesion molecule. The experimental approaches and the data obtained from them when taken together, make a convincing case for the participation of the endogenous lectins in the molecular mechanism(s) of intercellular recognition and adhesion. It should also be mentioned that a number of nonlectin-like proteins have also been implicated in slime mold cell cohesion and these have been reviewed elsewhere (12).

Despite considerable supporting evidence from a variety of more conventional experimental approaches, it has been the identification and isolation of a genetic variant (60) or mutant of *Dictyostelium discoideum* that has added the single strongest support for the functional participation of this lectin, discoidin, in cell recognition and adhesion. Ray and colleagues have isolated and characterized a defined mutant in which the lectin, discoidin, is present, is immunologically identical, yet has lost its carbohydrate-binding ability. This cell variant is noncohesive and fails to develop, but when the mutant is mixed with wild-type cells, the variant cells are carried through development. Also, revertants (60, 61, 63) that recover their carbohydrate-binding activity are capable of completing the differentiation cycle. This, in addition to strongly supporting the role of discoidin in cell recognition and adhesion, suggests that these processes are in fact mediated by the interactions of two complementary molecules on the surface of cells, the lock and key mechanism proposed by Weiss (69). It is this work in particular that points to the rationale, great strength, and potential in the selection and isolation of defined cell variants as an approach to the determination of animal lectin function.

These lectin-negative or lectin-function defective variant cell lines (clones) would be extremely useful in establishing lectin function in a definitive manner. It is the purpose of this chapter to demonstrate how chimeric toxins can be blueprinted and constructed to serve as effective selection reagents for selection and isolation of such lectin variants.

CHOICE OF A MODEL SYSTEM TO DEMONSTRATE THE UTILITY OF CHIMERIC TOXINS

Because of our interest in carbohydrate-mediated recognition phenomena and in the biological role(s) of vertebrate lectins, we chose the asialoglycoprotein receptor (3, 8), a hepatocyte-specific carbohydrate-binding protein or lectin, as a model system to demonstrate the utility of chimeric toxins as selection reagents, or tools to study animal lectin function.

For a ligand receptor system to serve as a model for studies aimed at the determination of lectin function(s), it should satisfy as many of the following criteria as possible. First, the ligand should be readily available in reasonable amounts. Second, the receptor, a lectin in this case, should be well characterized. Third, the receptor should be restricted to one or several cell types. Fourth, the pathway or mechanism by which the ligand is internalized should be well characterized. And last, a probe for the receptor other than the ligand itself should be available, a probe such as a monoclonal antibody to the receptor. It is also important, if one wishes to study endogenous lectin function by selecting and isolating lectin variant cell lines, employing toxic conjugates as selection reagents, to have available a *continuous* target cell line that expresses functional lectin on its cell surface.

The asialoglycoprotein receptor (3, 4, 8) presently satisfies all of these established criteria.

The asialoglycoprotein receptor protein has been sequenced, and monoclonal antibodies have been prepared (45). Binding of asialoglycoproteins to the large number of receptors on the surface of hepatocytes initiates a coated vesicle-mediated internalization of the lectin-glycoconjugate complex. The glycoconjugate is then transported to the lysosomes, where it is efficiently degraded, and the internalized asialoglycoprotein receptor is recycled and returned to the plasma membrane, where it is reutilized (3, 8).

The asialoglycoprotein receptor was chosen as a ideal model system to establish the usefulness of chimeric toxins in the study of lectin function for the following reasons: (a) asialoglycoproteins are easily obtained by the chemical or enzymatic desialylation of commercially available glycoproteins; (b) there are approximately a million asialoglycoprotein receptors per hepatocyte; (c) the receptor itself has been purified in large quantities and has been well characterized; (d) binding by the receptor is completely dependent on the presence of terminal galactose residues on the ligand; (e) the asialoglycoprotein receptor is found only on hepatocytes; (f) the ligand-receptor interaction initiates the internalization of the complex and subsequent degradation of the ligand; and (g) a human hepatoma cell line (46), Hep-G2, which expresses receptors, has recently been identified and characterized.

RESULTS

When these studies first began it was not possible to identify a continuous cell line that expressed asialoglycoprotein receptors and the Hep-G2 line had not yet been identified. Since we did not have a continuous target cell line we were forced to use primary cultures of hepatocytes in our studies. Rat hepatocytes were isolated and primary cultures were established (47).

It was then essential to demonstrate that our hepatocyte cultures were able to complete the cycle of asialoglycoprotein binding, internalization, and subsequent degradation. The results shown in Figure 12-3 suggest that our cells efficiently degraded iodinated ASF but not iodinated fetuin. Control cells, 3T3 cells, which do not have this receptor, were unable to degrade either labeled glycoprotein. The degradation of asialofetuin (ASF) by hepatocytes was linear for 20 hours, when the cells were able to degrade only 30% of the added ligand. The rate of degradation was measured to be 1-2 fmoles of iodinated ASF/min/10^6 hepatocytes. It was clear from these experiments that we were able to isolate and culture hepatocytes with active asialoglycoprotein receptors.

PROTEIN PREPARATIONS

Diphtheria toxin A chain (DTA) prepared from diphtheria toxin (48) was a gift from R. J. Collier. Ricin was prepared from castor beans (31). RTA was isolated from Sepharose-bound ricin by elution of the A chain with mercaptoethanol (36). ASF was prepared from commercial fetuin by enzymatic desialylation (31) with neuraminidase from *Clostridium perfringens.* The glycoproteins orosomucoid, asialo-orosomucoid, and asialogalacto-orosomucoid were gifts from Drs. M. Wickerhauser, J . Whitehead, and J. Paulson, respectively.

PREPARATION OF TOXIC CONJUGATES

The primary amino groups of asialofetuin were derivitized with N-succinimidyl-3(2-pyridyldithio)propionate (SPDP) by incubation with excess reagent. The 2-pyridyldithiopropionate derivative N-succinimidyl-3(2-pyridyldithio) pro-

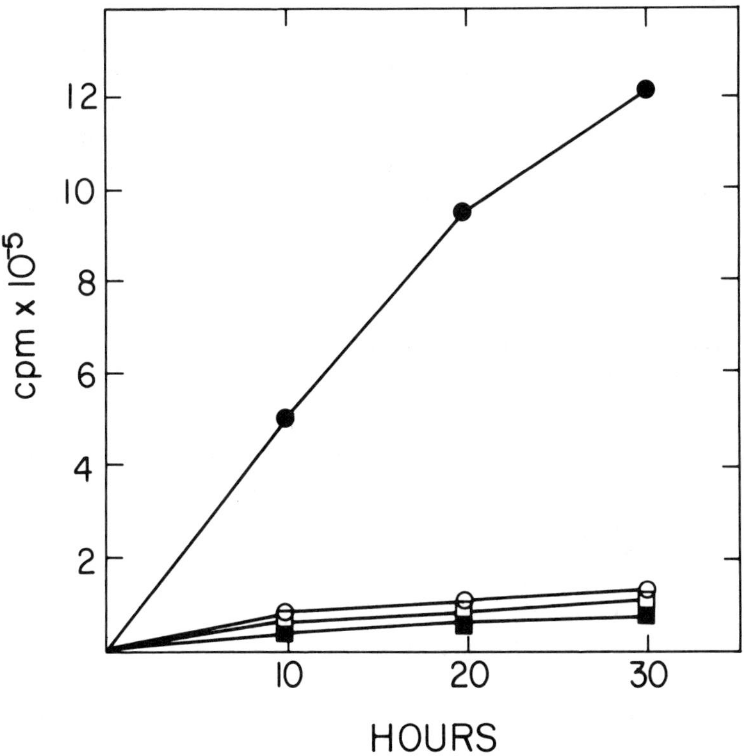

Figure 12-3. Degradation of asialofetuin by hepatocytes. Degradation of ^{125}I-ASF and ^{125}I-labeled fetuin by rat hepatocytes in culture. Cultures were 24 hr old. Approximately 1.25×10^6 cpm of ^{125}I-labeled fetuin or 0.9×10^6 cpm of ^{125}I-labeled fetuin or 0.9×10^6 cpm of ^{125}I-labeled ASF were applied to culture dishes in 2 ml of medium. All points were assayed in triplicate, and the data shown are the averages of these points. (●) Hepatocytes plus ^{125}I-labeled ASF; (○) hepatocytes plus ^{125}I-labeled fetuin; (■) 3T3 cells plus ^{125}I-labeled ASF; (□) 3T3 cells plus ^{125}I-labeled fetuin.

pionate (ASF-PDP) was separated from unreacted SPDP by chromatography on Sephadex G-25. Immediately prior to conjugation, RTA and DTA were reduced in dithiothreitol and were quickly separated from excess reducing reagent by Sephadex G-25 chromatography. Details of the conjugate synthesis have been described elsewhere (31, 32, 36). The two conjugates, ASF-DTA and ASF-RTA, were separated from the reactants and purified by gel-exclusion chromatography. The conjugates were characterized by polyacrylamide gel electrophoresis in the presence and absence of reducing agent.

CHARACTERIZATION OF THE NEWLY SYNTHESIZED CONJUGATES

The ASF-RTA conjugate obtained after purification on two columns of Sephadex G-100 is seen in Figure 12-4. Lanes 1, 2, and 3 of the gel are ASF (45,000), RTA (33,000 and 30,000), and purified ASF-RTA (presumed 75,000), respectively. After reduction with 2-mercaptoethanol, the disulfide-linked ASF-RTA conjugate (lane 6) is cleaved completely to ASF (lane 4) and RTA (lane 5). The ASF-DTA conjugate characterized in a similar fashion (not shown) has been described previously (36). We also prepared conjugates of ASF and native fetuin to control for non-receptor-mediated uptake.

TOXICITY OF THE ASF-RTA AND ASF-DTA CONJUGATES

When it was established that the hybrid protein conjugates can be cleaved into their respective ligand (ASF) and toxin A chain (DTA or RTA), we then tested for the toxicity of the conjugates on the target and control cells compared to the native toxin itself and the free A chain(s).

It is clear from an initial qualitative experiment seen in Figure 12-5 that the ASF-RTA conjugate is extremely toxic to cultured hepatocytes. Fetuin, the native glycoprotein, is unable to block the killing action of the conjugate, while excess ASF can compete with the toxin and protect cells against its cytotoxic action. When hepatocytes were exposed to control concentrations of ASF and fetuin under standard conditions, the cultures were indistinguishable from the control cultures.

To quantitatively assess the relative toxicity of the conjugates, native toxins, and A chains, we examined their effects on protein synthesis since the A chains are, in fact, inhibitors of protein synthesis. Ricin (Figure 12-6) was, as expected, a potent toxin on hepatocytes, with an $E.D._{50}$ in a 24-hour period of 5×10^{-12} M. (The $E.D._{50}$ is the concentration of reagent required to

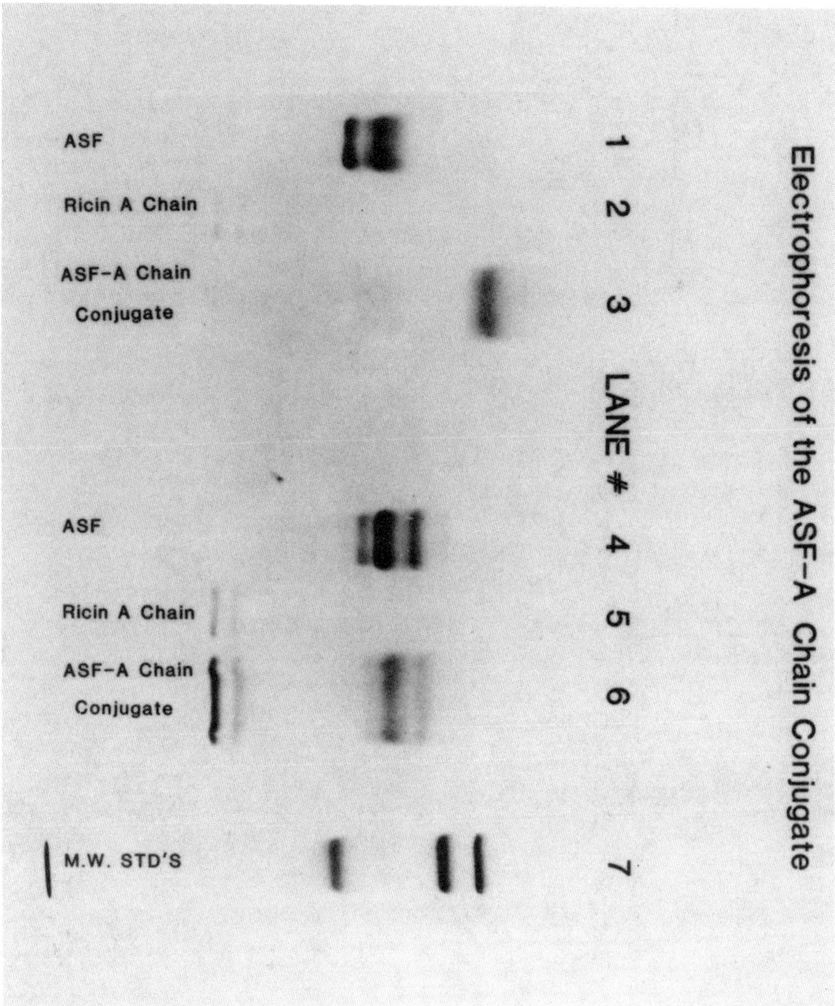

Figure 12-4. Electrophoresis of the ASF-RTA chain conjugate. Gel electrophoresis of the ASF-RTA conjugate. Proteins were denatured in 1% SDS prior to electrophoresis. The samples applied to lanes 1–3 were not reduced compared to those applied to lanes 4–7, which were reduced in 150 mM mercaptoethanol. The acrylamide concentration was 12.5% in the gel. Lanes (1, 4) ASF, 15 μg; (2, 5) RTA, 5 μg; (3. 6) ASF–RTA, 10 μg; (7) molecular weight standards (BSA, Mr 67,000; catalase, Mr 60,000; ovalbumin, Mr 44,000; cytochrome c, Mr 13,500). The two forms of pure RTA (Mr 30,000 and 33,000) are apparent in lanes 2, 5, and 6.

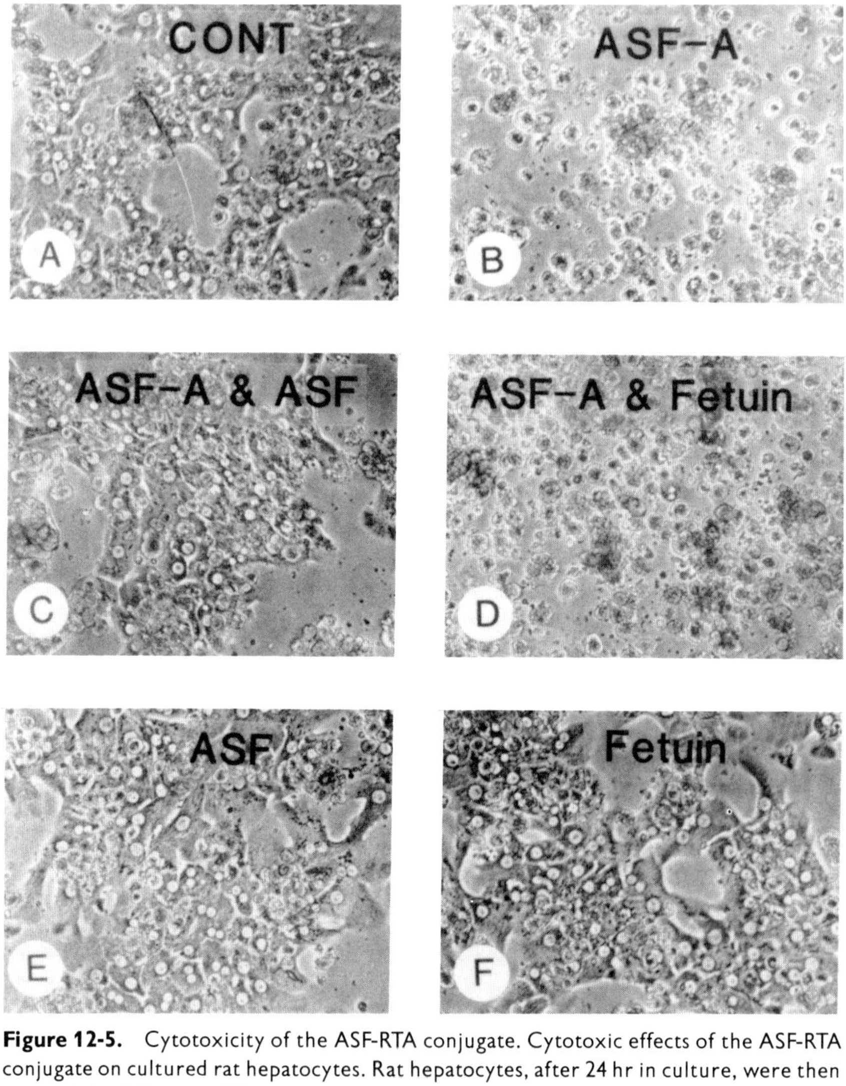

Figure 12-5. Cytotoxicity of the ASF-RTA conjugate. Cytotoxic effects of the ASF-RTA conjugate on cultured rat hepatocytes. Rat hepatocytes, after 24 hr in culture, were then exposed for 24 hr at 37°C to growth medium with no addition (A); ASF-RTA, 2.5 μg/ml (B); ASF-RTA, 2.5 μg/ml, plus ASF, 650 μg/ml (C); ASF-RTA, 2.5 μg/ml, plus fetuin, 650 μg/ml, (D); ASF, 650 μg/ml, (E); and fetuin, 650 μg/ml (F).

give 50% inhibition of protein synthesis in a 24-hr period.) On the other hand, diphtheria toxin, known to be noncytotoxic to rodent cells, was far less toxic. The purified RTA and DTA chains (Figure 12-6) were also considerably less toxic than ricin. It is evident that both the ASF-DTA and ASF-RTA conjugates are extremely effective toxins for hepatocytes, with an $E.D._{50}$ of about one order of magnitude greater than that of native ricin. In a control experiment (not shown), equimolar unconjugated mixtures of ASF and either RTA or DTA were essentially nontoxic. These results confirmed that the ASF must be joined covalently to the toxin A chain to confer toxic activity to cells.

We then took advantage of a fetuin-DTA conjugate to assess the upper limit of hepatocyte toxicity that is not asialoglycoprotein receptor-mediated.

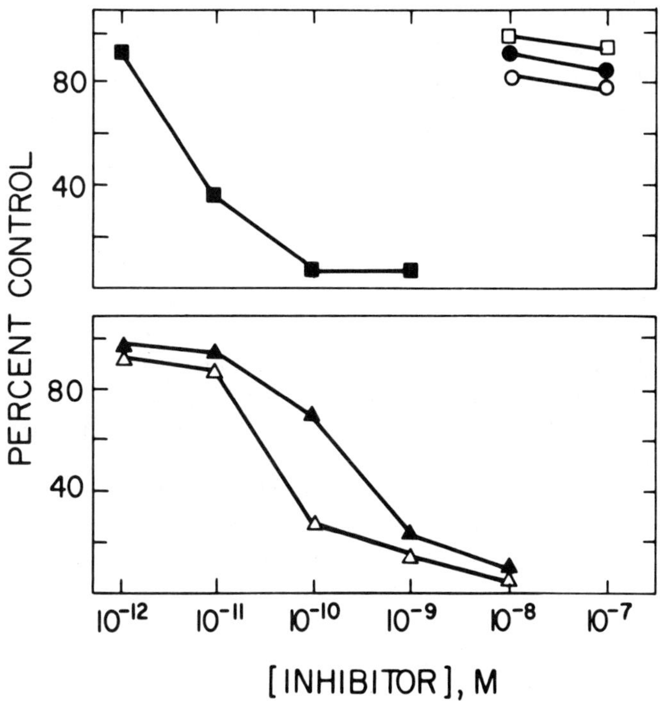

Figure 12-6. Protein synthesis inhibition by ASF-RTA and ASF-DTA conjugates. Inhibition of protein synthesis in cultured hepatocytes by ASF-RTA and ASF-DTA. Protein synthesis was assayed by measuring the incorporation of ^{3}H-leucine into protein by hepatocytes in a leucine-free medium. Protein synthesis was measured after 24 hr of exposure to reagents and compared to control heptocytes with no additions. Values are expressed as percent of control protein synthesis. Upper panel: ricin (■); RTA (□); diphtheria toxin (●); DTA (○). Lower panel: ASF-DTA (△); ASF-RTA (▲).

The control fetuin-DTA conjugate and the ASF-DTA conjugate differ only by the presence of terminal sialic acid residues on the oligosaccharide units of the native serum glycoprotein, fetuin. Fetuin should not be recognized and bound by the hepatocyte lectin due to the absence of terminal galactose moieties. As expected, the toxicity of the fetuin-DTA conjugate was 300-fold lower than that of the ASF-DTA conjugate. Competition binding studies with ASF suggested that the degree of toxicity, which is less than 1% of that observed for the ASF-DTA conjugate, reflects primarily non-asialoglycoprotein receptor-mediated uptake, often referred to as fluid or bulk phase uptake.

After establishing the toxicity of the ASF-RTA and DTA conjugates, we then turned our attention to the assessment of the degree of receptor specificity. In these experiments we tested a variety of potential antagonists for their ability to block the ASF-RTA or ASF-DTA mediated inhibition of protein synthesis. To do this, we simultaneously added a number of glycoprotein derivatives at varying concentrations, along with ASF-DTA or ASF-RTA at a concentration of 10^{-9} M, which inhibited protein synthesis by more than 95%. We then determined the reduction in the toxic effect of conjugate at 10^{-9} M at various concentrations of the glycoprotein derivatives.

Fetuin (Figure 12-7) at concentrations as high as 10^{-5} M had no antagonistic activity. In contrast, ASF blocked both the ASF-RTA and ASF-DTA inhibition of protein synthesis in a linear and dose-dependent fashion.

To show that this effect was not fetuin/asialofetuin-specific, another serum glycoprotein, orosomucoid, or alpha-1-acid glycoprotein was employed; like fetuin, this glycoprotein has oligosaccharide chains that terminate in sialic acid. Treatment of orosomucoid with neuraminidase yields the galactose-terminated derivative asialo-orosomucoid. To generate a second derivative, the terminal galactose residues of asialo-orosomucoid can be effectively removed with an appropriate galactosidase to give asialoagalacto-orosomucoid. The terminal monosaccharide of asialoagalacto-orosomucoid is N-acetylglucosamine, which is not recognized by the asialoglycoprotein receptor.

Neither native orosomucoid nor asialoagalacto-orosomucoid should interfere with the toxicity of the ASF-A chain conjugates, since neither glycocoprotein derivative has terminal galactose residues. As with asialofetuin, asialo-orosomucoid, with its galactosyl terminal oligosaccharide chains should be an effective antagonist of the ASF-DTA and ASF-RTA.

The predictions, based on the structures of the various glycoprotein derivatives, were confirmed in a series of experiments both qualitatively and quantitatively. Orosomucoid (Figure 12-7), like native fetuin, had no inhibitory activity at high concentrations. In contrast, asialo-orosomucoid, like ASF, was a potent antagonist of the toxicity mediated by the ASF conjugates. Asialoagalacto-orosomucoid was ineffective at concentrations as high as 10^{-6} M (not shown) and was only slightly active at 10^{-6} M. It should be noted that at this concentration, 10,000 molecules of potential inhibitor are present for each molecule to toxic conjugate. This means that this level of inhibition by the asialogalacto-orosomucoid derivative would be observed if only 0.25% of the asialo-orosomucoid were not converted to the agalacto derivative by the galactosidase.

To establish further the receptor specificity of the ASF-RTA conjugate, we also tested a variety of reagents for their ability to block the inhibition of protein synthesis caused by ricin or by the ASF-RTA conjugate. Ricin (0.5 μg/ml) effectively reduced protein synthesis in hepatocyte cultures to 1% of control values. Anti-ricin B chain immunoglobulin G (IgG) effectively blocked ricin's toxicity as well as lactose (5 mM), although lactose at higher concentrations proved to be somewhat cytotoxic. Anti-asialoglycoprotein receptor,

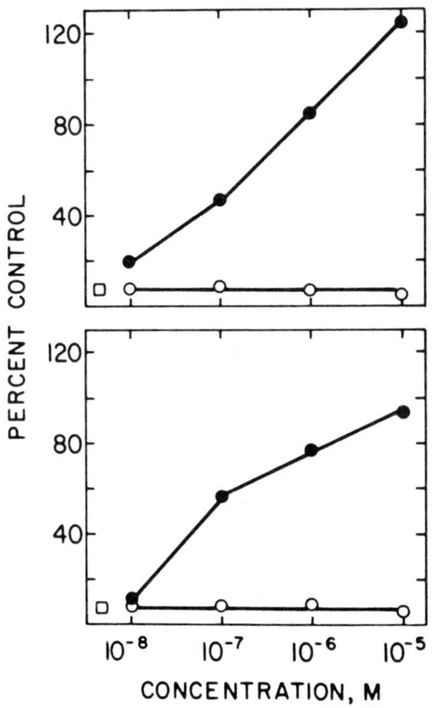

Figure 12-7. Effects of glycoprotein and glycoprotein derivatives on ASF-RTA–mediated toxicity. Either 10^{-9} M ASF-RTA alone or 10^{-9} M ASF-RTA plus the indicated concentrations of the potential antagonists of toxicity were added to cultured heptocytes. Protein synthesis was measured after 24 hr. Upper Panel: ASF-RTA (□); ASF-RTA plus fetuin (○); ASF-RTA plus ASF (●). Lower panel: ASF-RTA plus orosomucoid (○); ASF-RTA plus asialoorosomucoid (●).

which was capable of blocking the degradation of I^{125}-ASF, had no effect on the activity of ricin.

In experiments employing ASF-RTA as the cytotoxic reagent, in contrast to the protective effects against ricin, anti-B chain IgG and lactose did not significantly protect against the effect of the ASF-RTA conjugate. Nonimmune IgG had no effect on the activity of ricin and the ASF-RTA conjugate. Asialofetuin and anti-asialoglycoprotein receptor IgG were quite effective antagonists of the ASF-RTA hybrid. These experiments demonstrate that the native toxin, ricin, and the ASF-RTA conjugate bind to different cell-surface receptors on hepatocytes. Anti-A chain IgG can completely block the action of both ricin and the ASF-RTA conjugate, demonstrating that the toxicity of both the native toxin and the hybrid toxin is due to the action of the A chain.

Additional support for the receptor specificity of these toxic conjugates was obtained by demonstrating the differential effect of anti-rat liver lectin IgG and anti-chick liver lectin IgG with respect to their ability to protect hepatocytes against conjugate toxicity. The results shown in Figure 12-8 indicate that IgG derived from antisera raised to the purified rat liver lectin is

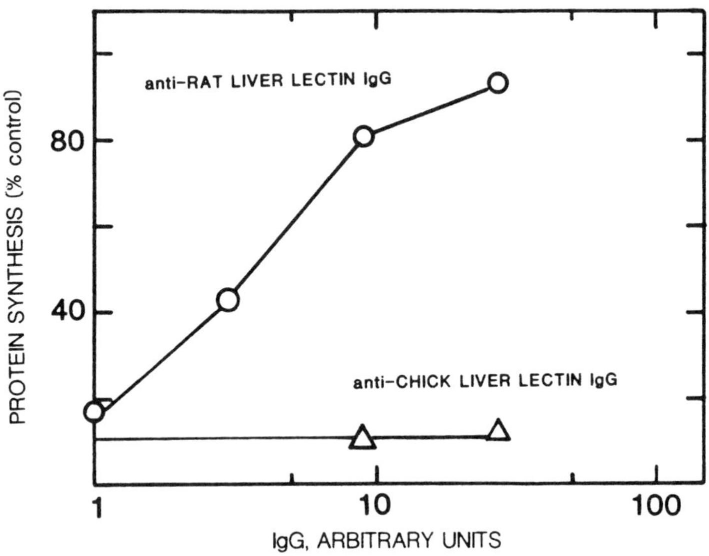

Figure 12-8. Effect of anti-rat liver lectin and anti-chick liver lectin IgG on conjugate-mediated toxicity. Either 10^{-9} M ASF-RTA alone or 10^{-9} M ASF-RTA plus the indicated arbitrary units of IgG as potential antagonists were added to cultured hepatocytes. Protein synthesis was measured after 24 hr. Anti-rat liver lectin IgG (○); Anti-chick liver lectin IgG (△).

able to protect hepatocytes in a dose-dependent fashion against a dose of ASF-DTA conjugate (10^{-9} M) sufficient to kill greater than 95% of the cells. On the other hand, IgG derived from antisera raised to the N-acetylglucosamine-binding lectin present in the hepatocytes of birds and reptiles had no protective effect whatsoever. This experiment demonstrates once again that the toxicity of the hybrid conjugate is mediated by galactose-specific mammalian liver lectin, the asialoglycoprotein receptor.

CELL-TYPE SPECIFICITY OF THE TOXIC CONJUGATES

Once it was established that the conjugates ASF-RTA and ASF-DTA were highly receptor-specific and that the toxicity was mediated by the asialoglycoprotein receptor, attention was turned to evaluating the cell type or target cell selectivity.

To ascertain the degree of cell type specificity, three continuous cell lines,—3T3, VERO, and CHO—and two primary cell cultures from heart and aortic endothelium were tested for their sensitivity to ASF-DTA inhibition of protein synthesis. None of these cell lines or primary cell cultures exhibited any inhibition of protein synthesis when exposed to ASF-DTA at concentrations as high as 10^{-8} M. In striking contrast, the $E.D._{50}$ of the ASF-DTA and the ASF-RTA conjugates on primary cultures of hepatocytes were on the order of 5×10^{-11} M, making them nearly as toxic as the native toxin, ricin. These experiments confirmed that the conjugate was not only receptor-specific but was also cell type-specific or hepatocyte-specific.

IDENTIFICATION OF A CONTINUOUS CELL LINE EXPRESSING FUNCTIONAL RECEPTOR FOR ASIALOGLYCOPROTEINS

Recently, the identification and characterization of a human hepatocellular carcinoma cell line referred to as Hep-G2 (64, 65) has made the isolation of asialoglycoprotein receptor variant cell lines possible. This cell line expresses functional receptors for asialoglycoproteins, although they are reduced in numbers compared to isolated hepatocytes. The design and synthesis of extremely toxic receptor-specific and cell type-selective chimeric toxins such as the ASF-DTA and ASF-RTA conjugates should make this task considerably easier.

The general experimental approach would be to use the receptor-specific ASF-DTA and ASF-RTA conjugates to select variant lines, clones, that are resistant to killing by the hybrid toxins. These selections could be made from a normal or a mutagenized population of Hep-G2 cells. The phenotype of variants that one would expect to isolate by using the toxic hybrids as selection reagents are of two general types. The variants would be either receptor-negative or at least have greatly reduced numbers of receptors. Other variants that would survive conjugate killing would be those that have receptors—it would be hoped a full complement—yet are defective in their ability to effectively internalize the conjugate-receptor complex. These would be termed receptor-positive internalization-negative variants. A third type that may be possible but has never been identified is a variant whose ribosomes are resistant to the effects of the toxin A chains. While the isolation of such a variant is not probable, it would certainly be welcome. Isolated clones of the first two phenotypes can be distinguished from one another and characterized by straightforward receptor-binding assays. Receptor-negative variants would not bind ^{125}I-ASF. Variants that are receptor-positive and internalization negative can be identified by their ability to bind ^{125}I-ASF and by their inability to down-regulate their asialoglycoprotein receptors in the presence of excess ASF.

SUMMARY

Certainly, the most provocative use of chimeric, or hybrid-protein, toxins is that of cancer chemotherapeutic agents. While the potential of this "magic bullet" approach is profound, and while most in vitro tissue culture studies have been quite promising, the in vivo, or animal studies, while also promising, have revealed certain inherent limitations.

Only recently have the isolated catalytic subunits of native toxins been used in conjunction with alternative binding or targeting subunit replacements (11). While the design and synthesis of many of these hybrid toxins have met with marked success, many others have proved to be relatively ineffective toxins on intended target cells. General strategies have not evolved (72) whereby one can design and construct a chimeric toxin with the assurance that it will be both highly specific and extremely toxic. The advent of such predictive design awaits a better understanding of the mechanisms by which the native toxin subunits traverse the plasma membrane (71, 72) and how the enzymatic subunit escapes, presumably, from endocytic vesicles and gains access to the cytosol subsequent to internalization. Chimeric toxins themselves will serve as useful reagents to elucidate the mechanisms and the pathways involved in the productive internalization of native toxins.

In addition to cancer, other clinical applications that hold promise are the use of chimeric toxins to combat viral, bacterial, and parasitic infections. They may also be useful as reagents to control birth. While the future of the clinical use of these reagents in the treatment of cancer is somewhat uncertain, their utility as experimental tools or reagents in biomedical research is without question.

As reagents of value to the experimental biologist, chimeric toxins have been shown to be a great value in the study of ligand-receptor interactions (24, 25, 72) and in understanding the mechanism of action of hormones. They will prove invaluable in elucidating the functions of cell surface antigens, as selection reagents to isolate useful cell variants, and as probes in molecular parasitology or as neuroanatomical mapping reagents. Hybrid protein toxins can also be used to eliminate or select certain cell types of subpopulations of cells during development and differentiation, particularly in the immune system. Another potential application is their use in the purification of primary cultures of neural cells by taking advantage of the differential expression of receptors for certain hormones; for example, EGF receptors (70). These hormone receptor-targeted toxins will also be of value in studying the interdependent relationships of EGF receptor-bearing cells and neuron in co-cultures of neural cells. They will serve to define the biological role or function(s) of developmental or differentiation-specific antigens in a variety of systems by the selective elimination of antigen-responsive cells during development and histogenesis.

In this chapter, the potential use of chimeric toxins in the elucidation of the function(s) of a large number of developmentally regulated endogenous cell surface animal lectins has been proposed and illustrated.

ACKNOWLEDGMENTS

This work was done in the laboratory of Dr. Harvey R. Herschman, Department of Biological Chemistry and Laboratory of Biomedical and Environmental Sciences, UCLA School of Medicine, Los Angeles, California 90024. The work was done in conjunction with Dr. Daniel B. Cawley, a postdoctoral fellow in Dr. Herschman's laboratory. This work was supported by a contract between the Regents of the University of California and the Department of Energy (DE-AM03 76-SF0012), and an NIH grant (GMS 24797) awarded to Dr. Herschman and Dr. Cawley held a National Research Service Award (F32 GM07489). The cost of the preparation of this article was supported by an NIH grant (NS 15490 NLS) awarded to David L. Simpson. Appreciation is extended to Mr. Salvador A. Casco for his expertise in illustrating this chapter.

REFERENCES

1. A. G. Morell, R. A. Irvine, I. Sternlieb, I. H. Scheinberg, and G. Ashwell, 1968, Physical and chemical studies of ceruloplasmin, *J. Biol. Chem.* **243:**155-159.
2. R. J. Stockert, A. G. Morrell, and I. H. Scheinberg, 1974, Mammalian hepatic lectin, *Science* **186:**365-366.

3. G. Ashwell, and J. Harford, 1982, Carbohydrate specific receptors of the liver, *Annu. Rev. Biochem.* **51:**531-544.
4. D. L. Simpson, D. R. Thorne, and H. H. Loh, 1978, Lectins, endogenous carbohydrate-binding proteins from vertebrate tissues: Functional role in recognition processes?, *Life Sci.* **22:**727-748.
5a. S. H. Barondes, 1981, Lectins: Their multiple endogenous cellular functions, *Annu. Rev. Biochem.* **50:**207-221.
5b. S. H. Barondes, 1984, Soluble lectins: A new class of extracellular proteins, *Science* **223:**1259-1264.
6. K. Olden, J. B. Parent, and S. L. White, 1982, Carbohydrate moieties of glycoproteins: A re-evaluation of their function, *Biochim. Biophys. Acta* **650:**209-232.
7. I. J. Goldstein, and C. E. Hayes, 1978, The lectins: Carbohydrate-binding proteins of plants and animals, *Adv. Carbohydr. Chem.* **35:**127-340.
8. E. F. Neufeld, and G. Ashwell, 1980, Carbohydrate recognition systems for receptor-mediated pinocytosis, in *The Biochemistry of Glycoproteins and Proteoglycans,* W. J. Lennarz, ed., Plenum Press, New York, pp. 241-266.
9. E. F. Neufeld, T. W. Lim, and L. J. Shapiro, 1975, Inherited disorders of lysosomal metabolism, *Annu. Rev. Biochem.* **44:**357-376.
10. W. S. Sly, and P. Stahl, 1980, Receptor-mediated uptake of lysosomal enzymes, in *Transport of Macromolecules in Cellular Systems,* S. C. Silverstein, ed., Dahlein Konferenzem, Berlin, pp. 229-244.
11. S. Olsnes, and A. Pihl, 1982, Chimeric toxins, *Pharm. Ther.* **15:**355-381.
12. J. R. Bartles, W. A. Frazier, and S. D. Rosen, 1982, Slime mold lectins, *Int. Rev. Cytol.* **75:**61-99.
13. J. C. Brown, and R. C. Hunt, 1978, Lectins, *Int. Rev. Cytol.* **52:**277-319.
14. E. F. Neufeld, T. W. Lim, and L. J. Shapiro, 1975, Inherited disorders of lysosomal metabolism, *Annu. Rev. Biochem.* **44:**357-376.
15. G. Levi, 1984, Some considerations on the possible role of animal lectins in processes of cell recognition, *Chim. Ind. Med.* **65:**727-732.
16. S. Olsnes, and A. Pihl, 1982, Chimeric toxins, in *Pharmacology of Bacterial Toxins,* J. Drews and F. Dorner, eds., Pergamon Press, New York, pp. 51-105.
17. S. Olsnes, and A. Pihl, 1976, Abrin, ricin and their associated agglutinins, *Specificity of Action of Animal, Bacterial and Plant Toxins,* Receptors and Recognition, series B, vol. 1, P. Cuatrecasas, ed., Chapman and Hall, London, pp. 129-173.
18. M. L. Hedblom, D. B. Cowley, S. Boguslawski, and L. L. Houston, 1978, Binding of ricin A chain to rat liver ribosomes: Relationship to ribosome inactivation, *J. Supramol. Struct.* **9:**252-269.
19. S. C. Benson, 1977, Correlation of inhibition of ribosomal GTPase and protein synthesis by Abrin and ricin, *Biochem. Biophys. Res. Commun.* **75:**845-850.
20. D. B. Cawley, M. L. Hedblom, and L. L. Houston, 1978, Homology between ricin and ricinus communis agglutinin: Amino terminal sequence analysis and protein synthesis studies, *Arch. Biochem. Biophys.* **190:**744-755.
21. R. J. Collier, 1975, Diphtheria toxin: Mode of action and structure, *Bacteriol. Rev.* **39:**54-85.
22. A. M. Pappenheimer, 1977, Diphtheria toxin, *Annu. Rev. Biochem.* **46:**69-94.
23. F. Stirpe, S. Olsnes, and A. Pihl, 1980, Gelonin, a new inhibitor of protein synthesis, nontoxic to intact cells, *J. Biol. Chem.* **255:**6947-6953.
24. H. R. Herschman, D. L. Simpson, and D. B. Cawley, 1982, Toxic ligand conjugates as tools in study of receptor ligand interactions, *J. Cell. Biochem.* **20:**163-176.
25. H. R. Herschman, D. B. Cawley, D. L. Simpson, 1982, in Growth of cells in hormonally defined Media, *Cold Spring Harbor Laboratories* **9:**381-395.
26. G. Moller, ed., 1982, Antibody carriers of drugs and toxins in tumor therapy, *Immunol. Rev.* **62:**1-216.

27. H. E. Blythman, P. Casellas, O. Gros, P. Gros, F. K. Jansen, F. Paolucci, B. Pau, and H. Vidal, 1981, Immunotoxins: Hybrid molecules of monoclonal antibodies and a toxin subunit specifically kill tumor cells, *Nature* **290:**145-146.
28. J. Collier, and D. A. Kaplan, 1984, Immunotoxins, *Sci. Am.* **251:**56-64.
29. D. C. Edwards, and P. E. Thrope, 1981, Targeting toxins—the retiarian approach to chemotherapy, *Trends Biochem. Sci.* **6:**313-316.
30. T. Davies, 1981, Magic bullets, *Nature* **289:**12-13.
31. D. B. Cawley, D. L. Simpson, and H. R. Herschman, 1981, Asialoglycoprotein receptor mediates the toxic effects of an asialofetuin-diphtheria toxin fragment A conjugate on cultured rat hepatocytes, *Proc. Natl. Acad. Sci. (USA)* **78:**3383-3387.
32. D. L. Simpson, D. B. Cawley, and H. R. Herschmann, 1982, Killing of cultured hepatocytes by conjugates of asialofetuin and EGF linked to the A chains of ricin or diphtheria toxin, *Cell* **29:**469-473.
33. T. Uchida, E. Mekada, and Y. Okada, 1980, Hybrid toxin of the A chain of ricin toxin and a subunit of *Wistaria floribunda* lectin, *J. Biol. Chem.* **255:**6687-6693.
34. D. G. Gillilan, R. J. Collier, J. M. Moehring, and T. J. Moehring, 1978, Chimeric toxins: Toxic, disulfide-linked conjugate of concanavalin A with fragment A of diphtheria toxin, *Proc. Natl. Acad. Sci. (USA)* **75:**5319-5323.
35. T. Yamaguchi, R. Kato, M. Beppu, T. Terao, Y. Inoue, Y. Ikawa, and T. Osawa, 1979, Preparation of concanavalin A-ricin A chain conjugate and its biological activity against various cultured cells, *J. Natl. Cancer Inst.* **62:**1387-1395.
36. D. B. Cawley, H. R. Herschman, D. G. Gilliland, and R. J. Collier, 1980, Epidermal growth factor-toxin A chain conjugates: EGF-ricin A is a potent toxin while EGF-diphtheria fragment A is nontoxic, *Cell* **22:**563-570.
37. W. K. Miskimins, and N. Shimizu, 1979, Synthesis of a cytotoxic insulin cross-linked to diphtheria toxin fragment A capable of recognizing insulin receptors, *Biochem. Biophys. Res. Commun.* **91:**143-151.
38. C. H. Hoffman, R. M. Lotan, W. W. Ku, and T. N. Oeltmann, 1983, Insulin-ricin B hybrid molecules mediate an insulin-associated effect on cells which do not bind insulin, *J. Biol. Chem.* **258:**11774-11779.
39. T. N. Oeltmann, and E. C. Heath, 1979, A hybrid protein containing toxic subunit of ricin and the cell-specific subunit of human chorionic gonodotropin, *J. Biol. Chem.* **254:**1022-1027.
40. T. N. Oeltmann, and E. C. Heath, 1979, A hybrid protein containing the toxin subunit of ricin and the cell-specific subunit of human chorionic gonadotropin, *J. Biol. Chem.* **254:**1028-1032.
41. T. M. Chang, A. Dazard, and D. M. Neville, 1977, Artificial hybrid protein containing a toxic protein fragment and a cell membrane receptor binding moiety in a disulfide-linked conjugate. Biochemical and biologic properties of diphtheria fragment A-S-S-human placental lactagen, *J. Biol. Chem.* **252:**1515-1522.
42. R. J. Youle, G. J. Murray, and D. M. Neville, 1979, Ricin linked to monophosphopentamannose binds to fibroblast lysosomal hydrolase receptors, resulting in cell-type specific toxin, *Proc. Natl. Acad. Sci. (USA)* **76:**5559-5562.
43. R. J. Youle, and D. M. Neville, 1980, Anti-thy 1.2 monoclonal antibody linked to ricin is a potent cell-type-specific toxin, *Proc. Natl. Acad. Sci. (USA)* **77:**5483-5486.
44. T. N. Oeltmann, and J. T. Forbes, 1981, Inhibition of mouse spleen cell function by diphtheria toxin fragment A coupled to anti-mouse thy-1.2 and by ricin A chain coupled to anti-mouse IgM, *Arch. Biochem. Biophys.* **209:**362-370.
45. S. H. Barondes, and S. D. Rosen, 1976, Cell surface carbohydrate-binding protein: Role in cell recognition, in *Neuronal Recognition,* S. Barondes, ed., Plenum Press, New York, pp. 331-356.
46. W. F. Loomis, ed., 1975, *Dictyostelium discoideum: A Developmental System,* Academic Press, New York.
47. S. D. Rosen, 1972, Ph.D. Thesis, Cornell University, Ithaca, New York.

48. S. D. Rosen, J. A. Kafka, D. L. Simpson, and S. H. Barondes, 1973, Developmentally-regulated carbohydrate-binding protein in Dictyostelium discoideum, *Proc. Natl. Acad. Sci. (USA)* **70:**2554-2558.
49. S. D. Rosen, D. L. Simpson, J. E. Rose, and S. H. Barondes, 1974, Carbohydrate-binding protein from Polyspondylium pallidum implicated in intercellular adhesion, *Nature* **252:**128, 149-151.
50. D. L. Simpson, S. D. Rosen, and S. H. Barondes, 1974, Discoidin, a developmentally regulated carbohydrate-binding protein from Dictyostelium discoideum: Purification and characterization, *Biochemistry* **13:**3487-3493.
51. W. A. Frazier, S. D. Rosen, R. W. Reitherman, and S. H. Barondes, 1975, Purification and comparison of two developmentally regulated lectins from Dictyostelium discoideum: Dicoidin I and II, *J. Biol. Chem.* **250:**9608-9615.
52. D. L. Simpson, S. D. Rosen, and S. H. Barondes, 1975, Pallidin: Purification and characterization of a carbohydrate binding protein from Polyspondylium pallidum implicated in intercellular adhesion, *Biochim. Biophys. Acta* **412:**109-119.
53. C. -M. Chang, S. D. Rosen, and S. H. Barondes, 1977, Cell surface localization of an endogenous lectin and its receptor in Polyspondylium pallidum, *Exp. Cell Res.* **104:**101-109.
54. D. R. Armant, and E. A. Berger, 1982, Immunochemical analysis of discoidins I and II at the cell surface of wild type and aggregation-defective mutants of Dictyostelium discoideum, *J. Cell. Biochem.* **18:**169-180.
55. C. -H. Sui, R. A. Lerner, G. Ma, R. A. Firtel, and W. F. Loomis, 1976, Developmentally regulated protein of the plasma membrane of Dictyostelium discoideum. The carbohydrate-binding protein, *J. Mol. Biol.* **100:**157-178.
56. R. W. Reitherman, S. D. Rosen, W. A. Frazier, and S. H. Barondes, 1975, Cell surface species-specific high affinity receptors for discoidin: Developmental regulation in Dictyostelium discoideum, *Proc. Natl. Acad. Sci. (USA)* **72:**3541-3545.
57. J. R. Bartles, and W. A. Frazier, 1980, Preparation of ^{125}I-discoidin I and the properties of its binding to Dictyostelium discoideum cells, *J. Biol. Chem.* **255:**30-38.
58. J. R. Bartles, N. J. Galvin, and W. A. Frazier, 1982, Discoidin I-membrane interactions. Discoidin I binds to two types of receptor on Dictyostelium discoideum cells, *Biochim. Biophys. Acta.* **687:**129-136.
59. S. D. Rosen, C. -M. Chang, and S. H. Barondes, 1977, Intercellular adhesion in the cellular slime mold Polyspondylium pallidum inhibited by interaction of asialofetuin or specific univalent antibody with endogenous cell surface lectin, *Dev. Biol.* **61:**202-213.
60. J. Ray, T. Shinnick, and R. Lerner, 1979, A mutation altering the function of a carbohydrate-binding protein blocks cell-cell cohesion in developing Dictyostelium discoideum, *Nature* **279:**215-221.
61. T. M. Shinnick, and R. A. Lerner, 1980, The cbpA gene: Role of the 26,000-dalton carbohydrate-binding protein in intercellular cohesion of developing Dictyostelium discoideum cells, *Proc. Natl. Acad. Sci. (USA)* **77:**4788-4792.
62. J. Ray, and R. A. Lerner, 1982, A biologically active receptor for the carbohydrate-binding protein(s) of Dictyostelium discoideum, *Cell* **28:**91-97.
63. S. Alexander, T. M. Shinnick, and R. A. Lerner, 1983, Mutants of Dictyostelium discoideum blocked in expression of all members of the developmentally regulated discoidin multigene family, *Cell* **34:**467-475.
64. A. L. Schwartz, A. Marshak-Rothstein, D. Rup, and H. F. Lodish, 1981, Identification and quantification of the rat hepatocyte asialoglycoprotein receptor, *Proc. Natl. Acad. Sci. (USA)* **78:**3348-3352.
65. A. L. Schwartz, S. E. Fridovich, B. B. Knowles, and H. F. Lodish, 1981, Characterization of the asialoglycoprotein receptor in a continuous hepatoma line, *J. Biol. Chem.* **256:**8878-8881.

66. A. D. Attic, D. B. Weinstein, H. H. Freeze, R. C. Pittman, and D. Steinberg, 1979, Unaltered catabolism of desialylated low-density lipoprotein in the pig and in cultured rat heptocytes, *Biochem. J.* **180:**647-654.
67. D. N. Chung, and R. J. Collier, 1977, The mechanism of ADP-ribosylation of elongation factor 2, *Biochim. Biophys. Acta* **483:**248-257.
68. D. B. Cawley, M. L. Hedblom, and L. L. Houston, 1978, Homology between ricin and Ricinus communis agglutinin: Amino terminal sequence analysis and protein synthesis inhibition studies, *Arch. Biochem. Biophys.* **190:**744-755.
69. P. Weiss, 1947, The problem of specificity in growth and development, *Yale J. Biol. Med.* **19:**235-278.
70. D. L. Simpson, R. Morrison, J. de Vellis, and H. R. Herschman, 1982, Epidermal growth factor binding and mitogenic activity on purified populations of cells from the central nervous system, *J. Neurosci. Res.* **8:**453-462.
71. S. Olsnes, and K. Sandvig, 1982, Entry of polypeptide toxins into animal cells, in *Endocytosis,* I. Pastan and M. C. Willingham, eds., Plenum Press, New York, pp. 195-234.
72. H. R. Herschman, D. Cawley, and D. L. Simpson, 1985, Toxic conjugates of epidermal growth factor and asialofetuin, in *Receptor-Mediated Targeting of Drugs,* G. Gregoriadis, G. Poste, J. Senior, and A. Trouet, eds., Plenum Press, New York, pp. 27-51.

Index